Bioinspired Intelligent Nanostructured Interfacial Materials

Bioinspired Intelligent Nanostructured Interfacial Materials

Lei Jiang
The Chinese Academy of Sciences, China

Lin Feng
Tsinghua University, China

Chemical Industry Press

World Scientific

NEW JERSEY · LONDON · SINGAPORE · BEIJING · SHANGHAI · HONG KONG · TAIPEI · CHENNAI

Published by

World Scientific Publishing Co. Pte. Ltd.

5 Toh Tuck Link, Singapore 596224

USA office: 27 Warren Street, Suite 401-402, Hackensack, NJ 07601

UK office: 57 Shelton Street, Covent Garden, London WC2H 9HE

and

Chemical Industry Press
No. 13 Qingnianhu South Street
Dongcheng District
Beijing 100011
P.R. China

British Library Cataloguing-in-Publication Data
A catalogue record for this book is available from the British Library.

First published 2010 (Hardcover)
Reprinted 2016 (in paperback edition)
ISBN 978-981-3203-59-4

This edition is jointly published by World Scientific Publishing Co. Pte. Ltd. and Chemical Industry Press, P. R. China.

BIOINSPIRED INTELLIGENT NANOSTRUCTURED INTERFACIAL MATERIALS

ISBN-13 978-981-4280-31-0
ISBN-10 981-4280-31-3

Typeset by Stallion Press
Email: enquiries@stallionpress.com

Preface

Intelligent materials are emerging composite materials that have boomed since 1990s. These novel materials have gained outstanding achievements and influenced multidisciplinary fields in recent years. In the 21st century, the content of intelligent materials is continuously expanding and the domains of which are constantly being broadened. The outstanding features are the tight connection between fundamental researches and practical applications, as well as the tight coupling of biomimetic technology and nanotechnology. The intelligent material system is of great scientific significance; it combines studies that explore nature, mimic nature and surpass nature, involving a multitude of structures and functions. It provides new ideas, new theories, and new methodologies for the innovation of science and technology. After millions of years of evolution, plants and animals have almost completely and perfectly adapted to natural environments. Thus, mimicking the natural microstructures and functions of these creatures will build a bridge between biology and technology, which may provide inspirations for solving today's technological problems.

By summarizing the research findings on the synthesis, properties, and applications of photonic/electrical stimulative intelligent interfacial materials, we were the first to propose the concept of "Binary cooperative complementary micro/nanoscale interfacial materials". According to this design idea, the contact and coupling of heterogeneous materials will result in novel properties on the surface or interface of materials, which may create new functional materials and devices. For example, investigations on superamphiphilic surfaces (the material's surface is both superhydrophilic and superoleophilic), super-amphiphobic surfaces (the material's surface is both superhydrophobic and superoleophobic), and smart switchable superhydrophobic/ superhydrophilic materials will have important applications in daily life, environmental protection and a good number of other domains.

This book devotes to give a complementary introduction about biomimetic intelligent micro/nanoscale interfacial materials, paying attention to the intelligent materials with special wettabilities. The first chapter summarizes the definition of intelligent materials, the design ideas and representative examples of biomimetic intelligent micro/nanoscale interfacial materials. The second chapter introduces some natural creatures with special surface properties, including self-cleaning of lotus-leaf, walking-on-water of water strider, walking-on-wall of gecko, water-collection of desert beetle, special structural color of opal, wing of butterfly, feather of peacock and many others. In Chapter 3, we demonstrate the relationship between surface microstructures and special wettabilities in theory. The fourth chapter gives some typical manufacturing methods of biomimic superhy-drophobic surfaces; the fifth chapter introduces intelligent micro/ nanoscale interfacial materials with special wettabilities. In the last and sixth chapter, we give a conclusion of intelligent materials with some personal perspectives in this area.

In this text we have tried to strike a balance between specialty and popular science. This monograph has summarized a great deal of relative literature and our group's research findings on biomimetic intelligent micro/nanoscale interfacial materials in recent years. It can be used not only as a reference book for researchers in areas including chemistry, materials science and biology, but also for science enthusi-asts. We would be very gratified if this book could spur readers' interests in biomimetic intelligent micro/nanoscale interfacial materials.

This book is the collective effort and creative results of our research group. It would not be possible without the help and sup-port from many colleagues. We would like to give our most sincere appreciation to the following colleagues: Prof. Jian Xu, Prof. Zhenzhong Yang, Prof. Lin Li, Prof. Yanlin Song, Prof. Meixiang

Wan, Prof. Yuliang Li, Prof. Yunqi Liu, Prof. Jin Zhai at the Institute of Chemistry, Chinese Academy of Sciences, Prof. Xi Zhang, Prof. Quanshui Zheng, Prof. Xiqiao Feng, Prof. Ziniu Wu at Tsinghua University, Prof. Guangzhao Zhang at the University of Science and Technology of China, Prof. Bai Yang at Jilin University, and Prof. Xianbao Wang at Hubei University. Great thanks to academician Daoben Zhu, academician Chunli Bai, academician Jiannian Yao, academician Zhenhe Tong, academician Qi Wu, academician Hongyuan Chen, and academician Long Jiang of the Chinese Academy of Sciences. Thanks for the hard work of graduates and post-docs of our group: Huanjun Li, Jian Jin, Yingshun Li, Taolei Sun, Xinjian Feng, Shuhong Li, Huan Liu, Chaowei Guo, Mei Li, Shutao Wang, Ying Zhu, Jingxia Wang, Yong Zhao, Xuefeng Gao, Meihua Jin, Yongmei Zheng, Xianbao Wang, Fan Xia, Weiqin Zhu, Hongli Ge, Wenlong Song, and others.

Particular thanks go to teacher Guilan Wang, Dr. Jinming Xi, Dongliang Tian, Nv Wang, Hui Zhang, Dan Wang, Wenfang Yuan, Liang Xu, Tianyi Zhao, Chunling Yu, Jihua Zhang, and Huiqiong Zhou for their efforts on images treatment, literature collection and proofreading of the manuscript.

Lei Jiang and Lin Feng
Beijing, China

About the Authors

Lei Jiang, born in 1965, is currently an academician of the Chinese Academy of Sciences (CAS). He is now a professor at both the Institute of Chemistry, CAS and School of Chemistry and Environment, Beihang University. He received his B.S. degree (1987), M.S. degree (1990) and Ph.D. degree (1994) from Jilin University, China (Tiejin Li's group). He then worked as a postdoctoral fellow in Prof. Akira Fujishima's group at Tokyo University. In 1996, he worked as a senior researcher in Kanagawa Academy of Sciences and Technology. He joined the Institute of Chemistry, CAS as part of the Hundred Talents Program in 1999. His research interests include bioinspired surface and interfacial materials.

Lin Feng, born in 1976, received her Ph.D. degree in Chemistry from the Institute of Chemistry, CAS (Prof. Lei Jiang's group) in 2003. She is now an associate professor in the Department of Chemistry, Tsinghua University, China. Her research interests are biomimetic superhydrophobic nanoscale interfacial materials. She has won many awards such as "Excellent Youth Prize" of Chinese Chemical Society (2004), and "National Excellent Doctor Degree Dissertation" (2005). She was a visiting scholar at University of California, Berkeley (Peidong Yang's group), USA from 2008 to 2009.

Contents

Chapter 1

Summary of Biomimetic Smart Nanoscale Interfacial Materials

In a sense, the progress of human society is based on the development of materials. Materials with excellent performance accelerate the development of human society; the development of society also desires more new materials. In the 21st century, the target of materials sciences has changed from research on relationship between chemical composition, structure and performance of materials to the systemic investigation of the materials, including the fabrication and processing of the materials. It is characterized remarkably by the fact that materials sciences and materials technology have become very closely connected, while the core issue of the development of materials science and technology is to discover and develop novel advanced materials. It has been a guiding concept in many research fields that novel materials and devices can be created by controlling the structures of materials and the peculiar functions of interface generated by contacting and amalgamating heterogeneous materials.

Smart materials are novel composite materials which have rapidly developed since 1990. They have attracted much attention for their wide applications and the evident influences on a variety of fields. For example, in modern medicine, smart materials can be used to prepare artificial muscle, artificial skin, and drug delivery system; in military field, they can be used in the construction of warships to restrict the travel of noise, which is advantageous as it enhances the stealth performance of submarines and warships; in daily life, they can be used to enhance the performance and comfort of cars, and also used to change the color of houses at your pleasure. With the development of surface and interface sciences in mesoscopic scale, it is advantageous to the interaction of many significant subjects such as physics, chemistry, and materials science, and also to the amalgamation of burgeoning sciences such as nanotechnology, biotechnology, and information science, though some new scientific theories still need to be established. In the 21st century, the connotation and fields of smart materials are expanding gradually, the outstanding characteristic of which is the tight combination of basic research and applied research, biomimetic technology and nanotechnology. For instance, self-cleaning materials have been fabricated by mimicking microstructures and properties of lotus leaves; tires with more security have been produced by imitating the function of cat's forepaw pad combined with supple structure and properties of spider web; low-energy-loss airplane coating has been manufactured by mimicking the ribbed microstructure of the scales of a shark; and "smart glass" has been designed by imitating color changing mechanism of animals such as squid and others. Therefore, research on smart materials has attracted extensive interests in many fields, such as the design of structures and the new fabrication strategies.

1.1 Definition of Smart Materials

Smart materials refer to those special materials which are characterized by their intelligence as they can respond appropriately to stimuli from internal and external environment. Specifically, smart materials need to have properties as follows:

(1) can perceive, that is to detect and identify the intensity of stimuli from external (or internal) world, such as electricity, light, heat, stress, strain, chemistry, environment, etc;
(2) can respond to external changes;
(3) can select and control the responses according to the designed ways;
(4) responses are sensitive, timely and appropriate; and
(5) can return to the original state quickly when external stimuli are removed.

The concept of smart materials derives from bio-mimicking. It is to develop some new "live" materials with diverse functions similar to those of organisms. Therefore, smart materials must be able to perceive the stimuli, respond according to stimuli and control the responding behavior, which are three basic factors of smart materials. However, existing materials are usually too simple to meet the requirements of smart materials. Thus smart materials generally refer to smart material systems which are composite materials of two or even more kinds of materials. As a result, studies on the design, manufacturing, processing, property and structural characteristics of smart materials all involve the forelands of materials sciences, which make smart materials the most active fields and the most advanced developing direction of materials sciences, that will play an important role in world economy, society, national security, and the development of science and technology.

Stimuli-responsive smart materials possess perception, response and control functions when stimulated by external fields such as electricity, light, heat, stress, strain, chemistry, nuclear radiation etc. Studies on smart materials and development of smart devices have attracted closed attention of many scientists in the whole world. For example, Hu *et al.*[1] reported that the surface patterns on Poly(N-isopropylacryamine) hydrogel could be controlled and modulated by diverse environmental stimuli. Beebe *et al.*[2] introduced active hydrogel into microfluidic channels to act as valves, such that local flow inside microchannels could be controlled by using external stumili to adjust the expanded and contracted state of hydrogel. This result is very important in the transport of microfluid. Russell[3] reported a surface-responsive material, the response mechanisms and rates of which could be easily manipulated by changing the length, chemical composition, architecture, and topology of the polymer chains. Livage[4] summarized the Stimuli-responsive materials — smart artificial muscles, and highlighted that the sheets made of entangled vanadium oxide nanofibers behaved like artificial muscles that contract reversibly on applying an electrical signal. Researchers from the Institute of Electron Physics, University of Stuttgart of Germany, invented a kind of smart solar clothes, which was a novel chemical fiber smart material using solar energy as energy resource, so that mobile telephone and electronic diary could be directly connected to the clothes.[5]

Biological smart interfacial materials are of great value in recognization and assembly of biomolecules and manipulation of complicated information. Kim and Abbott[6] reported that the specific binding of anti-biotin IgG to rubbed films of BSA functionalized by biotin could be detected by observing the optical appearances of liquid crystals supported onto these substrates. Liquid crystals were uniformly oriented when no antibody molecules were absorbed, while the alignment of liquid crystals was disrupted when complementary antibody molecules combined with antigens connected to BSA. Mason *et al.*[7] revealed that the excellent ability of identifying the sound sources of *Ormia* attributed to the special structures of its tympana(eardrums). Thereby, researchers prepared a tympana prototype which was suitable in ultrasonic range by using silicon to simulate the special tympana of *Ormia*. Furthermore, they invented a novel audiphone which could identify the direction of sound sources and enhance the effect of audiphones. This project had been recommended to the Ministry of Health and Department of Welfare of America as a demonstration project of results transforming by National Institutes of Health, USA in June, 2004.

Scientists from our country have done a lot of work with worldwide influence on fabrication, performances and theoretical studies of smart materials. Our group proposed a novel concept of "Binary Cooperative Complementary Interfacial Materials"[8] for the first time to guide the design and synthesis of new materials. We reported that the reversible transition between superhydrophobicity and superhydrophilicity on surfaces of nanoscale materials could be realized by stimulating with light, heat, etc.[9,10]

Professor Gu and co-workers from Southeast University of China has achieved great advances in studies on structural color and photonic crystals. Through thermo-processing method, they fabricated reverse opal photonic crystals by assembling composite nanospheres of polystyrene and silicon dioxide. Interestingly, the color of materials could be controlled conveniently by changing the structural periods.[11–13]

Professor Yang and co-workers from Jilin University of China reported that the colloidal crystals on PDMS stamp surface could serve as "ink" and be transferred onto the polymer-coated solid substrates by a modified microcontact-printing (μcp) technique. Polymer coatings were introduced to ensure effective interactions between colloidal crystals and substrates.[14–16]

Professor Yu and co-workers from the University of Science and Technology of China have done a lot of innovated studies on fabrication of biomimetic multiscale structures. They fabricated many kinds of inorganic nanoscale materials by chemical synthesis combined with imitation of the process of biomineralization. They reported that helices of achiral inorganic nanocrystals could be fabricated in the presence of a racemic polymer at normal temperature and pressure, which provides theoretical and experimental basis to build inorganic materials with complex special structures.[17–20]

Academician Yao and co-workers from the Institute of Chemistry, Chinese Academy of Sciences (CAS) systematically studied novel optical functional materials from basic research to their applications, and have achieved a lot of innovative results.[21–24] They introduced the concept of supramolecular chemistry and molecular assembly technology into the design and building of inorganic photochromic materials for the first time, and photochromic ultra-thin films with nanoscale thickness which is much lower than the traditional microscale ones had been built. Moreover, the photochromic properties of these films could be regulated by changing the chemical micro-environments of building blocks, and the response time had been shortened from milliseconds to nanoseconds.

Liu *et al.* from the Center for National Nanosciences and Nanotechnology made important breakthrough in studies on building molecular devices by biological process. They built a pH-driven nucleic acid motor and explored its application, and the work provided theoretical and experimental basis to building novel artificial muscles.[25–27]

Professor Jiang and co-workers from the Institute of Chemistry of CAS found out that functionalized Polydiacetylene vesicles could identify *Escherichia coli*, antibody-antigen, protein, etc. rapidly and easily as bionic cell identification devices, which may be further used to assemble bio-chips.[28,29]

Professor Chen and Prof. Xia from Nanjing University have done many groundbreaking works in the fields of nano-biological electrochemical analysis and microfluidic analysis system.[30,31] They used many chemical and physical means, such as supramolecular self-assembly, covalent bonding, electrochemical polymerization and adsorption, combined with methods of biology, physics and materials sciences, to design and assemble a series of novel, reliable nanoscale molecular electronic devices with recognition properties. They also developed modern analysis methods and detection technology, established new principles in the detection of proteins and enzymes, as well as microfluidic analysis systems, and proposed a number of important biomolecular sensors.

1.2 Designing Concept of Bioinspired Smart Interfacial Materials

The design concept of bionic smart interfacial materials has five levels: (i) Designing concept of biological intellectualization; (ii) Coordination of multiscale structure effects; (iii) Design and synthesis of target smart stimuli-responsive molecules; (iv) Design of heterogeneous interfaces; and (v) Cooperative complimentary effect of weak interactions and bistable states.

1.2.1 *The bioinspired concept*

Lives in nature have achieved all processes of intelligence through more than four billion years of evolution. Therefore, learning from nature should be the eternal theme for the development of smart materials.[32] Inspired by organisms in nature, we can mimic special functions of organisms from certain aspect to design smart materials, and we also combine the microscopic and macroscopic world. As a result, we may surpass the nature in some aspect eventually.[33]

1.2.1.1 *Scientific problems related to intelligence in life*

Scientific problems related to intelligence in life exist everywhere, and the organs involved can be found in all parts of the body. Both the responses to external stimuli and various life phenomena are based on the changes of intermolecular or intramolecular interactions of bio-macromolecules. For example, the organs involved in collecting, transferring, and storing information include nerve, brain, eyes, nose, tongue, ears, skin, etc.; those involved in loading, transporting, and transferring matters include haematocytes, vessels, alveoli, etc.; those involved in transforming, transferring, and storing energy include cardiac muscles and other muscles. In addition, some organisms also have organs with particular functions, such as antennae and compound eyes of some insects. Some special functions, such as walking on surface of water, breathing under water, collection of freshwater, and photosynthesis, also result from the intellectualization of organisms.

All of the senses of living organisms, including olfaction, vision, hearing, etc., are the results of transfer of external stimuli (smell, light, sound, etc.). These stimuli form a certain interface between the environment and the nervous system of organisms, making the organisms respond to the special stimulus.

Olfaction

How does olfaction come into being? The mechanism has been considered to be the most difficult puzzle among the different senses of human beings. Two American scientists, L. Buck and R. Axel, who were awarded the Nobel Prize for Physiology or Medicine, discovered that the generation of olfaction was based on the responsive changes of biomolecules to external stimuli, as a result of which human beings could distinguish and remember about ten thousand different odors.[34] First, the odorant molecules in air combine with odorant receptors which are distributed in olfactory receptor cells of nasal epithelium. When odorant receptors are exited by odorant molecules, nerve signals are generated and transported to a microstructure called "glomerulus" in the olfactory bulb of brain, and then transported to other areas of brain. Therefore, human beings can perceive certain odor consciously, and remember this smell at another instance.

Hearing

Ear consists of concha, middle ear, and inner ear. When a sound starts, it is transferred to tympanum through concha first, and then transferred to inner ear which was filled with liquid through middle ear, and finally transformed into corresponding signal by helical cochlea (Fig. 1.1). As a part of inner ear, cochlea is a hollow helical tube which looks like a snail, and a special device which generates hearing, called basilar membrane in medicine, exists in it. There are twenty-four thousand nerve fibers on the basilar membrane, and they can only be seen using scanning electron microscope (SEM). These nerve fibers are not of the same length and have many hearing cells loaded on them. When basilar membrane receives acoustic vibration transferred from middle ear, it transfers the vibration to hearing cell immediately to generate nerve impulses, which are transported to brain, subsequently forming hearing. That is the whole process of how we can hear the sound from external environment.

Vision

Eyes of insects consist of single eye and compound eye, through which insect can respond to changes of external light. A compound eye consists of thousands of helical small eyes [Fig. 1.2(a)].[36] The structure of a small eye is ingenious; it has a helical convex lens on the outside called cornea to collect light, a tapered crystal called crystalline cone in the middle, and a vision nerve called rhabdom at the innermost. From SEM image, it can be seen that many tubular microvilli, each 1–2 μm long and ~60 nm in diameter, are tightly packed together forming a 100-μm–long rhabdomere [Fig. 1.2(b)]. When stimulated by light, light point is transferred to nerve photoreceptor to generate a "point image", many point images of small eyes which interact with each other constitute an "image".[37]

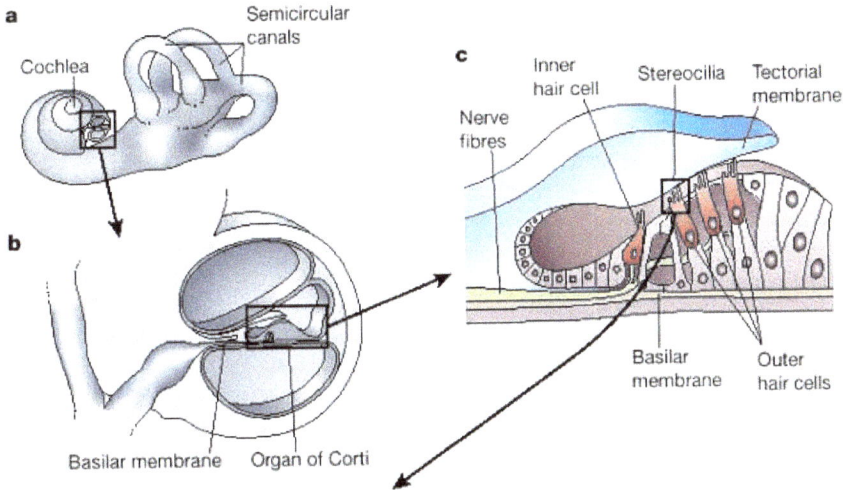

Fig. 1.1 The mechanism of hearing. (a) Skeleton map of part of inner ear. (b) Cross-section skeleton map of cochlea. (c) Sound causes the vibration of basilar membrane. (d) SEM image of nerve fibers.[35] (Reprinted by permission from Macmillan Publishers Ltd: *Nature*, copyright 2001.)

All the above-mentioned senses are the most familiar to human beings (olfaction, hearing, and vision) and they illustrate the intellectualization of special functions in organisms. It should be noted that microstructure is vital for the accomplishment of these special

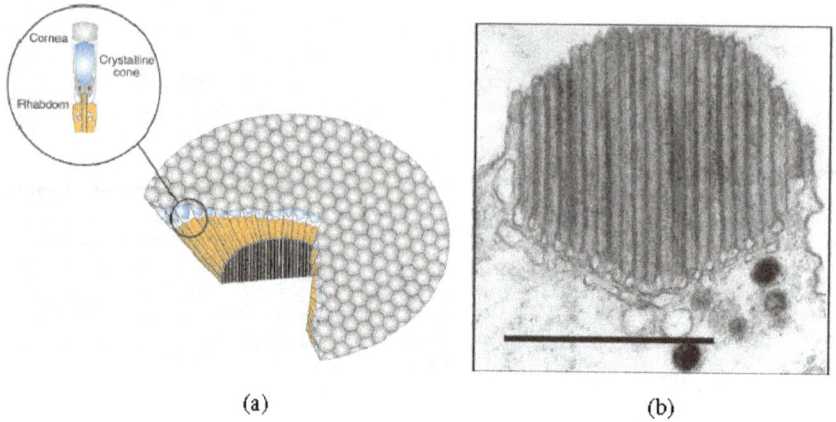

Fig. 1.2   (a) Schematics of compound eye of insect.[36] (b) Electron micrograph of one rhabdomere in compound eye of *Drosophila* (scale bar: 1 μm).[37] (Reprinted with permission from AAAS; Reprinted by permission from Macmillan Publishers Ltd: *Nature*, copyright 2001.)

functions, such as glomerulus in olfactory bulb of brain, nerve fibers on the basilar membrane of cochlea, and the rhabdomere in photoreceptive membrane of insect compound eye; all of them have micro/nano-composite structures and also can specifically respond to external stimuli. Therefore, it is inspired that most of intellectualization of special functions in organisms are closely related to the elaborate microscopic structures. If we can learn from nature, the unity of microstructure and macroscopic performances should be achieved inevitably by recognizing and mimicking the relationship between special microstructures and excellent performance of organisms.

1.2.1.2 *The unity of microstructure and macroscopic performance in nature*

Generally, the macroscopic performance of materials is codetermined by chemical composition and structure. As the chemical component are inherent in a material, therefore studies on the structure of material become very important. The structure of materials can be divided into macroscopic structure, mesoscopic (microscopic) structure, and molecular structure according to the levels of scale.

Generally, macroscopic structure can be seen by naked eye, with diameter above millimeter, centimeter, decimeter, and even meter. Accordingly, the simulation of macroscopic structure is to simulate the macroscopic appearance of subject. For example, the invention of modern aircraft is a result from the initial attempt to mimic the flying postures of birds. In fact, human beings made a mistake in trying to fly like a bird: they wanted to fly with their own strength like insects and birds did. Many machines which could fly had been made in ancient China, and Leonardo da Vinci in Renaissance had already drawn the blueprints of ornithopter and helicopters. A real flying machine must be based on physics, mathematics, and materials sciences. With the development of science, airplane industry has developed rapidly in recent 100 years after modern aircraft appeared. The speed, height, and consecutive flying ability far exceed the birds. Actually, the flying principle of airplane is absolutely different from that of birds. It depends on aerodynamics and hydrodynamics, which requires more strict mathematical calculation in theory, and more excellent technology and materials in the manufacturing process. Inspired from the flight of birds at first, humans tried to mimic birds, and finally airplane was made. This process reflects precisely the whole biomimic intellectualization process which starts from learning from nature, then mimicking nature, and finally surpassing nature (Fig. 1.3).

The simulation of molecular and atomic structure of materials is on the molecular-level. For example, the biochip technique which has aroused widespread attention in recent years is just this type of simulation. Biochip technique including cell chips, protein chips (biomolecules chips) and gene chips (DNA chips) has been the frontier technology of biomedical engineering, which is of great significance in theory and application.

Here, we focus on the mesoscopic structures in organisms, including microstructures and nanostructures. It is disclosed that when the structure size of materials is comparable to some physical constants, including the electron mean free path, electron/hole diffusion radius, molecule

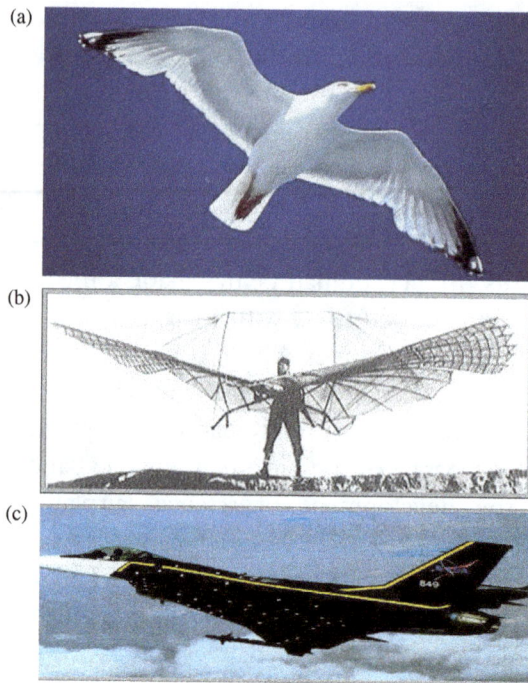

Fig. 1.3   Modern aircraft (b) and airplane (c) which mimic the posture of flying bird (a).

dipole interaction radius, exciton Bohr radius, and wavelength of light, etc., some new effects can be generated. We will illustrate some new physical properties resulting from the microstructure of materials next.

Eigler and co-workers[38] built an artificial atom structure to comfine electrons in order to study their quantum confinement behaviors (see Fig. 1.4). They individually positioned 48 iron atoms onto copper (111) surface to assemble a circular closed corral with radius 7.13 nm using the tip of scanning tunneling microscope (STM). Since surface state electrons on copper (111) surface were strongly scattered by Fe adatoms, part of electrons could be confined in the corral to form discrete quantum state, which provides a mini "lab" for scientists to study quantum phenomena. Furthermore, they enlarged the quantum corral to 78 iron atoms forming an elliptical artificial atom

Fig. 1.4   Corral of iron atoms on copper (111) surface. (From Ref. 38. Reprinted with permission from AAAS.)

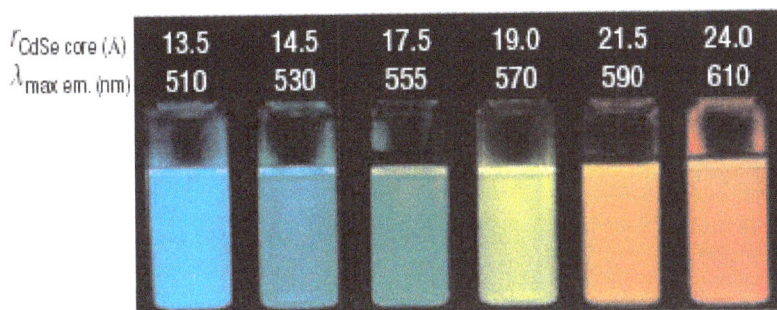

| CdSe core (Å) | 13.5 | 14.5 | 17.5 | 19.0 | 21.5 | 24.0 |
| λ max em. (nm) | 510 | 530 | 555 | 570 | 590 | 610 |

Fig. 1.5   Photo demonstrating the size-tunable fluorescence properties of the six QD with different size (1 Å = 0.1 nm).[40] (Reprinted by permission from Macmillan Publishers Ltd: *Nature Materials*, copyright 2005.)

structure.[39] They discovered that electrons in the corral fluctuated symmetrically encircling the two centers of ellipse, which is similar to the electronic wave field of biatomic molecules (hydrogen molecule, oxygen molecule). Therefore, they extended the confinement environment of quantum corral from the atomic to molecular field, building a molecule corral.

In another example, some physical properties of quantum dots (QD) will change if their sizes are different. Figure 1.5 indicates that with the radius of CdSe QD becoming bigger, the fluorescent color of them changes a lot when exited by 365 nm UV light.[40]

Special surface properties

The special micro/nanostructure of organisms can genarate special surface properties (Fig. 1.6). For example, the self-cleaning property of some leaves (lotus leaf) and inset wings (cicada, dragonfly, butterfly, etc.)[41–45] results from the special microstructures on the surface. On these surfaces, there is an air layer between the solid and water which can prevent water from wetting the surface, showing superhydrophobic property.[46,47] Theoretical studies show that the critical value of distance between two microstructures to generate superhydrophobicity is 100 nm, and this was confirmed by studies on superhydrophobic property of mosquito's compound eye. Moreover, some surfaces of organisms, such as rice leaves, feathers of duck and goose, and butterfly wings, etc., show anisotropic sliding property of water droplet. That is the rolling behavior of water droplet on parallel and vertical direction is quite different, which is determined by the arrangement of surface microstructure.[48,49] In addition, the antireflection property of some insect wings and eyes also results from the special microstructures of the surfaces.[50,51] Other specific functions of organisms — water striders can walk on water[52]; gecko, spiders, flies, and a kind of beetle (*Hemisphaerota cyanea*) are able to adsorb on smooth walls and glass, or even on the ceiling[53–57]; desert beetles can collect the fog[58]; beetles can change their color[59] — are also closely related to the special surface microstructures.

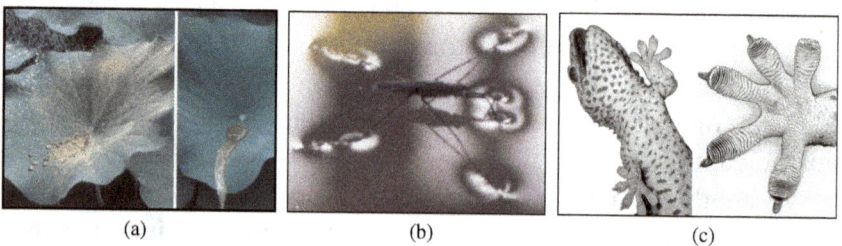

Fig. 1.6 Creatures with special surface properties. (a) Lotus leaf with self-cleaning property. (b) Water strider walking on water surface. (c) Gecko's feet with high adhesion force.

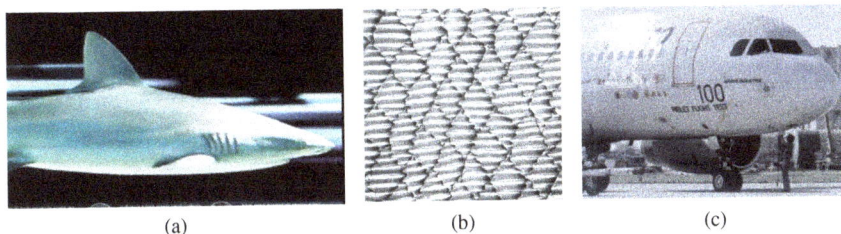

(a)   (b)   (c)

Fig. 1.7 (a) and (b): The riblets on shark skin provided inspiration for modeling studies of the drag reduction they confer. (c) Trials on an aircraft coated with a plastic film with this same microscopic texture.[60] (Reprinted by permission from Macmillan Publishers Ltd: *Nature*, copyright 1999.)

How can shark swim very fast in water? The secret lies in the ribbed texture [Fig. 1.7(a) and (b)] of the scales on its skin, which can provide aerodynamic efficiency relative to a smooth surface because of the way that the corrugations affect the viscous boundary layer of the water.[60] Inspired from this structure, drag-reducing coatings on airplane have been developed. A transparent plastic film with the same microscopic texture with shark skin helps to reduce aircraft drag by up to 8% — representing a fuel saving of about 1.5% [Fig. 1.7(c)]. Moreover, this textured film provides a self-cleaning function — the rough surface confers low wettability. When used as coating on ships, it can enhance the speed thus save energy by reducing the adhesion of marine organisms onto the hulk. Recently, the novel "shark skin" swimming suit has been favored by people.

Structural color of organisms

Organisms in nature have a variety of colors which are generated by the long years' evolution.[61] Color can be divided into chemical color and physical color according to the generating principle. Chemical colors are always generated by absorption of light by pigments in the body, while the reflection, scattering, interference or diffraction of light by submicron structures in organisms lead to physical color. Since the physical color is related to structures but not pigments, it is also called structural color. The colors on peacock feathers, some

Fig. 1.8 Organisms with structural color and their microstructures. (a) Peacock feathers. (b) Butterfly wings. (c) Natural opal. (Reprinted with permission.)

beetles, butterfly wings, natural opals, as well as pearls are all of this structural color (Fig. 1.8).[62–69] Color can also be divided into static and dynamic color according to whether it can change. Static color is non-random controlled color formed in the growth process of organisms, while dynamic color refers to the fact that color can change with the surrounding environment, such as the cases in chameleon, squid and octopus. It can be said that the color of animals is affected by various factors in environment, in particular by environmental background color, sunlight and temperature. If animals stay in an environment, with certain background color for a long time, the so-called "sematic color" can be formed. For example, if a crayfish is put in a white cup, it turns to white; when it is moved to a black cup, it turns black immediately. Some animals not only can change their color according to the background color, but also can respond to the patterns of background. For instance, if a halibut is put in a stripe-patterned environment, stripe patterns will appear on its body; if the environment is changed to dot pattern, dot patterns will appear on its body.

Most organisms with structural color or dynamic color contain photonic crystals,[70–74] but the molecular structure, micro/nanostructure, periodic structure and function of these photonic crystals are still unknown. Thus studies on color of organisms, particularly on the photonic crystals related to structural color and dynamic color, should play an important guiding role in the development of new generation of photonic materials, storage materials and display materials. Studying biological photonic materials should be an important approach to design and develop novel photonic materials. These bio-inspired photonic materials should find applications in many fields, such as electronics, environment, and materials, and so on.

Perfect biomineralization

Biomineralization materials are natural composite materials of bioceramics and biomacromolecules which are synthesized by the participation of life system, such as teeth and bones of human and animal, shells of mollusks, and so on. Compared to common natural materials and synthetic materials, biomineralization materials have special high-order structures and assembly methods, thereby possessing many almost perfect properties, such as high strength, excellent fracture toughness and wear resistance.[75–77]

There are many kinds of minerals in organisms, among which calcium carbonate is the most familiar one. Although calcium carbonate itself is very weak and fragile, its strength can be greatly enhanced by designing over multiple length scale. For example, the nacre of mollusk shells consists of more than 95% calcium carbonate,[78] but its fracture toughness is 3000 times higher than pure calcium carbonate. The attractive combination of high strength and excellent toughness stems from the fact that the organic matrices excreted by cells interact with inorganic crystals, forming advanced self-assembly structures.

Up to 95% of the hard inner layer in mollusk shells is fragile calcium carbonate, while the other 5% is biopolymer with good flexibility. They form a "brick and mortar" structure. Millions of ceramics-phase calcium carbonate platelets which are in the range of thousands of nanometers stack with each other like coins, and a thin layer of biopolymer encases all ceramics platelets on all sides. Moreover, each ceramics platelet has its own complex nanostructure.

Calcium phosphate plays an important role in the stability and hardness of biological hard tissue. For example, in enamel of rat incisor, ordered rods of calcium phosphate crystals are arranged in parallel along the long axis forming crystals bundle. The crystals bundles are then arranged in parallel forming enamel rod, and the enamel rods are arranged in parallel finally forming enamel. The extension direction of long axis in enamel rod is vertical to the surface of teeth. The spaces between enamel rods and spaces between crystals are filled with organic matrix. This highly ordered assembly makes the closed packing of minerals that account for 95% of the total quality available, leading to good mechanical properties.[79]

The iron-containing minerals (such as magnetite) in some organisms have special functions.[80] Many types of migratory birds, fish and animals such as sea turtles can use geomagnetic field to navigate and cruise. Magnetotactic bacteria can use the orderly arranged iron oxide of single magnetic domain size in its body to identify mobile directions [Fig. 1.9(a)]; and the iron oxide in the head of tuna and salmon can be used as a bio-magnetic navigation compass.

Some algae, fish scales, spicules of animal, and sponges have silicon-containing minerals.[81] Figure 1.9(b) indicates the layered structure of silicon dioxide in sponge spicules, which shows good optical and mechanical performance.

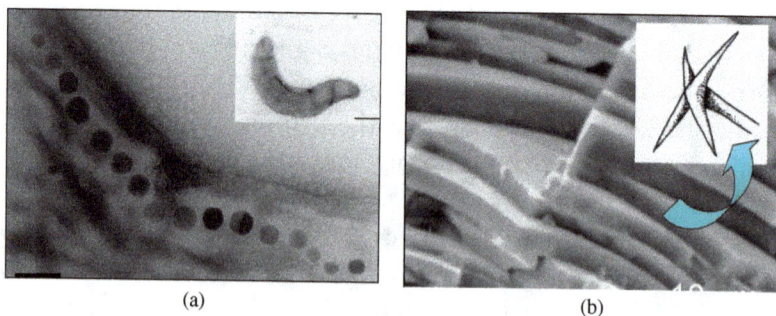

(a)   (b)

Fig. 1.9   Microstructures of biomineralization in nature. (a) The orderly arranged iron oxide in Magnetotactic bacteria.[80] (b) The layer structure of silicon dioxide in sponge spicule.[81] (Reprinted with permission.)

After the designing methods of biomineralization materials and functions of materials with the shortest size have been understood, human beings learn to build biomimic composite materials with high strength.[82–87] Using the mechanism of biomineralization to synthesize inorganic materials, that is using organic assemblies as templates to control the synthesis of inorganic materials, we can fabricate special microstructured materials which exhibit excellent physical and chemical performance. This is the so-called biomimic synthesis of materials or biomimic morphology fabrication method. For example, if we use materials with much better strength performance than materials that are used to construct shells, but adopt the same designing methods as shells, composite materials with much better strength performance could be fabricated, which may find applications in armor composite materials and aerofoil composite materials.

Advanced optical system

Polar bear is white from the appearance, but actually its skin is dark green.[88a] SEM image indicates that polar bear's fur is made of colorless hollow tubes, and the diameters of these tubes become bigger and bigger gradually from the top to the root [Fig. 1.10(a) and (b)]. The reason why the fur looks white is the rough inner surface of the thin tubes which can cause the diffuse reflection of light. When taking pictures of polar bears under natural light, clear images can be obtained; when infrared camera is used, only the face can be shown in the picture [Fig. 1.10(c) and (d)]. It indicates that the polar bear's fur has an excellent ability to absorb infrared light, thus exhibits excellent thermal insulation performance. This unique structure and color of polar bear result from its long-term struggle against cold. Therefore, scientists envisaged that if artificial hollow fibers with thermal insulation and energy-saving performance could be fabricated by mimicking the structure of polar bear's fur, the utilization of solar energy should be greatly improved, which is immeasurably beneficial to human beings.

Similar to polar bear's fur, feathers of many birds also have this extremely elaborate multichannel and multichamber tubular

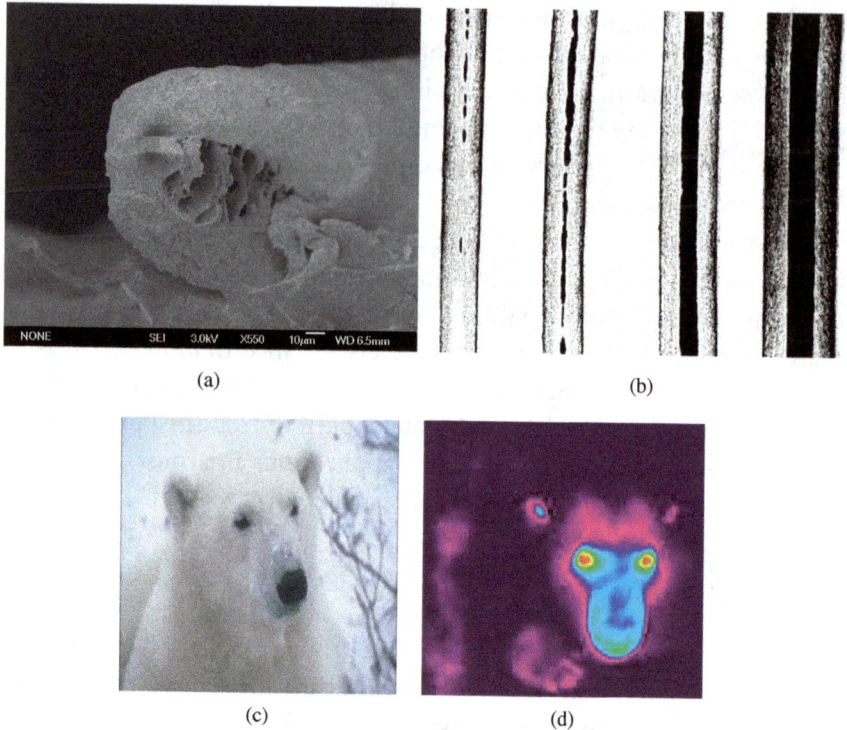

Fig. 1.10 Microstructure of the polar bear's hair: (a) Cross section image; (b) Vertical image.[88a] Pictures of polar bear under different light conditions: (c) Natural light; (d) Infrared night vision.

structure. This delicate structure can reduce the weight of feathers and serve as heat-shields on the basis of maintaining enough mechanical strength, so birds can fly in the sky with light feathers. Polar bears can survive in the extremely formidable polar environment also because of the multichamber structure of their fur. Many interests have been attracted to the fabrication of multichannel tubular structures for their outstanding performance, however, how to fabricate such multichannel micro/nanotubes is still a challenge. Recently, our group made a breakthrough in this area.[88b] We described a multifluid compound-jet electrospinning technique that could fabricate biomimic hierarchical multichannel TiO_2 microtubes in a facile way, and also the channel number can be controlled with ease. As a matter of

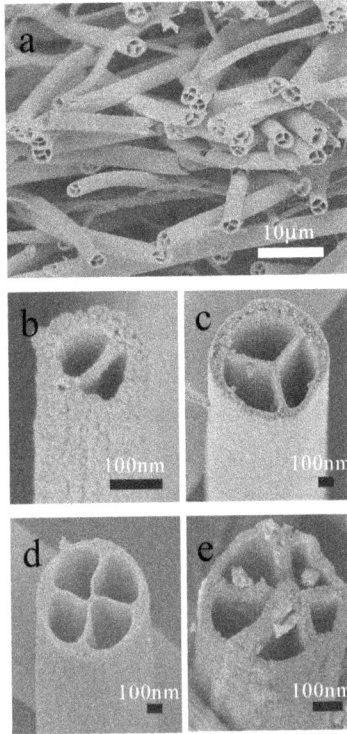

Fig. 1.11   SEM images of multichannel TiO_2 microtubes fabricated by multifluid compound-jet electrospinning technique. (a) SEM image of TiO_2 tri-channel microtube in large areas. Scale bar: 10 μm. (b)–(e): Magnified images of TiO_2 microtube with 2, 3, 4, 5 channels, respectively. Scale bar: 100 nm. (Reprinted with permission from Ref. 88(b). Copyright 2007 American Chemical Society.)

fact, tubes with two to five channels have been successfully fabricated (Fig. 1.11). These multichannel microtubes may find applications in many fields, such as in biomimic super-lightweight thermo-insulated textiles, high efficient filtering mesh, high efficient catalysts, and vessels for macro/nanofluid devices.

Generally, photosensitivity in most echinoderms attributes to "diffuse" dermal receptors. However, the latest research shows that the reason why brittlestarts (*Ophiocoma wendtii*) can change color from day to night is that the certain single calcite crystals used by brittlestars for skeletal construction are also components of specialized

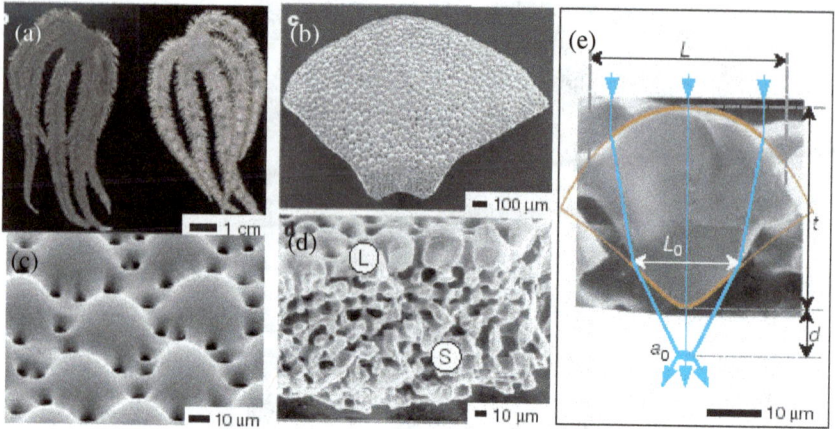

Fig. 1.12 Photosensitivity of *O. wendtii* brittlestars. (a) Ligtsensitive species *O. wendtii* brittlestars changes colour markedly from day (left) to night (right). (b)–(d): SEM of *O. wendtii* photoreceptor. (e) High-magnification SEM of the cross-section of an individual lens in *O. wendtii*.[89] (Reprinted by permission from Macmillan Publishers Ltd: *Nature*, copyright 2001.)

photosensory organs, conceivably with the function of a compound eye.[89] The analysis of arm ossicles in different kind of brittlestars showed that in light-sensitive species, the periphery of the labyrinth calcitic skeleton extends into a regular array of spherical microstructures that have a characteristic double-lens design [Fig. 1.12(b)]. SEM of the cross-section of a fractured dorsal arm plate (DAP) from *Ophiocoma wendtii* shows that the peripheral layer consists of calcitic stereom (S) and the enlarged lens structures (L) [Fig. 1.12(c) and (d)]. Further investigation of the cross-section of an individual lens shows that the operational part of the calcitic lens closely matches the profile of the compensated lens [Fig. 1.12(e)]. These results represent an example of a multifunctional biomaterial that fulfills both mechanical and optical functions.

Another special organism is a beetle named *Melanophila acuminate*, which can detect forest fires from 80 kilometers away.[90] Researches show that affected by temperature of forest fire, the special pit organs of larva can develop into infrared sensors during the growing process. These organs have about 50–100 sensors each with diameter 15 mm, and they can absorb infrared light, resulting in the expansion of surface organs. As a result, beetles can act properly to forest fire.

Bioinspired design of advanced materials and optical systems has become an emerging issue for scientist.[91] By simulating the advanced optical system of organisms and choosing practical polymers with desired optical properties, complex artificial systems can be produced quickly by soft lithography and three-dimensional microscale processing technology. For example, inspired from photosensitive brittlestars, microfluidic double-lens has been designed [Fig. 1.13].[92]

1.2.2 *Design and features of multiscale structures*

The multiscale effect of molecular structure, nanostructure and microstructure of material system is the inherent point to create new functions of materials (Fig. 1.14). The molecule is the most basic structural unit embodying the functions of materials, and the diversity of molecular structures determines the diverse functions and properties of materials. Micro/nanostructures are mesoscopic systems built according to certain laws using micro/nanoscale matter units, including one-dimensional, two-dimensional and three-dimensional structures. Due to special properties of nanostructure, there exist quantum size effect, small size effect, surface effect, etc.; new effects caused by the combination of nanostructures also exist, including

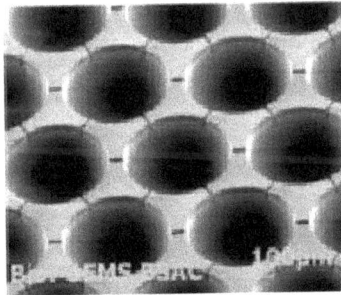

Fig. 1.13   SEM image of biomimic microfluid double-lens.[92] (Reprinted with permission.)

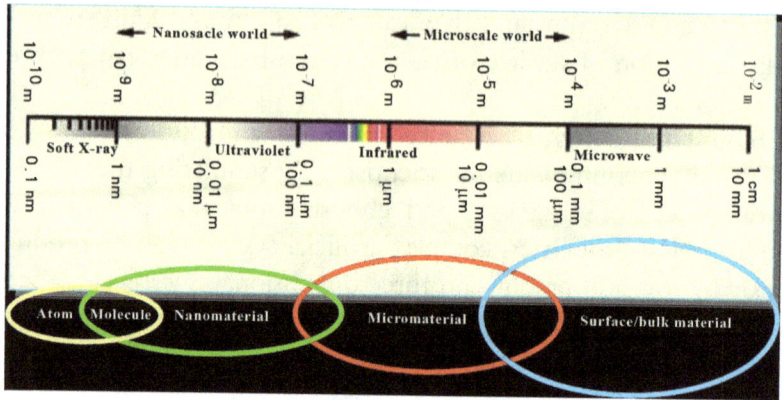

Fig. 1.14 Multiscale effects of material systems.

quantum coupling effect and synergistic effect. Therefore, the properties of nanostructures can be controlled easily through external fields (e.g. light, electric field, magnetic field). In conclusion, the design of material systems is not only confined to the bulk materials; furthermore, people can control the structures of materials over multiscale scope. That is, materials can exhibit peculiar macroscopic properties when the structures extend from the nanoworld to the macroworld.

1.2.2.1 *Multiscale structures of biological systems*

Many biological systems with outstanding mechanical properties have self-assembled hierachical structures over multiple scales, such as the shell structure mentioned above, bone and teeth, etc. Other systems, such as spider silk, wood, skin, tendon, also have multiscale structures, which lead to outstanding performance and versatility of biological systems. For instance, the complex and delicate multiscale hierarchical structures of bones lead to excellent mechanical properties. M. M. Stevens *et al.*[93] reported that the hierachical ordered structures in different scales of biological materials play an important role in cell culturing environment (Fig. 1.15). The multiscale structure of spider's silk confers excellent mechanical strength and good

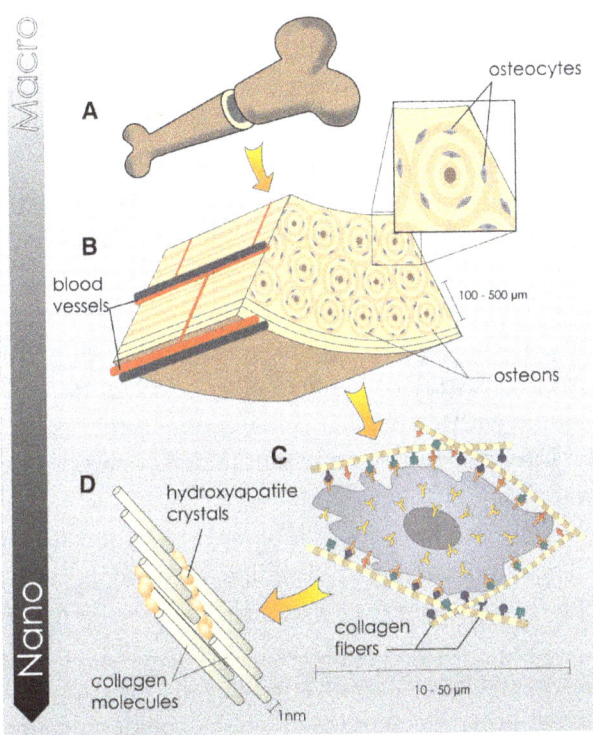

Fig. 1.15 Hierarchical organization of bone over different length scales. (A–D): The outer layer of bone is a strong calcified compact layer, which comprises many cylindrical osteons; the resident cells are coated in a forest of cell membrane receptors that respond to specific binding sites and the well-defined nanoarchitecture of the surrounding extracellular matrix. (From Ref. 93. Reprinted with permission from AAAS.)

flexibility, and this gives an example of using materials with the lightest weight to obtain the highest mechanical properties, e.g. 180 μg such material can be woven into a network of 100 cm² in order to catch flying insects.

On the other hand, these multiscale structures of biological materials also have special deformation characteristics. For example, controlled by multiscale structures, the deformation properties of tendon, wood, and bone exhibit special toughness and hardness. For bones, the multiscale concept lies in two aspects: (1) multiscale structures, that is, collagen molecules are linked by calcified inorganic particles from top to tail forming collagen fibers, and the spaces among collagen fibers are filled with interfacial polymer interfibrous matrix; (2) multiscale responses of deformation, that is, the smaller the inorganic particles' size in mineralized fibers, the greater the intensity of fibers; the polymer interfibrous matrix, similar to hydrogel, makes the tangential displacement of fibers possible, increasing the toughness of the material. Another example is wood (Fig. 1.16),[94,95] cellulose microfibrils twist helically against the tube-like inner wall of cells, allowing it to respond to stress liking a spring. Affected by polymer matrix consisting of hemicellulose and lignin, the hard cellulose fibers couple with each other in tangential direction, and the soft matrix at the interface controls the whole deformation process of the wood cell wall.

Nature has designed a series of interfacial polymers with interesting properties. These polymers can combine with fibers or layered film firmly to enhance the strength of materials, and also form hydrogel-like interfibrous matrix which can increase the flexibility and plastic shear deformation of the fibers, thus increasing the toughness of structures. These interfacial polymers should be good candidates to develop biomimic structures with strong strength and toughness.

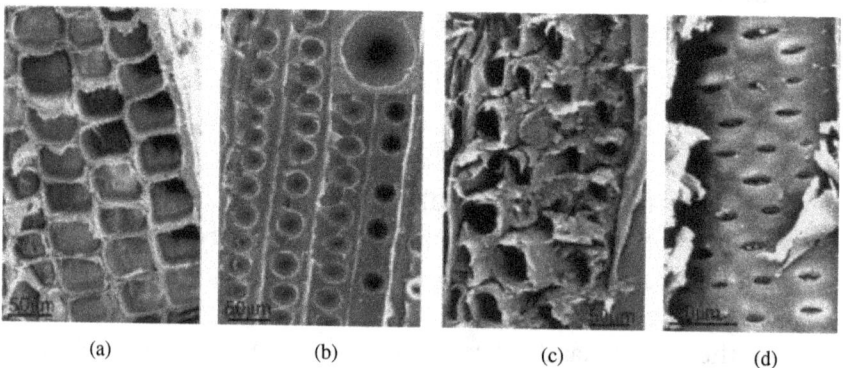

| (a) | (b) | (c) | (d) |

Fig. 1.16   Microstructure and multiscale structure in wood. (a) and (b): Cross-section and radial SEM images of pine. (c) and (d): Cross-section and radial SEM images of poplar.[94] (Copyright Wiley–VCH Verlag GmbH & Co. KGaA. Reproduced with permission.)

1.2.2.2 *Design of multiscale nanostructure*

Multiscale nanostructures can be built in diverse ways, among which the self-assembly technology is the most commonly-used method.[96] In self-assembly technology, building blocks (molecular or nanomaterials, micron or larger scale materials) organize spontaneously to form ordered structures,[97] i.e. a number of individuals associate with one another spontaneously and integrate into a close and ordered entity. In self-assembly process, building blocks organize or aggregate spontaneously because of the non-covalent bond interactions, forming a stable structure with regular geometric appearance. Self-assembly process is not just a simple stacking process of weak interactions between large numbers of atoms, ions and molecules, but is a complex and cooperative process of all forces. The molecular recognition in interface is crucial in this process; its achievement depends on the characteristics of building blocks, such as surface morphology, shape, surface functional groups, surface potential, and so on. The final organized structure has the lowest free energy. Research shows that the internal driving forces are the key point for self-assembly, including non-covalent interactions that act only at the molecular level, such as van der Waals force,[98,99] hydrogen bond,[100,101] electrostatic force,[102,103] and forces that act on the larger range, such as surface tension,[104] capillary force,[105] and so on.

Self-assembly systems over different scales from molecules to macroscopic objects have aroused much interest for a long time. Molecular self-assembly refers to the technology using non-bonded forces to organize molecules spontaneously forming stable supramolecular aggregates, among which the self-assembly of encoded protein of DNA double-helix structure and its complex derivatives has attracted much more attention. Recently, P. W. K. Rothemund reported the self-assembly of DNA molecules to create nanoscale shapes and patterns in a "down-top" way[106] by folding long, single-stranded DNA molecules into arbitrary two-dimensional shapes (Fig. 1.17). G. M. Whitesides *et al.* from Harvard University[107–112] did a lot of representative work in the self-assembly of macroscopic objects. Their researches provided a very simple and effective way to build micron,

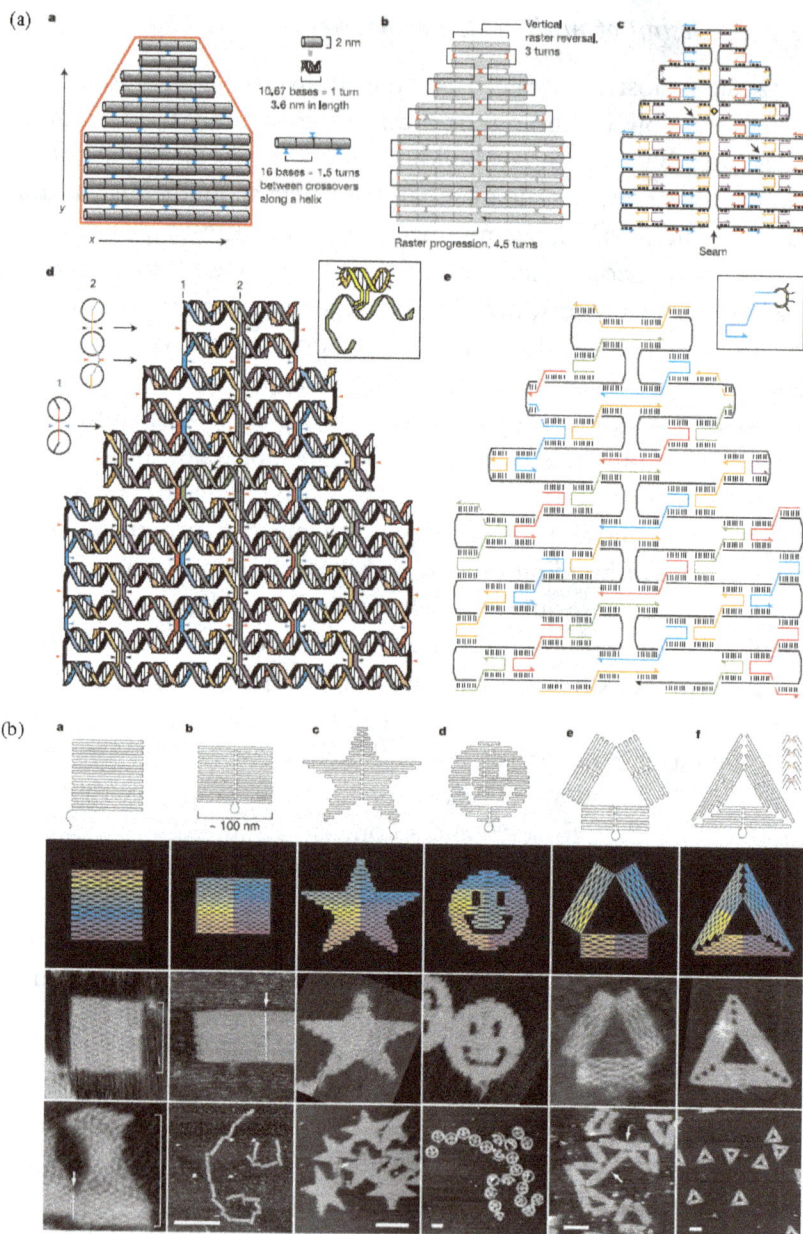

Fig. 1.17   Two-dimensional structures with diverse patterns can be obtained at wish by self-assembly of artificial DNA fragments with designed components.[106] (a) Building method. (b) Surfaces with a variety of patterns. (Reprinted by permission from Macmillan Publishers Ltd: *Nature*, copyright 2006.)

centimeter, and even larger-scale aggregates with regular geometric appearance.

The mesoscopic self-assembly between molecules and macroscopic objects is an emerging research hot spot in recent years. Using nanomaterials as building blocks to self-assemble diverse hierachically ordered organizations is of considerable significance in scientific research. It not only provides an effective way for us to organize functional materials into a highly ordered structure in an expected way, but also provides new opportunities for research on microdevices.[113–115]

Self-assembly using nanomaterials as building blocks can be divided mainly into the self-assembly of nanoparticles and self-assembly of one-dimensional materials. In addition, the simulation of biomineralization process to build multiscale structures is also an important aspect of this field, in which many scientists have done outstanding works.

Self-assembly of nanoparticles

In the self-assembly of nanoparticles, colloidal nanoparticles (metallic, non-metallic) stabilized by monolayer of molecules are ideal candidates to self-assemble various hierachical organizations. The most typical example is the self-assembly of gold or silver nanoparticles modified with thiol monolayer[116,117] induced by hydrogen bonds between the thiol molecules. In addition, our group[118] reported that Fe_3O_4 magnetic nanoparticles modified with 2-carboxyterthiophene (TTP-COOH) could self-assemble into uniform spherical aggregates through π–π interactions. Uniform spherical aggregates with a mean diameter of 148 ± 5 nm were obtained from Fe_3O_4 nanoparticles with a size of only 6.0 ± 1.3 nm [Fig. 1.18(a)]. The model for the self-assembly process is shown in Fig. 1.18(b). Aggregates resulted from interactions between nanoparticles have a uniform size distribution. However, the biggest disadvantage of this method is that the aggregates of nanoparticles modified by small molecules are too weak to be further used, though using macromolecules instead may resolve this problem. The self-assembly of inorganic nanoparticles modified with macromolecules has attracted extensive interests. The self-assembly

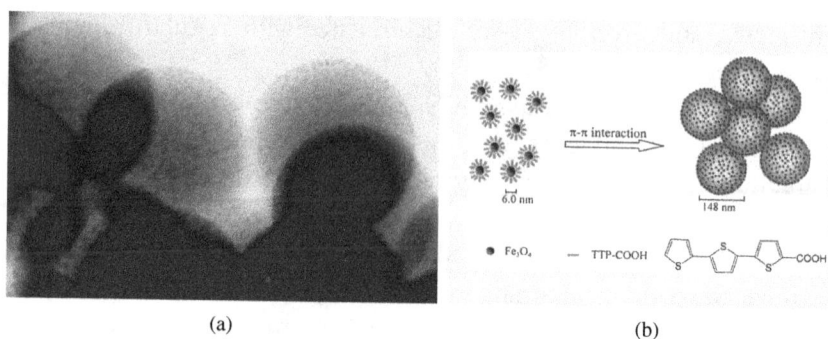

(a)   (b)

Fig. 1.18 Self-assembled microspheres from Fe_3O_4 magnetic nanoparticles through π–π interactions.[118] (a) Transmission electron microscope photographs. (b) Structure model proposed for the self-assembly process of individual nanoparticles. (Copyright Wiley–VCH Verlag GmbH & Co. KGaA. Reproduced with permission.)

process of inorganic nanoparticles modified with macromolecules depends strictly on the macromolecular chain in the functional groups, and assembling units recognize and aggregate with one another strictly through the interactions between macromolecules coated on nanoparticles.

It is worth mentioning that binary self-assembly of nanoparticles has attracted extensive attention in the self-assembly of multiscale structure in recent years.[119,120] The assembly of nanoparticles of two different materials into a binary nanoparticle superlattice (BNSL)[121,122] can provide a general and inexpensive way to obtain a large variety of materials (metamaterials) with precisely controlled chemical composition and tight placement of the components. For example, Shevchenko *et al.*[123] reported the self-assembly of more than 15 different BNSL structures, using combinations of semiconducting, metallic and magnetic nanoparticle building blocks (Au, Ag, Pd, Fe_2O_3, PbSe, PbS, LaF_3; dodecyldimethylammonium bromide DDAB and trioctylamine were used as stabilizer). For instance, NaCl-type superlattices can be assembled from Fe_2O_3 and Au; CuAu-type superlattices can be assembled from PbSe and Au nanoparticles (Fig. 1.19). This effective self-assembly method may play an important role in designing nanomaterials with new properties.

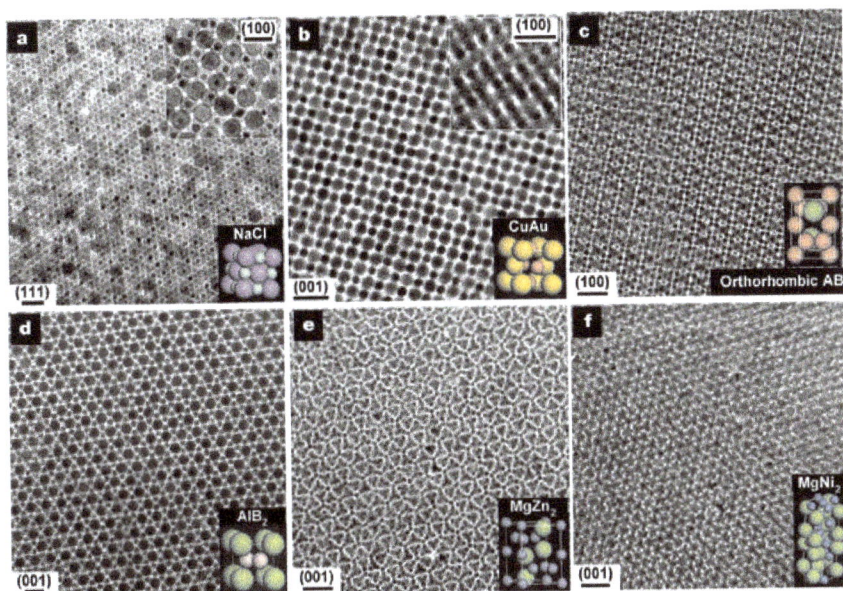

Fig. 1.19 TEM image of the characteristic projections of the binary superlattices self-assembled from different nanoparticles, and modeled unit cells of the corresponding three-dimensional structures (inset). The superlattices are assembled from (a) 13.4 nm g-Fe_2O_3 and 5.0 nm Au; (b) 7.6 nm PbSe and 5.0 nm Au; (c) 6.2 nm PbSe and 3.0 nm Pd; (d) 6.7 nm PbS and 3.0 nm Pd; (e) 6.2 nm PbSe and 3.0 nm Pd; (f) 5.8 nm PbSe and 3.0 nm Pd.[123] (Reprinted by permission from Macmillan Publishers Ltd: *Nature*, copyright 2006.)

Self-assembly of one-dimensional nanomaterials

Research on the self-assembly of one-dimensional nanomaterials has focused on the self-assembly assisted by liquid, i.e. using interfacial tension, capillary force of liquid, or different wetting properties of nanomaterials to induce self-assembly process. For example, our group[124] studied the behavior of water droplet on aligned carbon nanotube films in detail, and proposed a simple and effective water-spreading method to self-assemble carbon nanotube films into three-dimensional (3D) micropatterns. When a water droplet is placed on the fresh carbon nanotube film, it permeates slowly into the film and completely evaporates finally, as a result of which aligned

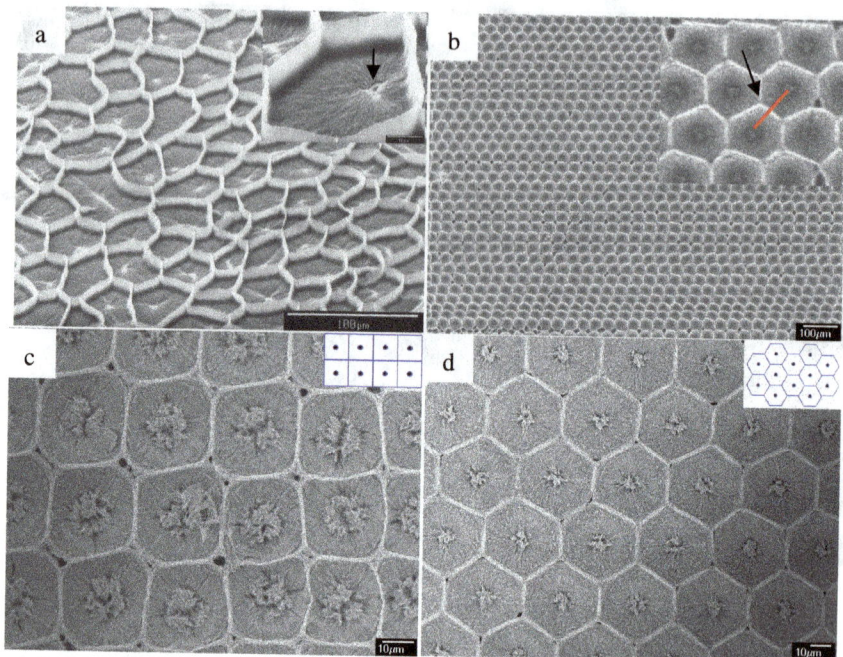

Fig. 1.20   Three-dimensional micropatterns self-assembled on aligned carbon nano-tube film by water-spreading method.[124] (a) Self-assembly of patterned surface without laser assistantance. (b)–(d): Self-assembly of patterned surface with laser assistantance. (Copyright Wiley–VCH Verlag GmbH & Co. KGaA. Reproduced with permission.)

carbon nanotube films assemble into honeycomb-like 3D patterns [Fig. 1.20(a)]. The driving force of this process is capillary force.

Self-assembly of one-dimensional materials can also be induced through templates, or different electrical properties of nanomaterials. Wang and co-workers[125,126] did a lot of innovative work in the self-assembly of one-dimensional materials induced by electrostatic force. Here the inorganic semiconductor materials are taken for example. They found that the electrical properties of ZnO nanobelts growing along the (001) direction were different on both ends, the zinc- and oxygen-terminated $\pm(0001)$ surfaces were positive and negative, respectively. As a result, right-handed helical nanostructures and nanorings were formed by rolling up single-crystal nanobelts induced by electrostatic forces. This structure was stable for its total energy was the lowest. Based on this result, our group disclosed another interesting phenomenon: the synthesized ZnO nanorods could self-assemble into flower aggregates. The synthesized ZnO nanorods grow along (001) direction, resulting in positively and negatively charged top and bottom surfaces, respectively. In order to decrease the total energy of the system, centric flower-like aggregates can be self-assembled, induced by electrostatic forces.

Liu and co-workers[127,128] used a solution-based approach to prepare complex hierachical ordered nanostructures by controlling the nucleation and growth of crystals [Fig. 1.21(a)–(c)]. Based on this result, they recently reported a multiscale patterned ZnO surfaces,[129] the growing process of which is: 1) growth of ZnO nanorods on substrate; 2) creation of new nucleation sites on exiting ZnO nanorods, in order to prepare hierachical structures. At the same time, a carboxyl patterned silver surface was fabricated using micro-contact printing technology. When the substrate was immerged into zinc nitrate aqueous solution subsequently, the highly ordered ZnO nanorods grew on the bare silver surface selectively, as a result of which patterned ZnO arrays can be obtained [Fig. 1.21(d) and (e)]. Similarly, if new nucleation sites had been created on the existing ZnO nanorods, patterned ZnO surfaces with hierachical nanostructures can be obtained.

Simulation of biomineralization

Using the mechanism of biomineralization to synthesize inorganic materials, that is using organic assemblies as templates to control the synthesis of organic materials, thus to fabricate special microstructured materials with excellent physical and chemical performance, is the so called biomimetic materials synthesis method or biomimetic morphology fabrication method.

Scientific researchers in China have achieved many significant results in the preparation of multiscale materials by simulating the process of biomineralization. For example, Yu *et al.*[130,131] from Chinese University of Science and Technology have been working for a long time on fabricating special materials with nanostructures, regulating effectively the size, microstructures and properties of nanomaterials in

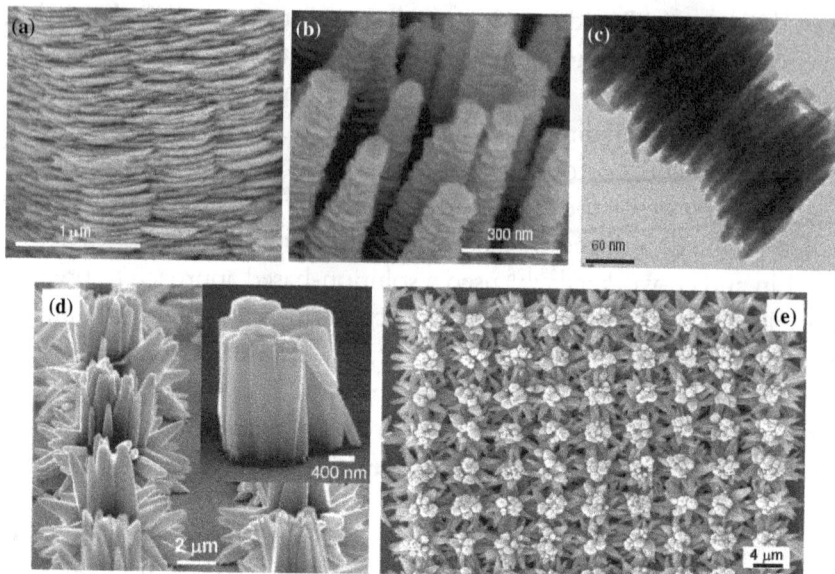

Fig. 1.21 (a)–(c): SEM and TEM images of oriented ZnO nanoplates and nanorods with hierachical structures.[128] (Reprinted by permission from Macmillan Publishers Ltd: *Nature Materials*, copyright 2003.) (d) and (e): Surface morphology of multiscale patterned ZnO nanostructure.[129] (Copyright Wiley–VCH Verlag GmbH & Co. KGaA. Reproduced with permission.)

order to design and build complex functional materials with special structures and shapes in natural conditions. They worked systematically on the simulation of biomineralization process and biomimetic synthesis of many kinds of inorganic materials such as minerals, and the nucleation, crystallization, oriented growth and assembly process of crystals regulated by organic molecules (biomolecules). They have achieved important results in biomineralization and biomimetic synthesis of particular nanostructured materials, and in developing new ways to synthesize nanomaterials chemically (Fig. 1.22). Recently, they reported a chemical route for the synthesis of well-defined concaved cuboctahedrons of copper sulfide crystals with 14 faces, which consist of six squares and eight triangles, and the overall structure shares 24 identical edges in a dymaxion way. The reaction of $Cu(NO_3)_2 \cdot 5H_2O$ and sulfur powders took place in ethylene glycol (EG) solution at

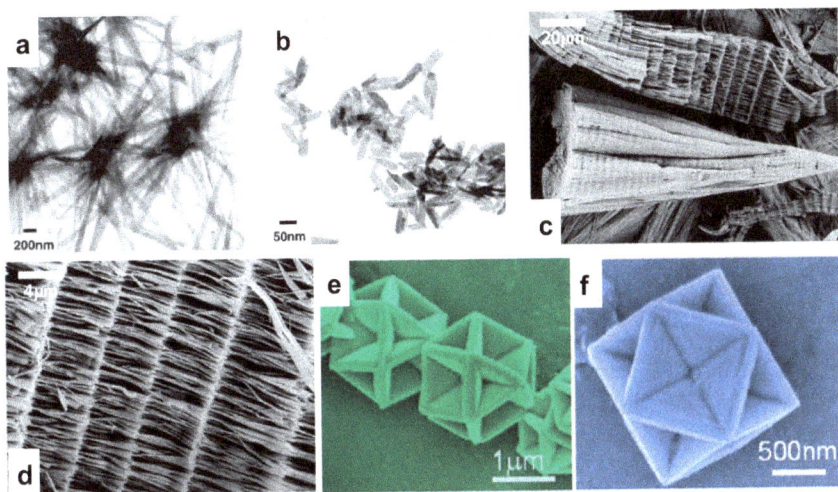

Fig. 1.22 Multiscale materials fabricated by simulating biomineralization process. (a)–(d): CdWO$_4$ and BaSO$_4$ nanofibers produced by adding organic molecules.[130,131] (e) and (f): SEM images of the typical caved cuboctahedron with 14 faces and its structural characteristics.[17b] [Figs. (a) and (b) (Copyright Wiley–VCH Verlag GmbH & Co. KGaA. Reproduced with permission. Figures (c)–(f) Copyright American Chemical Society.]

140°C in a Teflonlined stainless steel autoclave machine, and cuboctahedrons of copper sulfide crystals with 14 faces could be obtained by just centrifuging the mixture. It is amazing that microscopic objects with so elegant structures, which may be even impossible to create for an expert artisan, can be obtained by a very simple synthesis technique. These novel structured materials may find potential applications as a micrometer-sized construction element or a new kind of potential container for carrying, or encapsulating other materials.

Qi and co-workers from Peking University have also made important contributions to the research on simulating biomineralization process to fabricate hierachical ordered structures.[132–134] They used colloid chemistry methods and principles of biomimetic synthesis to control the fabrication of diverse inorganic or organic micro/nanostructures and highly ordered organization, and to explore effective, environmental friendly synthetic ways to prepare novel functional materials with specific size, morphology and structure (Fig. 1.23).

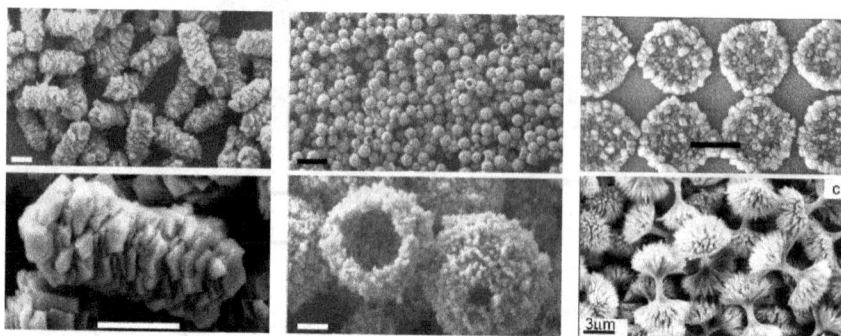

Fig. 1.23   CaCO$_3$ particles with multiscale structures fabricated by simulating the process of biomineralization.[132–134] (Reproduced with permission.)

Nature has given us a lot of inspiration. In organisms, molecules/biomacromolecules self-assemble into organelles/cells first, and then the cells recognize each other and aggregate into tissue, which forms organs subsequently, and finally an individual organism forms. Even the survival of individual organism also depends on certain recognition/self-organization/cooperation of individuals in a colony. So what nature has told us is that the realization of the most complex functions inevitably falls into a multiscale hierachical and ordered self-organization/cooperation process starting from small to big size. Scientists have devoted a lot of effort to obtain a variety of multiscale nanomaterial aggregates with regular geometric appearance through self-assembly method, hoping to achieve excellent physical and chemical properties which are different from monomers.

1.2.3 *Design, synthesis and application of stimuli-responsive molecules*

Design and synthesis of stimuli-responsive molecules make it possible for people to change the macroscopic interfacial properties through external stimuli. External stimuli can be classified into two types: physical and chemical stimuli. While the former includes light, electricity, magnetic field, heat, and stress, etc., the latter includes acids/bases, complexation, bond making/bond breaking, oxidizing/reducing

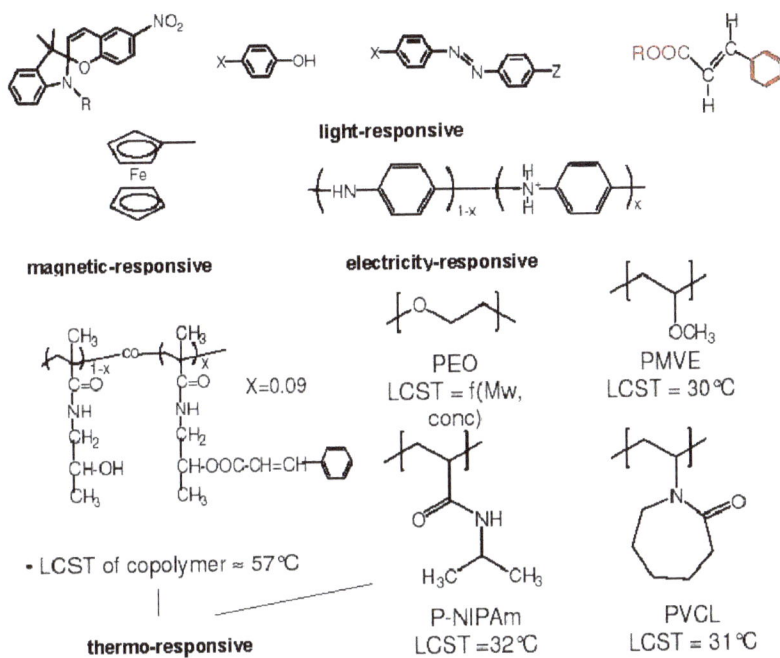

Fig. 1.24   Typical stimuli-responsive molecules. LCST: the lowest critical solving temperature.

agents, photochemistry, electrochemical forces, etc. So far many stimuli-responsive molecules have been developed, some typical ones[135] are listed in Fig. 1.24. If the polarity, conformation or functional group of an interfacial molecule can change when stimulated by external fields, physical properties of the stimuli-responsive molecules change accordingly.

Stimuli-responsive molecules have important applications in many fields including biology, physics, chemistry, and materials. Thomas and co-workers[136] reported a spiropyran-capped Au nanoparticle which could bind or release molecules via light mediation, and it offered intriguing possibilities for designing drug delivery systems with controlled release abilities. Kong and co-workers[137] reported a general strategy to control protein assembly on a switchable low-density self-assembly monolayer, which may open a new way to design functional biocomposite films for biosensors or protein chips. Willner and co-workers[138] reported that a non-dense long-chain tethered bipyridinium monolayer linked to an electrode support could move reversibly by the applied potential. Ito *et al.*[139] reported a nanometer-sized signal-responsive gate fabricated by self-assembly of ionizable polypeptide-poly (glutamic acid) brushes on a gold-coated porous membrane. The conformation of this polymer can change in response to pH. That is, in the region of low pH, polymer chain is protonated and folded to form an R-helical structure; in the region of high pH, it is deprotonated to form an extended random structure. As a result, water permeation through the membrane was reversibly regulated by pH. Stayton and co-workers[140] synthesized a low molecular weight copolymer of acrylic acid (AAc) and *N*-iso-propylacrylamide (NIPAAm) with reactive OH groups at one end, which displays both pH and temperature sensitivity. This copolymer can be conjugated to a specific cysteine-thiol site near the recognition site of streptavidin (SAv), forming a bioconjugate which can provide pH control of biotin-binding and -triggered release from the SAv. If stimuli-responsive molecules are combined with rough surfaces with micro/nanostructures, switch materials with controllable wettability responsing to external stimuli can be obtained.

1.2.4 *Design of heterogeneous interface — binary cooperative complementary nanometer-scale interfacial materials*

By summarizing the research results of fabrication of photo-/electro-controlled interfacial materials and their physical/chemical properties, we proposed a new concept of "binary cooperative complementary nanometer-scale interfacial materials" for the first time (Fig. 1.25).[8] With the development of surface sciences and interface sciences, it is known that a variety of interfaces can form between different objects, and studies on interfaces of metallic, inorganic, organic, semiconductor and biological materials disclose a lot of significant phenomena. Using peculiar surface and interface functions resulting from contact and amalgamation of heterogeneous materials to create novel materials and devices has been a leading concept of many research fields.

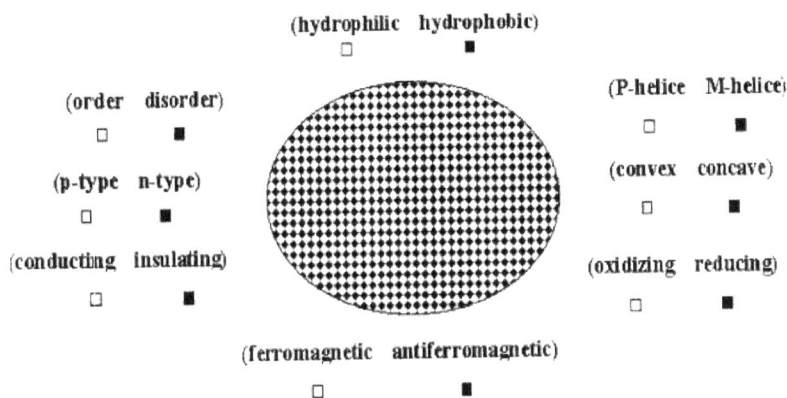

Fig. 1.25 Design of binary cooperative complementary nanometer-scale interfacial materials.[8]

From the viewpoint of physics, the symmetry and free energy of surface phase in condensed matter differ from bulk phase; when the size of a certain object is reduced from macroscopic scale to mesoscopic scale, the influence of surface phase on the physical properties of materials is not negligible. Therefore, design and control of surface phase should be a key point to study novel interfacial materials.

As far as binary property of material world was concerned, our ancestors had already used the logical thinking of the binary "yin and yang". The development of modern sciences also confirms that the matter world consists of a variety of binary cooperative complementary basic particles even in the smallest unit. Both the chemical composition and the properties of matter reflect the binary cooperative complementary property. People have consciously used the binary cooperative complementary property to develop new materials. For example, the following are used in molecular and supramolecular materials: surfactant (molecules with both hydrophilic and hydrophobic end groups), organic nonlinear optical material (functional molecules with both push and pull electron groups), as well as organic superconductor and organic strong field magnet, in which compound molecular crystals are generated by charge-transfer salts composed of electron donors and electron receptors. All of them indicate the important role of the binary cooperative complementary concept in designing novel materials.

"Binary cooperative complementary interfacial materials" extend the binary cooperative complementary property to nanoscale interface to study the new interfacial physical properties. It is known that the binary cooperative complementary property (physical property) is a universal concept, which is displayed in a variety of forms, e.g. hydrophilicity/hydrophobicity (oileophilicity/oileophobicity), conduction/insulation, oxidation/reduction, complementarity of surface geometric structure (convex/concave), stable/metastable structure, paramagnetism/antimagnetism, P type/N type semiconductor, left-handed chirality/right-handed chirality. Generally, surface phase and bulk phase of materials show a single characteristic. However, when the concept of binary cooperative complementary interfacial materials is used to build binary cooperative complementary interfaces in mesoscopic or even nanoscopic scale, the obtained interfaces usually show extraordinary interfacial physical properties. To achieve such binary cooperative complementary property, basic principles of the physics of soft condensed matter and nanosciences should be used to design target interfacial molecules.

Different from conventional simple bulk phase materials, the novel concept of "binary cooperative complementary nanometer-scale interfacial materials" is to build binary cooperative complementary nanometer-scale interfacial structures on the macroscopic surfaces of materials. The designing concept of such materials is: the synthesis of new materials is not necessary all the time, and the aim is to obtain two-dimensional surface phases with two different properties mixing with each other in mesoscopic scale after special surface processing has been applied; the area of each phase and the "interface" between two phases are both in nanometer scale. Studies indicate that nanoscale phases with different or even absolutely opposite physicochemical properties can cooperate with each other under certain conditions, as a result of which extraordinary interfacial physical properties emerge on macroscopic surfaces. Materials with such properties are the very binary cooperative complementary nanometer-scale interfacial materials.

Based on the designing concept of "binary cooperative complementary nanometer-scale interfacial materials", our group focuses on researches about super-amphiphilic interfacial materials (surfaces with both superhydrophilicity and superoileophilicity) and super-amphiphobic interfacial materials (surfaces with both superhydrophobicity and superoileophobicity), which are very important in daily life, depollution of environment, and others. It has been studied that binary cooperative complementary nanoscale interfaces of TiO_2 with hydrophilic and oileophilic nanoscale areas coexisting on the macroscopic surfaces can be fabricated by light irradiation.[141,142] As a result, macroscopic surfaces of TiO_2 show fantastic super-amphiphilicity. New materials fabricated according to this principle can be used to modify surface of glasses and construction materials, providing self-cleaning and fog-resisting functions.[143] Using epitaxial growth method which is a down-top, from-atom-to-molecule-to-aggregate, nanoscience method, nanoscale interfacial structures with complementary geometry (e.g. convex and concave) can be built on specific surfaces. Since gas molecules can adsorb onto nanoscale concave surfaces stably, which corresponds to a stable gas film existing on macroscopic surfaces, oil and water cannot contact with surfaces directly, leading to the extraordinary amphiphilicity of surfaces.

1.2.5 *Multiple weak interactions — cooperative complementary effect of system under nonequilibrium conditions*

Construction of bistable systems embodies truly the characteristics of intellectualization of materials. In the presence of external stimuli, the materials can perceive and distinguish them; when external stimuli disappear, the materials can recover to their original states quickly. The reversible switch between two (or even more) stable states can only be achieved through multiple weak interactions. Here, the weak interactions are crucial. Just because of these multiple weak interactions, lives become so colorful in nature. The weak interactions refer to short-/long-range synergistic actions of weak and less directional non-covalent bonds including hydrogen bonds, van der Waals forces, weak ionic bonds, molecular isomerization, capillary effect, and so on. For example, the oxygen carrying and releasing process of hemoglobin is a bistable reversible process resulting from the structural isomerization caused by cooperative complementary effect.[144]

As we have known, heme is a kind of pigment which makes blood red, and it is a complex of Fe^{2+} and porphyrin. When Fe^{2+} links with peptide chains of protein from both sides perpendicular to the porphyrin ring plane through nitrogen atoms of histidine with which Fe^{2+} coordinates, hemoglobin is formed. The hemoglobin has four subunits (designated as α and β subunits), each subunit containing a heme group which can bind reversibly with an oxygen molecule. It is a molecular machine with very high efficiency and its oxy-deoxy transformation can be regulated by moving and changing the conformation of the subunits. The combination of oxygen molecules with the four oxygen-binding sites on hemoglobin do not occur simultaneously. When blood flows through the lungs, ample oxygen molecules easily bind with the first subunits and subsequently drive the cooperative binding to other subunits. When blood circulates in body parts with poor oxygen but rich carbon dioxide, hemoglobin begins to release oxygen molecules. Once an oxygen molecule is released, the conformation of hemoglobin will change accordingly, causing the other three oxygen molecules to disengage from the protein quickly. In this way, hemoglobins can bind the maximum amount of oxygen from the lungs, and then transport them to the various organs. The cooperative complementary interaction of 16 hydrogen bonds in hemoglobin is the key point of this process.

When an oxygen molecule binds with the Fe^{2+} of heme in a subunit, the histidine at the bottom will be pulled up, which drives the movement of α subunit. The motion is transported to other subunits rapidly and they slide by each other. Therefore, the conformation of hemoglobin changes to form oxyhemoglobin. When oxygen molecules are released, the protein returns to the original conformation. Subsequently, hemoglobin binds carbon dioxide molecules through certain amido groups and then transports them to the lung. This process is repeated so that oxygen is transported by hemoglobin.

1.3 Typical Examples of Using Above-Mentioned Five Principles to Design Smart Materials

The designing concept of biomimetic smart materials derives from biological systems; the interfacial biomimetic design of materials opens a new avenue to create intellectualized materials. We combine interface, structure, function and multiple weak interactions flexibly, so as to obtain multiple binary cooperative complementary systems, i.e. to develop multiple smart systems on the premise of achieving the intellectualization of single material. Just like biological systems, it aims to make materials with more than two kinds of opposite physical properties, such as hydrophilicity/hydrophobicity, multiple responses to electricity, light, magnet, temperature etc., thereby to obtain advanced smart material systems.

Our group's studies on reversible superhydrophobicity/ superhydrophilicity switch embody a complete process of creating smart materials, which effectively combines the five-level of research including (1) designing concept of biomimetic intellectualization, (2) multiscale structure effects, (3) interfacial effects, (4) binary cooperative complementary opposite physical properties, and (5) multiple weak interactions. First of all, we found that the unique self-cleaning property of lotus leaves is based on the compound micro/ nanostructures and wax on the leaf surface.[45] This is considered as our designing concept of biomimetic intellectualized materials to construct smart responsive materials with special wettability. Hydrophilicity and hydrophobicity of materials are two opposite physical properties (binary cooperative complementary physical property). We chose a thermo-responsive polymer, which displays a reversible switch between weak hydrophilicity and weak hydrophobicity when stimulated by temperature, and the responses to temperature is based on multiple weak interactions (switch between multiple intermolecular and intramolecular hydrogen bonds), to construct new interface. We introduced these polymer molecules to a surface with compound micro/nanostructure forming a new interface. As a result, the weak hydrophilicity can be enlarged to be superhydrophilicity, and weak hydrophobicity to be superhydrophobicity,

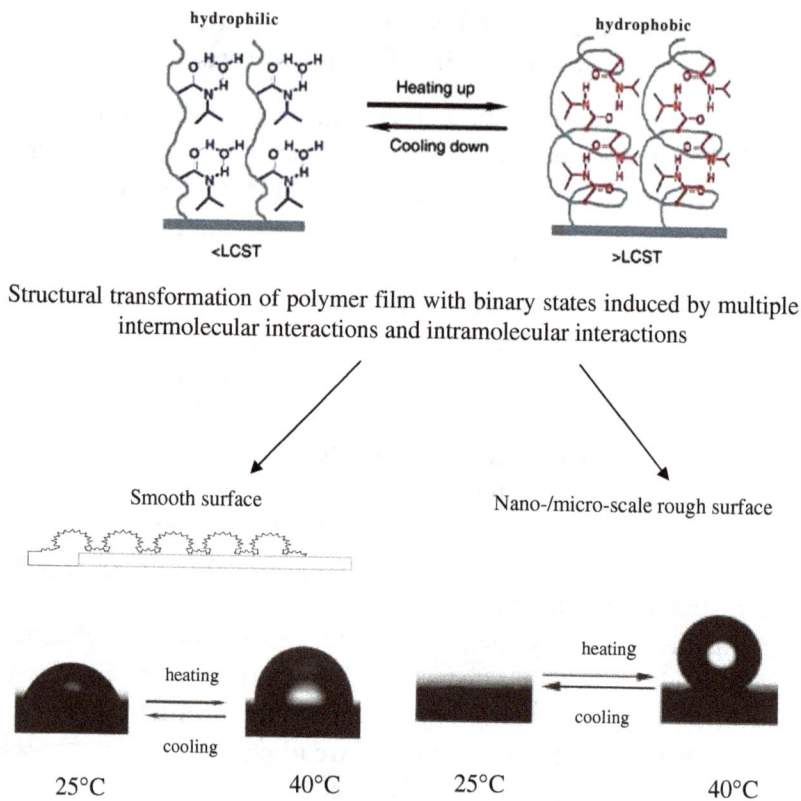

Structural transformation of polymer film with binary states induced by multiple intermolecular interactions and intramolecular interactions

Fig. 1.26 Superhydrophobic/superhydrophilic reversible switch under multiple weak interactions.[10] (Copyright Wiley–VCH Verlag GmbH & Co. KGaA. Reproduced with permission.)

hence reversible superhydrophibicity/superhydrophobicity switch can be obtained successfully (Fig. 1.26).[10]

After this work was accepted by *Angew. Chem. Int. Ed.*, it was immediately selected as VIP (very important paper) and cover paper. In the explanation, the editor said, "The yin and the yang represent the two opposing fundamental properties of nature and the universe in ancient Taiji Chinese philosophy. Superhydrophilicity and superhydrophobicity are fundamentally opposing properties of wettability. Reversible switching between these surface properties can be thermally induced through the combination of chemical modification of the surface, and surface roughness". The journal *Science* highlighted the work on thermo-responsive superhydrophobicity/superhydrophilicity reversible switch under the tile of "Super Switcher".[145] The designing concept of such switch materials was inspired from self-cleaning property of lotus leaf in nature, the superhydrophobicity of which is the combination of compound micro/nanoscale surface structure and materials with low surface energy. Meanwhile, the superhydrophobic/superhydrophilic reversible switch exceeds the single superhydrophobicity of lotus leaf. Therefore, biomimetic intellectualization, a process of learning from nature, finally exceeds nature. It is a typical example of fabricating biomimetic smart interfacial materials.

1.4 Intellectualized Design of Biomimetic Interfacial Materials

Nature has given us an important inspiration: by mimicking the smart property of organisms, simulating a certain aspect of their special functions, and choosing proper materials to create micro/nanostructures, excellent macroscopic performances can be obtained; therefore the unity of microscopic and macroscopic world is achieved.

In this book, we discuss mainly the intellectualized design of biomimetic interfacial materials. In particular, we introduce biomimetic nanoscale interfacial materials with special wettability based on superhydrophobic surfaces, and also summarize the current research conditions and developing prospect of such materials.[146]

Our group has contributed a lot to the research on biomimetic superhydrophobic surfaces in recent years (Fig. 1.27). Guided by the concept of "binary cooperative complementary nanoscale interfacial materials", we have achieved a series of research results in "studies on biomimetic superhydrophobic nanoscale interfacial materials".[147] The work on "superhydrophobicity/superhydrophilicity reversible switch materials" has been especially evaluated as an example of "using 'the yin and the yang' Taiji ancient Chinese philosophy" to develop new materials.

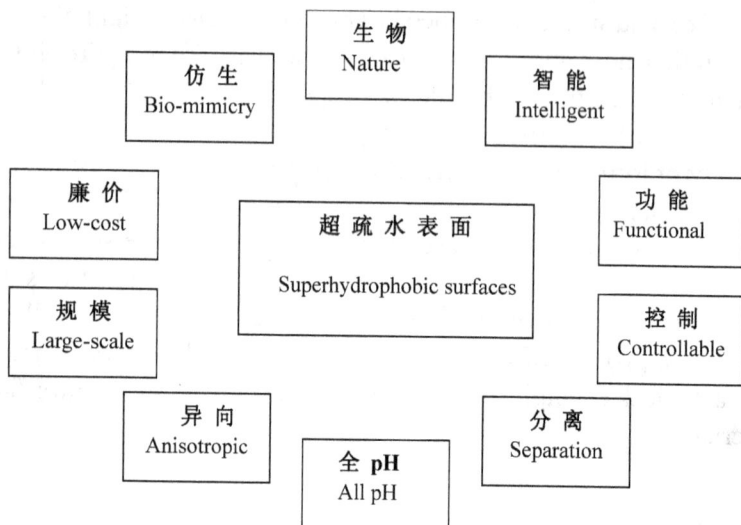

Fig. 1.27   Intellectualized design of superhydrophobic surfaces.

The major research findings of our group can be summarized as follows:

(1) Several typical surfaces of organisms with superhydrophobicity were studied in detail, and the mechanism of superhydrophobicity was disclosed:

    (a) The key reason of superhydrophobicity of lotus leaf was the micro/nano-composite structures existing on the surface; the mechanism of superhydrophobic self-cleaning property of lotus leaf was thus explained. Biomimetic two-dimensional carbon nanotube film with micro/nano-composite structures was fabricated which confirmed that micro/nanostructures showed smaller sliding angle. Aligned carbon nanotube film with both superhydrophobicity and superoleophobicity, that is super-amphiphobicity, was also fabricated, which confirmed further that nanostructure was necessary for super-amphiphobicity.[148] These research results are very important for developing self-cleaning materials.

(b) Studies on the mechanism of sliding anisotropic behavior of droplet on rice leaf revealed that this phenomenon resulted from aligned arrangement of micro/nanostructures on the surface. Carbon nanotube film with biomimetic aligned micro/nanostructures was successfully fabricated, which confirmed that the moving trend of droplet was affected by the arrangement of microstructures.

(c) It was found that the stable aquatic characteristics of water strider were due to the special hierarchical structure of its legs, which were covered by large numbers of oriented tiny hairs (microsetae) with fine helical nanogrooves. Mechanical measurement of legs indicated that the maximal supporting force of a single leg was about 15 times the total body weight of the insect. This result may play a guiding role in designing dynamic superhydrophobic surfaces and developing novel aquatic vehicles.

(2) Inspired by the above research results, a variety of characteristic superhydrophobic surfaces were fabricated, which showed a lot of advantages, such as low cost, large scale, anisotropic property, wide pH range, separation of oil and water, controllable property, functionalization, and others.

(a) Super-amphiphobic aligned carbon nanotube films were fabricated.[148] Superhydrophobic polymer films were obtained by introducing microstructures to the surface, and this film showed superhydrophobicity without any modification by materials of low surface energy (e.g. organic fluorin, organic silicon), which opened a new way to make superhydrophobic materials.[149] Superhydrophobic film was also fabricated using amphiphilic polymer, which extended the selective range of materials to prepare superhydrophobic materials.[150] Using one-step casting process, super-amphiphobic polymer coatings with lotus-leaf–like hierachical micro/nano-composite structures were prepared.[151] In addition, by immersing copper plate into the solution of fatty acid, stable superhydrophobic surface without any modification or processing could be obtained, which was a breakthrough in simple fabrication of superhydrophobic surfaces under ordinary conditions.[152]

(b) Electrohydrodynamics (EHD) technique was used to fabricate biomimetic superhydrophobic polymer film with a lotus-leaf–like structure consisting of porous microspheres interlinked by nanofibers, which offered new possibilities for the continuous manufacture of superhydrophobic film materials.[153] By simulating the microscopic structures of cicada's wings, superhydrophobic polymer nanopillar film was prepared[154]; by mimicking the structure of keratinous hairs in gecko's feet, superhydrophobic aligned polymer nanotube film with high adhesive forces was fabricated, which could be used as a "mechanical hand" to transport liquid without any loss.[155]

(c) By imitating the arrangement of microstructures on rice's leaf, three-dimensional anisotropic aligned carbon nanotube was prepared, and different orientation of the aligned carbon nanotube was used to control the wettability of surfaces,[156] which showed the relationship between wettability and orientation of nanostructures, and this result could be applied in DNA chips and microfluid systems.

(d) Superhydrophobic nanostructured carbon films in a wide range of pH values were prepared, and these films could find wide applications in the chemical industry, microfluid systems, and others[157]; a composite film of polyaniline doped with azobenzenesulfonic acid (ABSA) blended with polystyrene with a lotus-leaf–like structure was prepared via a simple electrospinning method, and this film showed stable superhydrophobicity and conductivity, even in a wide pH range.[158]

(e) A novel coating mesh film with both superhydrophobic and superoleophilic properties was prepared by an easy and inexpensive spray-and-dry method, which can be used in the separation of oil and water. Separation and filtration equipments assembled from this mesh film can be used for pipe laying of vehicles such as ships to purify fuel and reduce accidents.[159]

(3) As far as smart materials were concerned, our group fabricated a compound micro/nanostructured copper film by electrochemical deposition (ECD) approach; the copper film was modified further with self-assembled monolayers of n-alkanoic acids with different length. By manipulating the deposition time and numbers of carbon atoms in fatty acid, superhydrophobicity and superhydrophilicity of film could be manipulated effectively.[160] Moreover, we had achieved important results in studies on external field–controlled superhyrophobicity/superhydrophilicity reversible "switch" materials. For example, thermo- and photo-responsive "switch" materials which can switch between superhyrophobicity and superhydrophilicity reversibly,[9,10,161] dual-responsive "switch" materials which can switch between superhyrophobicity and superhydrophilicity reversibly and are responsive to changes in both temperature and pH,[162] dual-responsive films that combine wettability conversion with photochromic behavior induced by irradiation,[163] and multiresponsive surfaces sensitive to pH, temperature and glucose,[164] had been prepared successfully.

Development of the above materials embodies the whole process of learning from nature, mimicking nature, and exceeding nature, which plays a guiding role in studies on biomimetic nanoscale interfacial materials. Related research works will be introduced in detail in the following chapters. Since the binary property of matter is endless, binary cooperative complementary interfacial materials should also have infinite number of permutations and combinations. A colorful world of novel advanced functional materials is waiting for us.

References

1. Hu Z, Chen Y, Wang C, Zheng Y, Li Y. *Nature*, 1998, **393**: 149.
2. Beebe D J, Moore J S, Bauer J M, Yu Q, Liu R H, Devadoss C. *Nature*, 2000, **404**: 588.
3. Russell TP. *Science*, 2002, **297**: 964.

4. Livage J. *Nature Mater.*, 2003, **2**: 297.
5. Graham-Rowe D. *New Scientists*, 2001, Apr. 11.
6. Kim S R, Abbott N L. *Adv. Mater.*, 2001, **13**: 1445.
7. Mason A C, Oshinsky M L, Hoy R R. *Nature*, 2001, **410**: 686.
8. Jiang L, Wang R, Yang B, Li T J, Tryk D A, Fujishima A, Hashimoto K, Zhu D B. *Pure Appl. Chem.*, 2000, **72**: 73.
9. Feng X J, Feng L, Jin M H, Zhai J, Jiang L, Zhu D B. *J. Am. Chem. Soc.* 2004, **126**: 62.
10. Sun T L, Wang G J, Feng L, Liu B Q, Ma Y M, Jiang L, Zhu D B. *Angew. Chem. Int. Ed.*, 2004, **43**: 357.
11. Gu Z Z, Fujishima A, Sato O. *Appl. Phys. Lett.*, 2004, **85**: 5067.
12. Gu Z Z, Horie R, Kubo S, Yamada Y, Fujishima A, Sato O. *Angew. Chem. Int. Ed.*, 2002, **41**: 1153.
13. Gu Z Z, Uetsuka H, Takahashi K, Nakajima R, Onishi H, Fujishima A, Sato O. *Angew. Chem. Int. Ed.*, 2003, **42**: 894.
14. Yan X, Yao J, Lu G, Chen X, Zhang K, Yang B. *J. Am. Chem. Soc.*, 2004, **126**: 10510.
15. Yan X, Yao J, Lu G, Li X, Zhang J, Han K, Yang B. *J. Am. Chem. Soc.*, 2005, **127**: 7688.
16. Yao J, Yan X, Lu G, Zhang K, Chen X, Jiang L, Yang B. *Adv. Mater.*, 2004, **16**: 81.
17. (a) Yu S H, Cölfen H, Tauer K, Antonietti M. *Nature Mater.* 2004, **5**: 51; (b) Wu C Y, Yu S H, Antonietti M. *Chem. Mater.*, 2006, **18**: 3599.
18. Luo L, Yu S H, Qian H, Zhou T. *J. Am. Chem. Soc.*, 2005, **127**: 2822.
19. Liu B, Yu S H, Li L, Zhang Q, Zhang F, Jiang K. *Angew. Chem. Int. Ed.*, 2004, **43**: 4745.
20. Chen S, Yu S H, Wang T, Jiang J, Cölfen H, Hu B, Yu B. *Adv. Mater.*, 2005, **17**: 1461.
21. Yao J N, Hashimoto K, Fujishima A. *Nature*, 1992, **355**: 6361.
22. Xiao D, Xi L, Yang W, Fu H, Shuai Z, Fang Y, Yao J. *J. Am. Chem. Soc.*, 2003, **25**: 6740.
23. Tian Z, Chen Y, Yang W, Yao J, Zhu L, Shuai Z. *Angew Chem. Int. Ed.*, 2004, **43**: 4060.
24. Xiao D, Yang W, Yao J, Xi L, Yang X, Shuai Z. *J. Am. Chem. Soc.*, 2004, **126**: 15439.
25. Liu D, Bruckbauer A, Abell C, Balasubramanian S, Kang D J, Klenerman D, Zhou D. *J. Am. Chem. Soc.*, 2006, **128**: 2067.

26. Liu D, Balasubramanian S. *Angew. Chem. Int. Ed.*, 2003, **42**: 5734.

27. Shu W, Liu D, Watari M, Riener C K, Strunz T, Welland M E, Balasubramanian S, McKendry R A. *J. Am. Chem. Soc.*, 2005, **127**: 17054.

28. Ma Z, Li J, Liu M, Cao J, Zhou Y, Tu J, Jiang L. *J. Am. Chem. Soc.*, 1998, **120**: 12678.

29. Lin L, Zhao H, Li J, Tang J, Duan M, Jiang L. *Biochem. Biophys. Res. Commun.*, 2000, **274**: 817.

30. Qiu X, Burda C, Fu R, Pu L, Chen H, Zhu J. *J.Am. Chem. Soc.*, 2004, **126**: 16276.

31. Xu J, Bao N, Xia X, Peng Y, Chen H. *Anal. Chem.*, 2004, 76: 6902.

32. Douglas T. *Science*, 2003, **299**: 1192.

33. Taton T A. *Nature Mater.*, 2003, **2**: 73.

34. Linda B, Richard A. *Cell*, 1991, **65**: 175.

35. Gillespie P G, Walker R G. *Nature*, 2001, **413**: 194.

36. Lee L P, Szema R. *Science*, 2005, **310**: 1148.

37. Hardie R C, Raghu P. *Nature*, 2001, **413**: 186.

38. Crommie M F, Lutz C P, Eiger D M. *Science*, 1993, **262**: 218.

39. Heller E J, Crommie M F, Lutz C P, Eiger D M. *Science*, 1994, **369**: 464.

40. Medintz I L, Uyeda H T, Goldman E R, Mattoussi H. *Nature Mater.*, 2005, **4**: 435.

41. Barthlott W, Neinhuis C. *Planta*, 1997, **202**: 1.

42. Neinhuis C, Barthlott W. *Annals of Botany*, 1997, **79**: 667.

43. Wagner T, Neinhuis C, Barthlott W. *Acta Zoologica*, 1996, 77: 213.

44. Gorba S N, Keselb A, Berger J. *Arthr. Struc. Develop.*, 2002, **29**: 129.

45. Feng L, Li S H, Li Y S, Li H J, Zhang L J, Zhai J, Song Y L, Liu B Q, Jiang L, Zhu D B. *Adv. Mater.*, 2002, **14**: 1857.

46. Lum K, Chandler D, Weeks J D. *J. Phys. Chem. B*, 1999, **103**: 4750.

47. Gao X. PhD Dissertation, Institute of Chemistry, Chinese Academy of Sciences, 2006.

48. Kennedy R J. *Nature*, 1970, **227**: 736.

49. Zheng Y M, Gao X F, Jiang L. *Soft Matter*, 2007, **3**: 178.

50. Watson G S, Watson J A. *Appl. Surf. Sci.*, 2004, **235**: 139.

51. Stavenga D G, Foletti S, Palasantzas G, Arikawa K. *Proc. R. Soc. B.*, 2006, **273**: 661.

52. Gao X F, Jiang L. *Nature*, 2004, **432**: 36.

53. Autumn K, Liang Y A, Hsieh S T, Zesch W, Chan W P, Kenny T W, Fearing R, Full R J. *Nature*, 2000, **405**: 681.
54. Autumn K, Peattie A M. *Integr. Comp. Biol.*, 2002, **42**: 1081.
55. Kesel A B, Martin A, Seidl T. *Smart Mater. Struct.*, 2004, **13**: 512.
56. Eisner T, Aneshansley D J. *PNAS*, 2000, **97**: 6568.
57. Yang F, Yang W. *Mech. Eng.*, 2004, **26**: 12.
58. Parker A R, Lawrence C R. *Nature*, 2001, **414**: 33.
59. Hadley N F. *Science*, 1979, **203**: 367.
60. Ball P. *Nature*, 1999, **400**: 507.
61. Vukusic P, Sambles J R. *Nature*, 2003, **424**: 852.
62. Sanders J V. *Nature*, 1964, **204**: 1161.
63. Zi J, Yu X, Li Y, Hu X, Xu C, Wang X, Liu X, Fu R. *PNAS*, 2003, **100**: 12576.
64. Srinivasarao M. *Chem. Rev.*, 1999, **99**: 1935.
65. Vukusic P, Sambles J R, Lawrence C R, Wootton R J. *Nature*, 2001, **410**: 36.
66. Vukusic P, Sambles J R, Lawrence C R. *Nature*, 2000, **404**: 457.
67. Vukusic P, Hooper I. *Science*, 2005, **310**: 1151.
68. Parker A, Welch V L, Driver D, Martini N. *Nature*, 2003, **426**: 786.
69. Wong S, Kitaev V, Ozin G A. *J. Am. Chem. Soc.*, 2003, **125**: 15589.
70. Yablonovitch E. *Phys. Rev. Lett.*, 1987, **58**: 2059.
71. John S. *Phys. Rev. Lett.*, 1987, **58**: 2486.
72. Zakhidov A A, Baughman R H, Iqbal Z, Cui C, Khayrullin I, Dantas S O, Marti J, Ralchenko V G. *Science*, 1998, **282**: 897.
73. Vlasov Y A, Bo X, Sturm J C, Norris D J. *Nature*, 2001, **414**: 289.
74. Campbell M, Sharp D N, Harrison M T, Denning R G, Turberfield A J. *Nature*, 2000, **404**: 53.
75. Sarikaya M, Tamerler C, Jen A Y, Schulten K, Baneyx F. *Nature Mater.*, 2003, **2**: 577.
76. Ouyang J. *Matrix Manipulation of Biomineralization and Its Appilications.* 1st Ed., Beijing: Chemical Industry Press, 2006.
77. Cui F, Zheng C. *Biomimetic materials.* 1st Ed., Beijing: Chemical Industry Press, 2004.
78. Mayer G. *Science*, 2005, **310**: 1144.
79. Dorozhkin S V, Epple M. *Angew. Chem. Int. Ed.*, 2002, **41**: 3130.
80. Schüler D, Frankel R B. *Appl. Microbiol. Biotechnol.*, 1999, **52**: 464.
81. Sarikaya M, Fong H, Sunderland N, Flinn B D, Mayer G, Mescher A, Gaino E J. *Mater. Res.*, 2001, **16**: 1420.

82. Sarikaya M. *PNAS*, 1999, **96**: 14183.
83. Mayer G, Sarikaya M. *Exp. Mech.*, 2002, **42**: 395.
84. Naik R R, Stringer S J, Agarwal G, Jones S, Stone M O. *Nature Mater.*, 2002, **1**: 169.
85. Naik R R, Brott L L, Rodriguez F, Agarwal G, Kirkpatrick S M, Stone M O. *Prog. Org. Coatings*, 2003, **47**: 249.
86. Aizenberg J, Muller D A, Grazul J L, Hamann D R. *Science*, 2003, **299**: 1205.
87. Braga D. *Angew. Chem. Int. Ed.*, 2003, **42**: 5546.
88. (a) Grojean R E, Sousa J A, Henry M C. *Appl. Opt.*, 1980, **19**: 339; (b) Zhao Y, Cao X Y, Jiang L. *J. Am. Chem. Soc.*, 2007, **129**: 764.
89. Aizenberg J, Tkachenko A, Weiner S, Addadl L, Hendler G. *Nature*, 2001, **412**: 819.
90. Sowards L, Schmitz H, Tomlin D, Naik R, Stone M. *Ann. Entomol. Soc. Am.*, 2001, **94**: 686.
91. (a) Cha J N, Stucky G D, Morse D E, Dming T J. *Nature*, 2000, **403**: 289.
92. Jeong K, Liu G L, Chronis N, Lee L P. *Opt. Express.*, 2004, **12**: 2494.
93. Stevens M M, George J H. *Science*, 2005, **310**: 1135.
94. Shin Y, Wang C, Exarhos G J. *Adv. Mater.*, 2005, **17**: 73.
95. Keckes J, Burgert I, Fruhmann K, Muller M, Kolln K, Hamilton M, Burghammer M, Roth S V, Stanzl-Tschegg S, Fratzl P. *Nat. Mater.*, 2003, **2**: 810.
96. Liu H, Zhai J, Jiang L. *Chin. J. Inorg. Chem.*, 2006, **22**: 597.
97. Kim F, Kwan S, Akana J, Yang P D. *J. Am. Chem. Soc.*, 2001, **123**: 4360.
98. Andres R P, Bielefeld J D, Henderson J I, Janes D B, Kolagunta V R, Kubiak C P, Mahoney W J, Osifchin R G. *Science*, 1996, **273**: 1690.
99. Patil V, Mayya K S, Pradhan S D, Sastry M. *J. Am. Chem. Soc.*, 1997, **119**: 9281.
100. Mirkin C A, Letsinger R L, Mucic R C, Storhoff J J. *Nature*, 1996, **382**: 607.
101. Boal A K, Rotello V M. *J. Am. Chem. Soc.*, 2000, **122**: 734.
102. Caruso F, Susha A S, Giersig M, Möhwald H. *Adv. Mater.*, 1999, **11**: 950.
103. Caruso F, Caruso R, Möhwald H. *Science*, 1998, **282**: 1111.
104. Bowden N B, Weck M, Choi I S, Whitesides G M. *Acc. Chem. Res.*, 2001, **34**: 231.

105. Bico J, Roman B, Moulin L, Boudaoud A. *Nature*, 2004, **432**: 690.
106. Rothemund P W K. *Nature*, 2006, **440**: 297.
107. Wolfe D B, Snead A, Mao C, Bowden N B, Whitesides G M. *Langmuir*, 2003, **19**: 2206.
108. Choi I S, Weck M, Xu B, Jeon N L, Whitesides G M. *Langmuir*, 2000, **16**: 2997.
109. Boncheva M, Ferrigno R, Bruzewicz D A, Whitesides G M. *Angew. Chem. Int. Ed.*, 2003, **42**: 3368.
110. Choi I S, Bowden N, Whitesides G M. *Angew. Chem. Int. Ed.*, 1999, **38**: 3078.
111. Bowden N, Terfort A, Carbeck J, Whitesides G M. *Science*, 1997, **276**: 233.
112. Bowden N, Oliver S R J, Whitesides G M. *J. Phys. Chem. B*, 2000, **104**: 2714.
113. Duan X, Huang Y, Cui Y, Wang J, Lieber C M. *Nature*, 2001, **409**: 66.
114. Huang Y, Duan X, Wei Q, Lieber C M. *Science*, 2001, **291**: 630.
115. Paul S, Pearson C, Molloy A, Cousins M A, Green M, Kolliopoulou S, Dimitrakis P, Normand P, Tsoukalas D, Petty M C. *Nano Lett.*, 2003, **3**: 533.
116. Brust M, Walker M, Bethell D, Schiffrin D J, Whyman R. *Chem. Commun.*, 1994, **7**: 801.
117. Hostetler M J, Templeton A C, Murray R W. *Langmuir*, 1999, **15**: 3782.
118. Jin J, Iyoda T, Cao C, Song Y, Jiang L, Li T J, Zhu D B. *Angew. Chem. Int. Ed.*, 2001, **40**: 2135.
119. Kiely C J, Fink J, Brust M, Bethell D, Schiffrin D J. *Nature*, 1998, **396**: 444.
120. Shevchenko E V, Talapin D V, Rogach A L, Kornowski A, Haase M, Weller H. *J. Am. Chem. Soc.*, 2002, **124**: 11480.
121. Redl F X, Cho K S, Murray C B, Brien S. *Nature*, 2003, **423**: 968.
122. Shevchenko E V, Talapin D V, O'Brien S, Murray C B. *J. Am. Chem. Soc.*, 2005, **127**: 8741.
123. Shevchenko E V, Talapin D V, Kotov N A, O'Brien S, Murray C B. *Nature*, 2006, **439**: 55.
124. Liu H, Li S, Zhai J, Li H, Zheng Q, Jiang L, Zhu D. *Angew. Chem. Int. Ed.*, 2004, **43**: 1146.
125. Kong X Y, Wang Z L. *Nano Lett.*, 2003, **3**: 1625.

126. Gao P X, Ding Y, Mai W, Hughes W L, Lao C, Wang Z L. *Science*, 2005, **309**: 1700.

127. Tian Z R, Voigt J A, Liu J, Mckenzie B, Mcdermott M J. *J. Am. Chem. Soc.*, 2002, **124**: 12954.

128. Tian Z R, Voigt J A, Liu J, Mckenzie B, Mcdermott M J, Rodriguez M A, Konishi H, Xu H F. *Nat. Mater.*, 2003, **2**: 821.

129. Sounart T L, Liu J, James A, Voigt J A, Hsu J W P, Spoerke E D, Tian Z, Jiang Y B. *Adv. Funct. Mater.*, 2006, **16**: 335.

130. Yu S H, Antonietti M, Cölfen H, Giersig M. *Angew. Chem. Int. Ed.*, 2002, **41**: 2356.

131. Yu S H, Antonietti M, Cölfen H, Hartmann J. *Nano Lett.*, 2003, **3**: 379.

132. Qi L M, Li J, Ma J M. *Adv. Mater.*, 2002, **14**: 300.

133. Lu C H, Qi L M, Ma J M, Cheng H M, Zhang M F, Cao W X. *Langmuir*, 2004, **20**: 7378.

134. Zhang D B, Qi L M, Ma J M, Cheng H M. *CrystEngComm.*, 2002, **4**: 536.

135. Gil E S, Hudson S M. *Prog. Polym. Sci.*, 2004, **29**: 1173.

136. Ipe B I, Mahima S, Thomas K G. *J. Am. Chem. Soc.*, 2003, **125**: 7174.

137. Liu Y, Mu L, Liu B, Zhang S, Yang P, Kong J. *Chem. Commun.*, 2004, 1194.

138. Wang X, Kharitonov A B, Katz E, Willner I. *Chem. Commun.*, 2003, 1542.

139. Ito Y, Park Y S, Imanishi Y. *Langmuir*, 2000, **16**: 5376.

140. Bulmus V, Ding Z, Long C J, Stayton P S, Hoffman A S. *Bioconjugate Chem.*, 2000, **11**: 78.

141. Wang R, Hashimoto K, Fujishima A, Chikuni M, Kojima E, Kitamura A, Shimohigoshi M, Watanabe T. *Nature*, 1997, **388**: 431.

142. Wang R, Hashimoto K, Fujishima A, Chikuni M, Kojima E, Kitamura A, Shimohigoshi M, Watanabe T. *Adv. Mater.*, 1998, **10**: 135.

143. Fujishima A, Rao T N, Tryk D A. *J. Photochem. Photobio. C.*, 2000, **1**: 1.

144. Voet D, Voet J G. *Biochemistry.* 3rd Edn., John Wiley & Sons, Inc. 2004.

145. Chin G. *Science*, 2004, **303**: 287.

146. Feng X J, Jiang L. *Adv. Mater.*, 2006, **18**: 3063.

147. Sun T L, Feng L, Gao X F, Jiang L. *Acc. Chem. Res.*, 2005, **38**: 644.

148. Li H J, Wang X B, Song Y L, Liu Y Q, Li Q S, Jiang L, Zhu D B. *Angew. Chem. Int. Ed.*, 2001, **40**: 1743.

149. Feng L, Li S H, Li H J, Zhai J, Song Y L, Jiang L, Zhu D B. *Angew. Chem. Int. Ed.*, 2002, **41**: 1221.
150. Feng L, Song Y L, Zhai J, Liu B Q, Xu J, Jiang L, Zhu D B. *Angew. Chem. Int. Ed.*, 2003, **42**: 800.
151. Xie Q D, Xu J, Feng L, Jiang L, Tang W H, Luo X, Han C C. *Adv. Mater.*, 2004, **16**: 302.
152. Wang S T, Feng L, Jiang L. *Adv. Mater.*, 2006, **18**: 767.
153. Jiang L, Zhao Y, Zhai J. *Angew. Chem. Int. Ed.*, 2004, **43**: 4338.
154. Guo C W, Feng L, Zhai J, Wang G J, Song Y L, Jiang L, Zhu D B. *Chem Phys Chem*, 2004, **5**: 750.
155. Jin M H, Feng X J, Feng L, Sun T L, Zhai J, Li T J, Jiang L. *Adv. Mater.*, 2005, **17**: 1977.
156. Sun T L, Wang G. J, Liu H, Feng L, Jiang L, Zhu D B. *J. Am. Chem. Soc.*, 2004, **125**: 14996.
157. Feng L, Yang Z L, Zhai J, Song Y L, Liu B Q, Ma Y M, Yang Z Z, Jiang L, Zhu D B. *Angew. Chem. Int. Ed.*, 2003, **42**: 4217.
158. Zhu Y, Zhang J C, Zheng Y M, Huang Z B, Feng L, Jiang L. *Adv. Funct. Mater.*, 2006, **16**: 568.
159. Feng L, Zhang Z Y, Mai Z H, Ma Y M, Liu B Q, Jiang L, Zhu D B. *Angew. Chem. Int. Ed.*, 2004, **43**: 2012.
160. Wang S T, Feng L, Liu H, Sun T L, Zhang X, Jiang L, Zhu D B. *Chem Phys Chem*, 2005, **6**: 1475.
161. Feng X J, Zhai J, Jiang L. *Angew. Chem. Int. Ed.*, 2005, **44**: 5115.
162. Xia F, Feng L, Wang S T, Sun T L, Song W L, Jiang W H, Jiang L. *Adv. Mater.*, 2006, **18**: 432.
163. Wang S T, Feng X J, Yao J N, Jiang L. *Angew. Chem. Int. Ed.*, 2006, **45**: 1264.
164. Xia F, Ge H, Hou Y, Sun T L, Zhang G Z, Jiang L. *Adv. Mater.*, 2007, **19**: 2520–2524.

Chapter 2

Living Organisms with Special Surface Performance

Organisms in nature have realized the so-called *intelligent control* through evolution. Through the process of natural selection (survival of the fittest), various features of living species have reached a level of perfection.

This chapter starts with an introduction of some living organisms with special surface properties, including the self-cleaning and anisotropic sliding/rolling properties of the surfaces of plant leaves, the self-cleaning, anti-reflection and structural color of insect wings, the superhydrophobic legs of water strider, the gecko adhesive setae, and the water collection and metachrosis of a desert beetle. This is followed by a discussion on the fundamental causes of these special properties; the relationship between the macroscopic properties and the nanoscale structures is revealed. Also, based on these living organisms with different structures and properties, artificial simulations of these microstructures and their special properties are presented.

2.1 Self-Cleaning Property of the Surfaces of Plant Leaves

2.1.1 *Surface roughness — Lotus effect*

Many leaf surfaces of terrestrial plants are water-repellent, and display a high self-cleaning ability. Contaminating particles adhere to the surface of the coming water droplet and are removed from the leaf when the droplet rolls. This effect is impressively demonstrated in the leaves of *N. nucifera*, and is, therefore, called the "Lotus Effect". Based on the observation of around 300 kinds of plant leaf surfaces, it is reported firstly by Barthlott and Neinhuis that this unique property is based on the coexistence of surface roughness caused by micrometer-scale papillae and the epicuticular hydrophobic wax (Fig. 2.1(a)).[1,2] It is well known that these surfaces show superhydrophobic properties with contact angles greater than 150° and are almost completely free of contamination (Fig. 2.1(b)). These are the two main points that characterize the self-cleaning surfaces.

2.1.1.1   Self-cleaning property of lotus leaf with the micro/nanostructures

Lotus leaf has the reputation of living in the silt but staying pure since ancient times in China. Studies by the authors' groups in 2002 revealed that a superhydrophobic surface with both a large contact angle (CA) and small sliding angle (α) needs the cooperation of nano and microscale structures, and the arrangement of the microstructures on this surface can influence the way that a water droplet tends to move.[4] Figure 2.2(a) shows a typical large-scale scanning electron microscopy (SEM) image of the surface of a lotus leaf. Many papillae were found on this surface in a random distribution with diameters

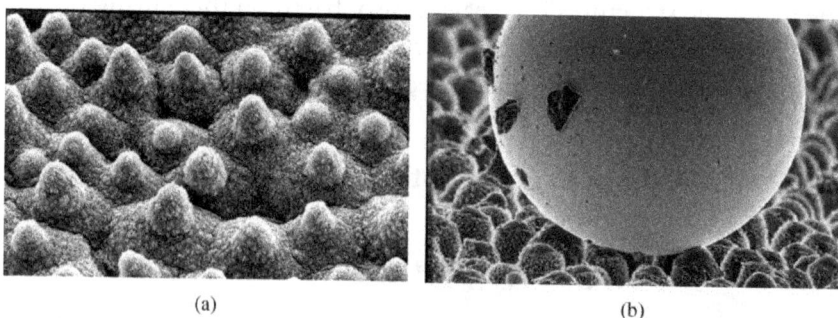

(a)   (b)

Fig. 2.1   Lotus effect. (a) The self-cleaning surfaces of Lotus leaf. (b) Contaminating particles are removed from the leaf when the droplet rolls. (Reprinted with permission from Ref. 3. Copyright 2004 American Chemical Society.)

ranging from 5 to 9 μm. CA and α on the surface of this lotus leaf were about 161.0° ± 2.7° and 2°, respectively. Figure 2.2(b) shows a high-resolution SEM (HRSEM) image of a single papilla, on which hierarchical structures in the form of branch-like nanostructures with average diameters of 124.3 ± 3.2 nm are clearly observed. Nanostructures which would effectively prevent the underside of the leaf from being wetted were also found on the lower surface of the lotus leaf (Fig. 2.2(c)).

A model of such a surface with superhydrophobicity and multiscale character can be inferred as described by Adamson and Gast.[5] The surface roughness was considered on the top of the papillae. Since the hierarchical structures of the lotus leaf are very similar to the description of the triadic Koch curve in fractal geometry,[6] the fractal formula was used to calculate the roughness factor. By changing the

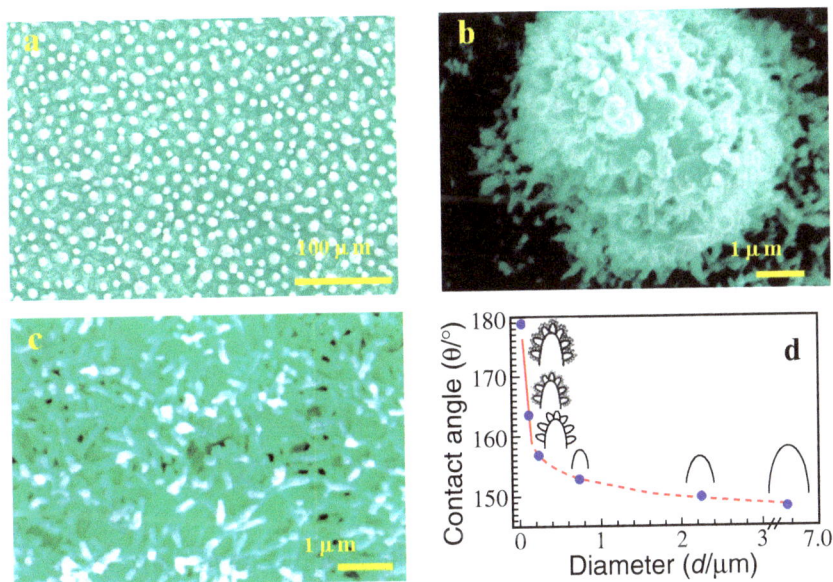

Fig. 2.2 SEM images of the lotus leaves. (a) Large-area SEM image of a lotus leaf. (b) Enlarged view of a single papilla from (a). (c) SEM image of the lower surface of the lotus leaf. (d) The fitted curve based on calculated data (contact angle against the mean outer diameter of protruding structures).[4] (Copyright Wiley–VCH Verlag GmbH & Co. KGaA. Reproduced with permission.)

roughness factor, a mathematical model was evolved as follows to describe the relationship between CA on a rough surface (θ_f) and that on a responding smooth surface (θ):

$$\cos\theta_f = f_s \left(\frac{L}{l}\right)^{D-2} \cos\theta - f_v. \tag{2.1}$$

Here, $(L/l)^{D-2}$ is roughness factor, where L and l are the upper and lower limit scales of the fractal behavior of the surface, respectively, and D is the fractal dimension. For the surface of lotus leaf, L and l are defined with the size of the diameters of the papillae and branch-like nanostructures, respectively. From the curve model, the value of D in three-dimensional space was found to be about 2.2618, and the value of L/l equal to 3^n. The value of n is determined by the object acting on the fractal structures. The surface roughness factor becomes larger with increasing n value. Therefore, the larger the value of n, the smaller that of l when the upper limit scale L is certain. In Eq. (2.1), f_s and f_v are the fractions of the surface under the water droplet occupied by solid material and air, respectively; ($f_s + f_v = 1$). f_s and f_v in Fig. 2.2(a) were evaluated as 0.2056 and 0.7944, respectively, and the value of θ was $104.6 \pm 0.5°$ measured from the primary reference.[7] Therefore, when n equals 0, 1, 2, 3 and 4, respectively, the values of θ_f can be calculated as 147.8°, 149.7°, 152.4°, 156.5° and 163.4° (Fig. 2.2(d)). It was indicated on the curve that the diameter was about 128 nm when CA was about 160°.

2.1.1.2 *Biomimetic self-cleaning surface — Lotus leaf–type structure of aligned carbon nanotube film*

As mentioned earlier, the micro-geometry of a lotus leaf surface with self-cleaning ability is actually a kind of hierarchical micro/nano-composite structure. However, the mere nanoscale structures can lead to the superhydrophobic phenomenon. In order to study the role between micro/nano-composite and pure nanostructure, the densely-packed aligned carbon nanotube (ACNT) films with lotus-like structure and pure nanostructure were prepared respectively.[8] First, we used laser etching methods to adjust the distribution of catalyst on the surface, and prepared the lotus-like surface structure.[4] Figure 2.3(a) shows the SEM image of the top view of the lotus-like ACNT films; the average diameter of the papillae were about 3 μm. The HRSEM image of one single micropapilla with nanostructures is shown in Fig. 2.3(b). CA on these ACNT films was about 166°, and α was about 3°. The lotus-like ACNT films have both a larger CA and a lower α, that is, anti-adhesion to water. In comparison, Fig. 2.3(c) and (d) show SEM images of the top and cross-sectional views of the densely packed ACNT films, respectively. The CA on these nano-structured films

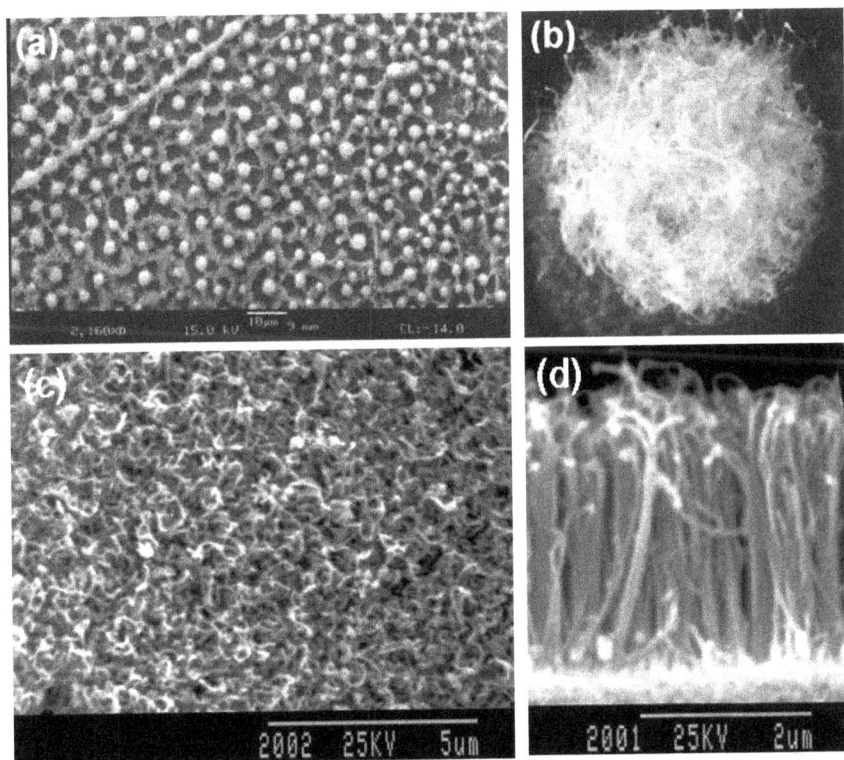

Fig. 2.3 SEM images of ACNT films. (a) Top view of the lotus-like ACNT films.[4] (b) Enlarged view of a single papilla from (a).[4] (c) and (d): Top and cross-sectional view of densely packed ACNT films.[8] (Copyright Wiley–VCH Verlag GmbH & Co. KGaA. Reproduced with permission.)

was about 158°. However, the value of α, the difference between advancing and receding CAs was higher than 30°.

It was indicated that the hierarchical composite structures would make the contact angle increase to a certain extent and enable the rolling angle to reduce greatly, that is, water droplets on the surface can significantly reduce the adhesion. Detailed theoretical studies have shown that complex hierarchical structure can greatly reduce the extent of contact of a droplet with a solid surface, thus affecting the outline, perimeter and continuity of the three-phase line and so on, leading to a decline in the rolling angle.[9–11] This will be discussed in more details in Chapter 3.

In short, superhydrophobic ability can be induced by simple micro- or nanostructure, but it is difficult for water droplets on the surface to roll down. Only the complex hierarchical micro/nano-structure can form the self-cleaning surface both with larger contact angle and smaller rolling angle (Fig. 2.4).

2.1.1.3 *Artificial superhydrophobic surfaces with self-cleaning properties*

Self-cleaning ability can be induced by the microstructure of leaf surface, which provides us with the inspiration of building artificial superhydrophobic surface. The leaf itself is a good template, and it is expected to be similar to the leaves with microstructure and self-cleaning property by copying its structure. The wetting and the self-cleaning properties of three types of superhydrophobic surfaces have been investigated by Fürstner and Barthlott[12]: silicon wafer specimens with different periodic arrays of spikes hydrophobicized by chemical treatment, replicates of water-repellent leaves of plants, and

(a)   (b)

Fig. 2.4 Schematic drawings of (a) simple microstructure and (b) hierarchical micro/nanostructure.

industrially fabricated metal foils additionally hydrophobicized by means of a fluorinated agent. Ji *et al.* used a casting process to produce artificial Lotus leaf, which used poly (dimethylsiloxane) (PDMS), a soft material widely used for casting to replicate the surface structures.[13] Kwon and Lee recently reported a mass-production method of highly hydrophobic surfaces by simply replicating the highly hydrophobic plant leaf surfaces in two steps: the first step makes a nickel mould via electroforming and the second step replicates via a UV-nanoimprint lithography. When making a nickel mould, either a plant leaf or its negative polymer replica is used as a mandrel in electroforming, and final products become positive or negative polymer replicas of a plant leaf, respectively. It is found that the nickel-mould making using the plant leaf as a mandrel is quite successful and the final products in the form of a positive replica are better than those in the form of a negative replica in terms of replication quality and hydrophobicity. Contact angle values of the positive replicas are smaller than those of the natural leaves' surfaces by only $2°–5°$.[14]

2.1.2 *Hairy structures — Elastic effect*

The self-cleaning ability of lotus leaf which makes water bead roll off completely and thereby wash off contamination very effectively has been termed the "Lotus effect", although it is observed not only on the leaves of the Lotus, but also on many other species. Herminghaus *et al.*[15] recently proposed that water-repellent plant leaves could be divided into at least two distinctly different types. The first type of leaves, such as Lotus and Indian cress, looks macroscopically smooth. The mat appearance suggests the presence of some structure at small scales, which has to be considered to understand the wetting behavior. The second type is hair-covered leaves such as the Lady's Mantle (*Alchemilla vulgaris L.*). Figure 2.5(a) shows that a water droplet which is put onto the leaf carefully rests on the fur as a sphere with a contact angle close to $180°$.

The hair-covered leaves have superhydrophobic surface that is quite distinct from the lotus leaf surface. Because of the role of surface

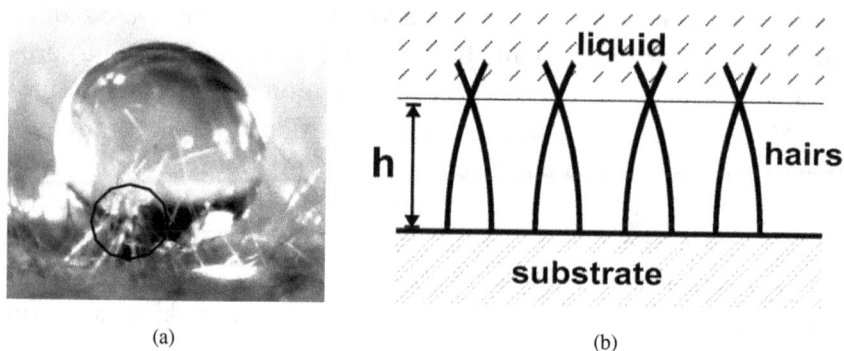

Fig. 2.5   Elastic effect of hairy structures. (a) Condensed hair cluster holding a water droplet. (b) Sketch of the elastic deformation of the hairs due to bundle formation at the liquid-air interface. (Reprinted with permission from Ref. 15. Copyright 2004 American Chemical Society.)

tension to force the hairs into bundles, most of the hairs have to bend, and this costs elastic energy.

At this point, water-air interface is maintained by hair, and relies on its flexibility to stop water droplets from wetting the surface. However, the formation of the superhydrophobic structure relying on this approach is not stable and it is just a steady-state kinetics, because the hair is hydrophilic and flexible. When water pressure is greater, the droplet can still contact the substrate, and the entire surface is wetted. At this time, the wetting comes back to the thermodynamic stability determined by the rough structure and composition of surface.

The contact angle, as obtained from the droplet meniscus at the hair, was found to be below 60° along the hair. The advancing contact angle, which was checked by moving the hair into the drop, was not found to be significantly larger. Another interesting observation was made by Herminghaus' group when they looked with the optical microscope at small droplets nucleating on a cooled Lady's Mantle leaf. The small droplets nucleated on the ground with a CA of more than 90°, but as soon as they made contact with the hairs, they were lifted from the cuticle into the brush, and stuck to the hairs. They seemed more hydrophilic than the cuticle and were thus energetically preferred. The droplets sticking to the hairs showed the known states of fiber wetting and coalesced when they got into contact with other droplets. SEM image confirmed that on a small scale, the leaves of the Lady's Mantle were covered with cuticular wax dandruffs. As a result, on a microscopic scale the droplets sensed the same hydrophobicity similar to the rough surface of the lotus leaf structure. During the growth of the droplets, they were easily moved to the brush of hydrophilic hairs, apparently because the latter were more hydrophilic than the cuticle.

To understand why a brush of hydrophilic hairs might be effectively hydrophobic, Herminghaus *et al.* also considered a bundle of hairs stuck into a liquid-air interface and calculated the elasticity of the hairs (Fig. 2.5(b)). If the hairs had a CA different from 90°, the liquid surface would deform around the hair according to Young's equation. The surface energy of a liquid was defined with $f(1 + |\Delta f^2|)^{1/2}$, where Δf represents deformation. Deformation of the liquid surface in either way would cost energy and results in an attractive potential between the hairs. It was noteworthy that this was the case for any microscopic CA different from 90°, either above or below. So it could be expected that a number of hairs would condense to individual bundles, such that the hairs met each other at the water-air interface. To force the hairs into bundles, most of them would cost elastic energy to bend. Supposing an infinite sample, the total energy of the system was found to be minimized when the hairs gathered into bundles of a particular size, which depended on the distance h between the substrate (cuticle) and the water-air interface. To move the interface closer to the substrate, on which the hairs were anchored, the hairs would bend sharply which can be seen in Fig. 2.5(b). Herminghaus *et al.* proposed the elastic energy contribution Γ:

$$\Gamma \propto \sqrt{K} h^{-1/2} \tag{2.2}$$

where K represented the elastic modulus of the hairs. If the mean density of hairs was high enough to form clusters and if the contact angle of water at the hairs differed significantly from 90°, a droplet would be kept away from the cuticle, as was observed on the macroscopic scale. Nevertheless, they also argued that this state could be overcome by forcing the liquid all the way to the cuticle, so that the hairs were completely immersed in water. In this state, the CA was only determined by the roughness of the cuticle.

To sum up, it is indicated that the water repellence of certain plant leaves can be explained by considering surface roughness on different length scales and sometimes additional elastic structures (hairs) covering the surface. For the hairy plant leaves, the elastic modulus of the hairs and the aerial coverage are important additional parameters to gain hydrophobicity with spherical droplets resting on the leaf's fur, even though the hairs are hydrophilic.

2.2 Surface Anisotropy

2.2.1 *The effect of the arrangement of microstructures on sliding anisotropy*

Anisotropic wetting and dewetting are very important properties on patterned surfaces that have recently attracted much interest. In plants, it is well known that a water droplet can roll freely in all directions on the surface of a lotus leaf. However, the authors' group recently found a sliding anisotropy tendency on the surface of a rice leaf.[4] Figure 2.6(a)

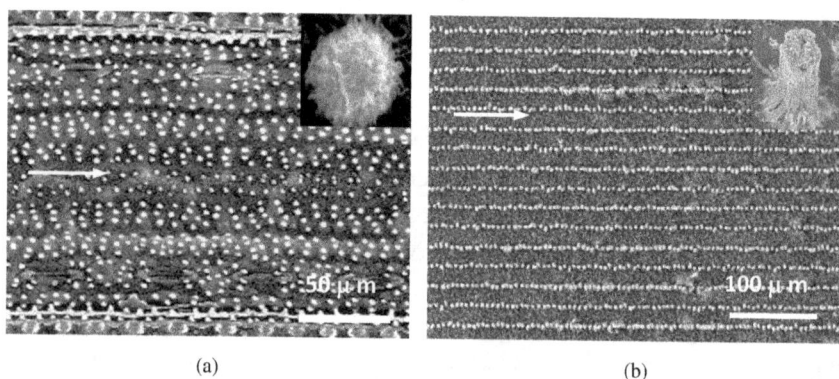

(a)   (b)

Fig. 2.6   Topography of the surface of a rice leaf and a rice-like ACNT. (a) SEM image of the surface of a rice leaf (the inset is a HRSEM image of an individual papilla). (b) SEM image of the top view of a rice-like ACNT film (the inset is a HRSEM image of an individual cluster of carbon nanotubes).[4] (Copyright Wiley–VCH Verlag GmbH & Co. KGaA. Reproduced with permission.)

shows a large-scale image of the surface of a rice leaf (the inset is a HRSEM image of an individual papilla), which has micro/nanostructures similar to those of the lotus leaf. However, the papillae are arranged in one-dimensional order parallel to the leaf edge (direction of arrow) and randomly in the other directions (perpendicular to the arrow). Correspondingly, the value of sliding angle was different for water rolling along the direction of the arrow (3°–5°) and perpendicular to the arrow (9°–15°). It is considered that the arrangement of micropapillae on the surface of a rice leaf could influence the motions of water droplets, causing a sliding anisotropy tendency on the surface. These results offered important information for designing controllable wettability on a solid surface. Accordingly, researchers prepared the rice-like ACNT films shown in Fig. 2.6(b), the ordered pattern of these films is in two dimensions with different spacings. The result shows that the water droplets rolled readily across the larger spacings (along arrow direction).

Besides the rice leaf, the feathers of some water bird, goose and duck also show a sliding anisotropy when water droplets roll off the feathers.[16] It is a common phenomenon that goose and duck frolic and look for food in water while keeping their feathers dry. Compared to dogs, they do not need to shake their bodies fiercely to get rid of water. This is because goose and duck feathers are waterproof. This waterproof function is attributed to the micro/submicro- well-ordered strip-shaped structures of their feathers (Fig. 2.7). This kind of structures endow feathers with good hydrophobicity and air permeability. Furthermore, such well-ordered microstructures have the function of directional water-repellency so that water droplets can roll along the outer side of strips easily.

Recently, the authors' group revealed that the surface of butterfly (*Morpho aega*)'s wings also exhibited the sliding anisotropy.[17] SEM observations reveal that the wings of the butterfly are covered by a large number of micro-scales. Each scale is consisted of well-ordered nano-stripes, each of which is stacked by tilted periodic lamellae (Fig. 2.8). This structure makes the surface of butterfly's wings exhibit the anisotropy of surface wettability, which means that the surface has different CA and SA in different directions.

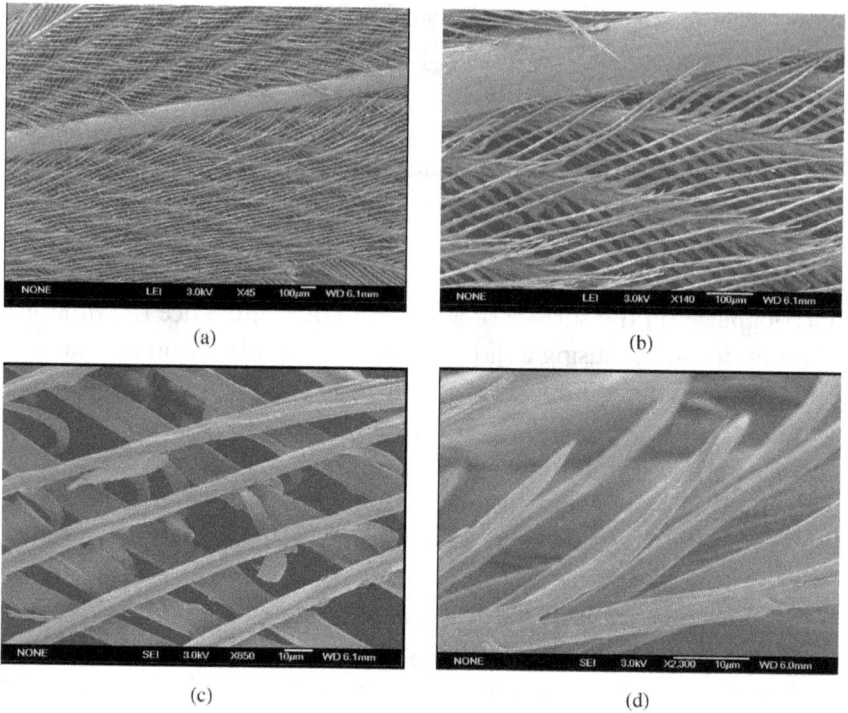

Fig. 2.7 SEM images of the duck feather. (a)–(d): Gradual magnifying images indicate the micron and sub-micron well-ordered strip-shaped structures.

The models proposed to elucidate the potential mechanism of the sliding anisotropy on the surface of butterfly's wings are shown in Fig. 2.9. When the wing is not tilted, the tips of the micro-scales and nano-stripes are in natural state so that the water droplet is in a metastable state. When the wing is tilted downward, the sliding angle (SA) is low, and a droplet easily rolls off the surface of the wings along the radial outward (RO) direction of the central axis of the body. But the water droplet is pinned tightly against the RO direction even when it is fully upright, which means that the water SA is high in this direction. While in the two vertical directions of RO, it is hard for both of water droplets to roll off the surface, so the water SAs are also high in these directions. The reason for the sliding anisotropy on the surface of butterfly's wing is that specific microstructure of the surface

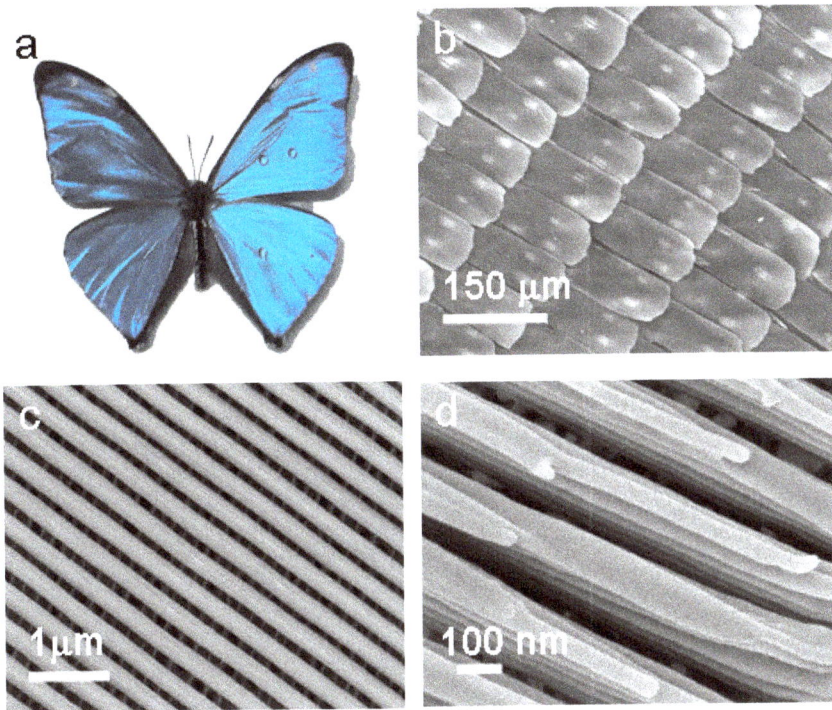

Fig. 2.8 Images of the surface of the butterfly's wings (*Morpho aega*). (a) Butterfly. (b)–(d): SEM images of the surface. (From Ref. 17 — Reproduced by permission of The Royal Society of Chemistry.)

causes the three-phase contact line (TCL) in each direction to become different. When the wing is tilted downward, the micro-scales with ridged nano-stripes are spatially separated from each other. In this case, air can be efficiently trapped between the water droplet and the wing. As a result, it forms the extremely discontinuous TCL. So the contact area between the water droplet and the surface is small and the droplet easily rolls off the wings along the RO direction. However, when the wing was tilted upward, the flexible nano-stripes and micro-scales take on a close arrangement so that there is little air existing between the water droplet and the wing. As a result, a quasi-continuous TCL is formed, and the contact area between the water droplet and the surface is large. So the droplet pins tightly on the wing as it is tilted upward.

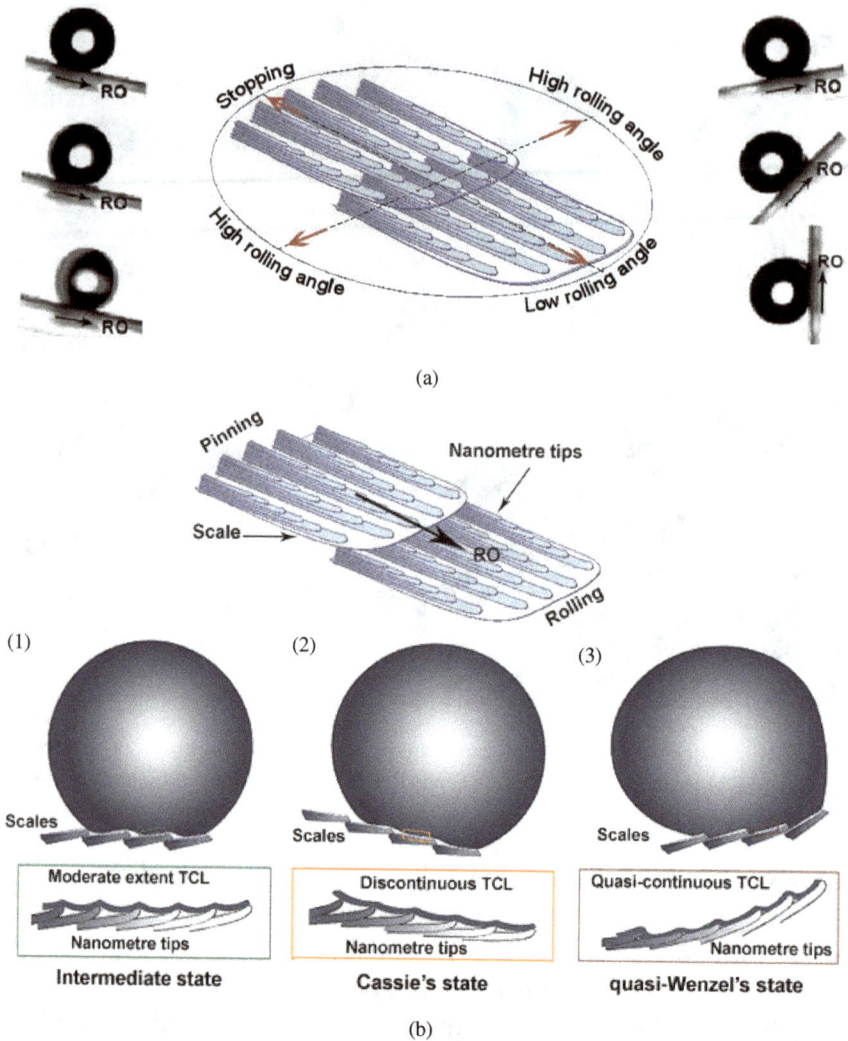

(a)

(b)

Fig. 2.9   The models proposed for elucidating the potential mechanism of the sliding anisotropy on the surface of butterfly's wing. (a) The sliding anisotropy: (1) Roll off easily along the RO; (2) Different sliding anisotropy in different direction; (3) Pin on the wing against the RO. (b) Three states of superhydrophobicity caused by the micro-scales and nano-strips on the wing: (1) metastable state; (2) composite contact state; (3) wetting contact state. (From Ref. 17 — Reproduced by permission of The Royal Society of Chemistry.)

The arrangement of surface microstructures has an important effect on the sliding anisotropy. The TCL is considered as the main factor of affecting water droplets' rolling behavior. For example, on a rice leaf surface, the TCL is determined by micropapillae. In contrast, on the surface of lotus leaf, the distribution of micropapillae is uniform, thus in all directions the TCLs are similar. Multi-composite structures on the surfaces of duck's feathers and butterfly's wings also confer themselves sliding anisotropy. This feature not only represents the organism's own properties and specific functions, but also provides people with new inspiration: the thought of varying parameters to synthesize practical superhydrophobic surface can be applied in the microfluid area based on the nanotube technology.

2.2.2 *Anisotropic aligned carbon nanotubes (ACNTs) — the changing of wettability induced by surface microstructures*

The authors' group also studied the influence of three-dimensional anisotropic microstructures on the surface wettability.[18] We fabricated three-dimensional anisotropic ACNT films by chemical vapor deposition (CVD) method on silicon templates with well-defined structures, and studied its influence on wettability. The results showed that superhydrophobicity with a water contact angle (CA) larger than 150° and hydrophilicity with a water CA smaller than 30° could both be achieved on the anisotropic ACNT films by simply varying the structural parameters without alteration of surface free energy. Studies indicated that this effect could be explained by the coexistence of the horizontal and vertical ACNT arrays. This work develops a new method of controlling surface wettability by utilizing surface anisotropic geometric patterns.

Figure 2.10(a)–(c) show SEM images of periodic patterns of ACNT microstructures on silicon templates with spacings between quadrate pillars of 20 μm, 15 μm, and 10 μm, respectively. Mutually orthogonal ACNT microstructures can be clearly seen in the higher magnification image (Fig. 2.10(d)). The ACNTs were densely packed

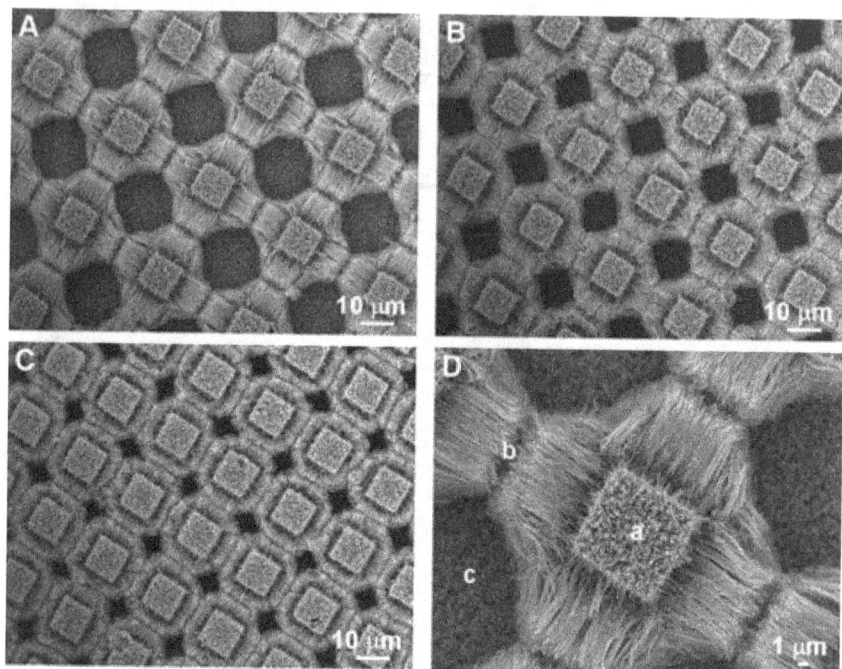

Fig. 2.10 SEM images of three-dimensional anisotropic ACNT films. (a)–(c): Periodic ACNT arrays with pillar spacings of 20 μm, 15 μm, and 10 μm, respectively. (d) Magnified image of (a). (Reprinted with permission from Ref. 18. Copyright 2004 American Chemical Society.)

with a fairly uniform length (about 10 μm) and diameters ranging from ~25 nm to ~50 nm. The areas of (1) and (2) respectively represent the vertical ACNTs on the top of quadrate silicon pillars and horizontal ACNTs from the side faces, while area (3) represents ACNTs grown from the bottom of the templates. When the pillar spacing is large, the horizontal ACNTs are rather straight. While the spacing reduces, the horizontal ACNTs bend upward owing to the space hindrance, and some extent of disorder of ACNT growth appears. The extent of bending and disorder increases with the decrease of pillar spacing.

Surface free energy is an important factor for the surface wettability. In order to contrast with the unmodified ACNT films, we modified the surface of ACNT films with different molecules which have different free energy: vinyltrimethoxysilane (VTMS) and (2-(perfluorooctyl)ethyl)trimethoxysilane (FETMS). The wettability of the untreated and VTMS-modified anisotropic ACNT films shows great dependency on the pillar spacing of templates. For untreated films with a pillar spacing of 20, 15, 10 and 6 μm, the equilibrium CAs are $22.2 \pm 4.1°$, $142.9 \pm 1.8°$, $25.5 \pm 2.7°$, and $10 \pm 1.5°$. The water CA of ACNT films on flat substrates is 158°. Figure 2.11 indicates the relationship between the water spreading behavior and pillar spacings when the water drops spread on the VTMS-modified anisotropic ACNT films. The water drops spread rapidly on films with small spacings. On films of 6 μm and 10 μm, the water CA reaches 20.8° in 0.2 s and 27.2° in 1.5 s, respectively. While for that of 20 μm, the water drop could stand on the film for several seconds before its rapid spreading. The final CA reaches 21.2°. Similar to the untreated film, the ACNT films grown from flat or proper spacing template both exhibit superhydrophobic properties (CA is 155°). When modified with FETMS, films with different structural parameters all exhibit superhydrophobic properties. Pillar spacings of 20, 15, 10 and 6 μm all show superhydrophobic properties with water CAs of $162.7 \pm 1.5°$,

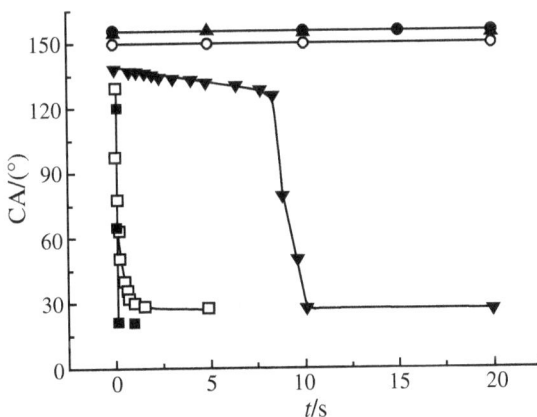

Fig. 2.11 Water spreading behavior on anisotropic ACNT films with pillar spacings of ■ 6 μm, □ 10 μm, ○ 13 μm, ● 15 μm, ▼ 20 μm and ▲ flat silicon substrate. (Reprinted with permission from Ref. 18. Copyright 2004 American Chemical Society.)

159.6 ± 0.9°, 154.2 ± 1.8°, and 153.8 ± 2.6°, respectively. These results indicate that the films can change from very hydrophilic to very hydrophobic by simply varying the structural parameters without changing surface free energy.

2.3 The Self-Cleaning and Anti-Reflection Functions of the Surfaces of Insect Wings

In some finned insects, including *odonata* (such as dragonfly), *hymenoptera* (such as bee), *trichoptera* (such as moth), *lepidoptera* (such as butterfly), *homoptera* (such as cicada), *coleoptera* (such as beetle), *diptera* (such as mosquito and fly) and so on, the surfaces of their wings also have self-cleaning performance, because all of them have different microstructures on their wings.[19,20]

The authors' group mainly studied the surface microstructures and wettability of cicada (*Tettigia orni*)'s wing. The result indicates that there are a number of nanopillar structures which can lead to superhydrophobicity to be uniformly distributed on the surface of cicada's wing (Fig. 2.12). By imitating the microstructure of cicada's wing surface, we prepared polycarbonate (PC) nanopillar film through template process.[21] The study of wettability on the film indicated that the smooth PC film was hydrophilic (contact angle was 85.7°); whereas the PC nanopillar film was strongly hydrophobic (contact angle was 145.6°). When the diameter of PC nanopillar increases, the contact angle of the film decreases (Fig. 2.13). Lee *et al.*[22] prepared a nanostructure polystyrene (PS) array film, which looked like the surface of cicada's wing and exhibited superhydrophobicity, using an analogous process. They then discussed the surface dynamic wetting behavior with different PS pillar diameters.

As mentioned in the first chapter, the periodical arranged microstructure on the scales of butterfly's wings can generate a specific structural color. Simultaneously, such a specific structure can also confer self-cleaning performance to the surface. The water CA of the surface is greater than 150°.[23] Based on the above research, Gu *et al.*[24] from Southeast University successfully obtained the biomimic nanomaterial with both structural color property and superhydrophobicity

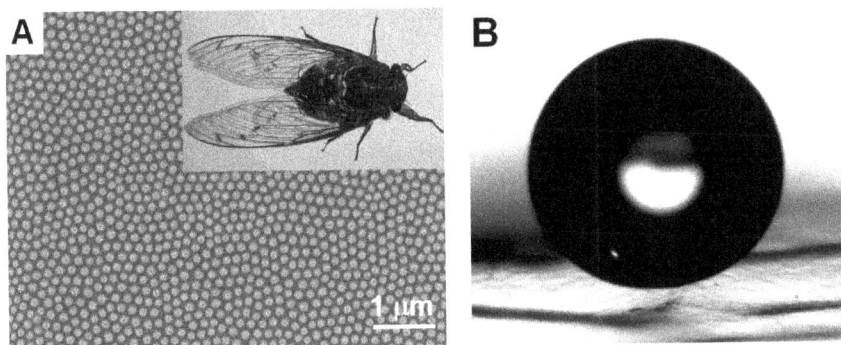

Fig. 2.12 Microstructures and wetting properties of the cicada's wing. (a) SEM of the self-cleaning surface of the cicada's wing. (b) Image of the water droplet on the surface of the cicada's wing.

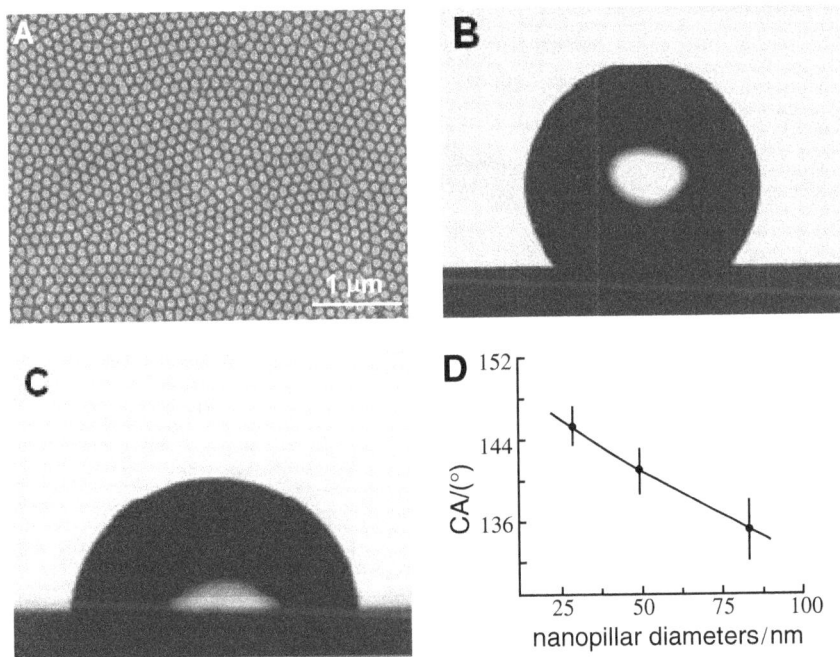

Fig. 2.13 Cicada's wing-like polystyrene (PS) nanopillar array.[21] (a) SEM image of polystyrene (PS) nanopillar array. (b) The water CA on the polystyrene (PS) nanopillar array film. (c) The water CA on the flat polystyrene (PS) film. (d) The relationship between nanopillar diameters and the water CAs. (Copyright Wiley–VCH Verlag GmbH & Co. KGaA. Reproduced with permission.)

by imitating the butterfly's (*Morpho*) structure and characteristics. They dispersed the polystyrene microspheres of different diameters and SiO$_2$ nanoparticles in water by ultrasonating polystyrene microspheres (diameter of several hundred micrometers) and SiO$_2$ nanoparticles. The concentrations of polystyrene with the diameter of several hundred micrometers and SiO$_2$ with diameter around 6 nm are 2% and 0.3%, respectively. Then they used glasses lifting and pulling in the suspension to form the film. In order to remove the polymer and bring about the curing process of SiO$_2$ nanoparticles, the formed film was dried in air and then calcined at 450°C. Figure 2.14 shows that the film surface has uniform, coarse nanostructure. Besides, the film exhibited superhydrophobicity (contact angle is 155°) after being modified with fluorosilane.

The special protrusive nanostructure of the cicada's wing surface makes it not only have the self-cleaning property, but also have reduced reflection function, so it can resist external aggression.[25] These two characteristics are the results of the long-term natural selection and evolution. Other insects' wings also have similar features.

Besides, the structure of some insects' compound eyes makes them have anti-reflecting characteristics. According to the former research, the compound eyes of moths consist of ordered hexagonal nanostructure arrays. These arrays form the homogeneous transparent layer of the corneal surface. Each protrusive nanostructure is one

Fig. 2.14   SEM image of an inverse opal PS film.[24] (a) A plan-view image. (b) The cross-sectional image. (Copyright Wiley–VCH Verlag GmbH & Co. KGaA. Reproduced with permission.)

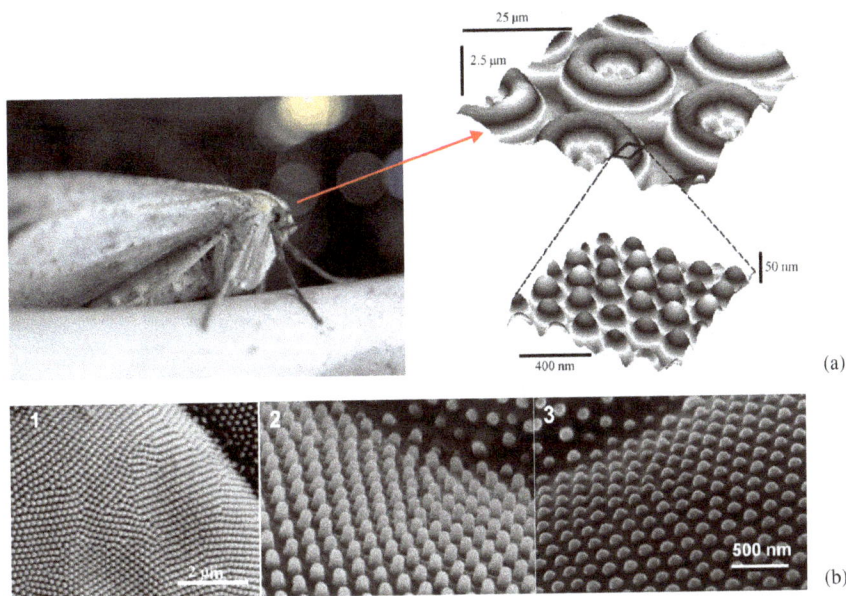

Fig. 2.15 (a) The microstructure of compound eyes and single eyes of the moth. (b) The different butterfly corneal nipple arrays: (1) *Inachis io*; (2) *nymphalid Polygonia c-aureum*; (3) *lycaenid Pseudozizeeria maha*. [(Figure (a) — Reprinted from Ref. 25, Copyright 2004, with permission from Elsevier. Figure (b) — Reprinted from Ref. 26, Copyright 2006, with permission from the Royal Society.]

anti-reflective unit (Fig. 2.15(a)). The low reflection generated by the protrusive nanostructure of the moth's eyes gives them abnormal inkiness and makes them less visible for predators even if the moth flies in the evening — this is called moth eye effect. Studies have found that butterfly's cornea also has similar structures (Fig. 2.15(b)).[26] Furthermore, different species of butterfly's compound eyes have different size, resulting in the difference of surface anti-reflective degree. The larger the height of nanostructure, the lower the surface reflectivity.

Recently, Yang group of Jilin University and the authors' group[27] co-discovered that the eyes of mosquito also had excellent super-hydrophobicity and antifogging property.[27] These characteristics allow mosquitoes to keep clear vision in damp environment. Both characteristics are generated by micronipples and the hexagonally close-packed nanostructures on the micronipples (Fig. 2.16(a)–(e)). We have also

Fig. 2.16   Mosquito compound eyes and the artificial counterpart. (a)–(e): Micro- and nanostructures of mosquito compound eyes. (f) Artificial compound-eye analogues.[27] (Copyright Wiley–VCH Verlag GmbH & Co. KGaA. Reproduced with permission.)

fabricated artificial compound eyes by soft-lithography methods and investigated the effects of micro/nanostructures on the properties (Fig. 2.16(f)). These research results will provide important scientific and technological bases for the design and development of new optical devices.

As described above, the surface microstructures may render organisms themselves multiple functions. For example, the nanopillar structures of cicada's wing surface and the hexagonally close-packed nanostructures of insect compound eyes give them self-cleaning and antireflecting properties; the layered periodic structures of butterfly's wing give it self-cleaning property and structural color. These results provide important foundation in both theory and practice for bionic functional materials.

2.4 Walking on Water — Water Strider

2.4.1 *The mechanism of water strider walking on water*

Water strider *Gerridae* is a small insect of about 1 cm in length that usually resides on the surface of ponds, slow streams, and marshes. Like other insects, it has six jointed legs, which are covered with numerous setae. Water strider is also known as pond skater, for it not only can walk on the water freely, but also can "fly" on the water surface like an ice-skater, while the legs do not pierce the water surface and are kept dry. How can the water strider stand on the water surface effortlessly, even move and jump rapidly? This has caught wild concerns in scientific field for a long time.

In the past, scientist considered that water striders on the water surface were supported by the surface tension generated by the curvature of the free surface, and they propelled themselves by driving their central pair of hydrophobic legs in a sculling motion.[28,29] Moreover, Prof. Bush and co-workers from MIT, USA have captured the walking process of different insects on water by high-speed video and particle-tracking system. They revealed that the strider transferred momentum to the underlying fluid primarily not through the capillary waves, but rather through hemispherical vortices shed by its driving hairy legs (Fig. 2.17). Analogously, some other creatures such as birds, bats, insects, and fish, also propel themselves across the water surface by a series of vortex as they move.[31,32]

Furthermore, based on the mechanism of insect's water walking, scientists have designed a mechanical water strider, Robostrider, constructed to mimic the motion of a water strider (Fig. 2.18). Robostrider was 9 cm long, weighed 0.35 g, and had proportions consistent with those of its natural counterpart. Its body was made from lightweight

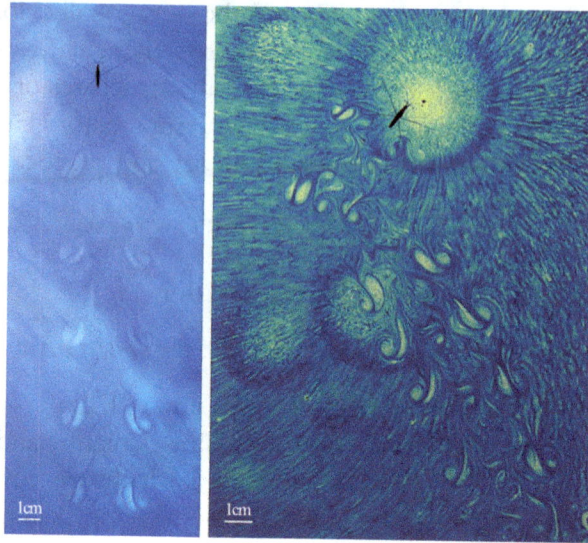

Fig. 2.17   The locomotion of water strider on water surface captured by high-speed video and particle-tracking system. The propulsion forward is primarily from the vortices beneath the surface of water shed by the strider's driving legs.[30] (Reprinted by permission from Macmillan Publishers Ltd: *Nature*, copyright 2003.)

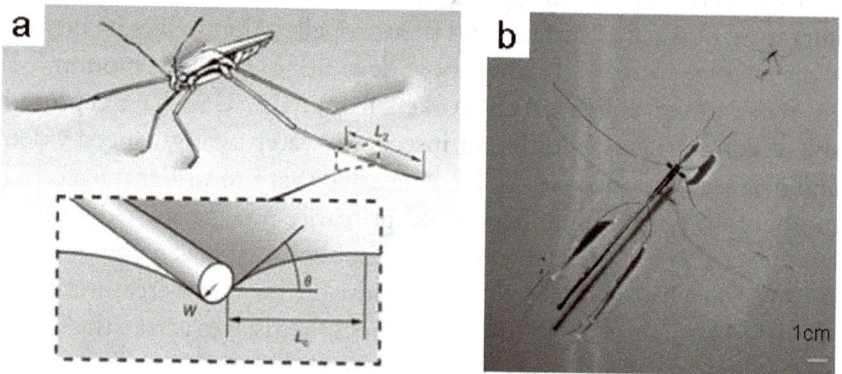

Fig. 2.18   A static water strider on the free surface and its artificial counterpart, Robostrider.[30] (a) Mechanical analysis of a static water strider on the free surface using a smooth structure model. (b) A water strider (upper-right corner, the small one) and Robostrider (lower-left corner, the big one). (Reprinted by permission from Macmillan Publishers Ltd: *Nature*, copyright 2003.)

aluminium, and two pairs of its legs were made of hydrophobic stainless steel wire, while another pair of legs were an elastic thread and a pulley. Robostrider supported itself on the water surface via the curvature force provided by its bent legs.

In 2004, the authors' group revealed the secret of water strider standing effortlessly and walking quickly on water surface in essence. We proposed that the special micro/nano-hierarchical structures of the legs are more important in inducing the superior water resistance (Fig. 2.19). High resolution SEM observation indicated that there were numerous tiny oriented micrometer-scale needle-shaped setae covering the legs (Fig. 2.19(b)). The diameter of human hair varies from 80 to 100 μm, while that of the water strider setae is less than 3 μm. Moreover, many elaborate helical nanoscale grooves were found on each microseta, forming the unique hierarchical structure (Fig. 2.19(c) and (d)). Air is trapped in the spaces of microsetae and nanogrooves to form an air cushion at the leg/water interface that prevents the legs from getting wet. Therefore, water strider can move quickly on the water surface without wetting the legs, and the superhydrophobicity of the legs is presented in macroscopic view

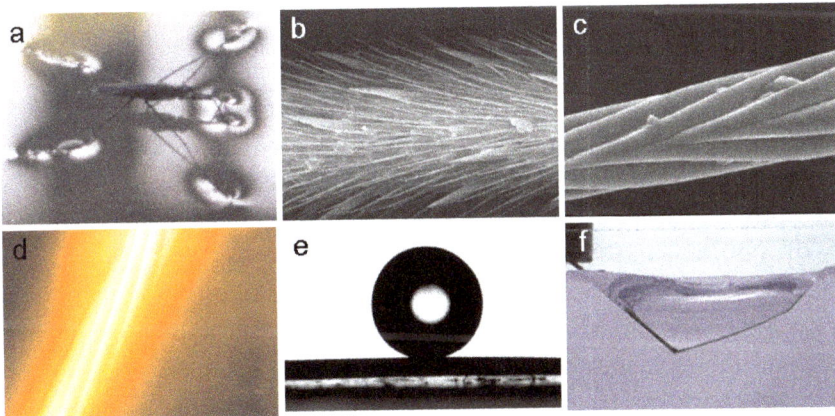

Fig. 2.19 (a) A water strider standing on water surface. (b) and (c): SEM images of the leg with mircosetae and nanogrooves. (d) AFM image of the nanogrooves on a seta. (e) A water droplet on a leg. (f) The maximal-depth dimple just before the leg pierces the water surface.[33]

(Fig. 2.19(e)). We have also found that 4 mm deep dimple could form when water strider's leg strikes the water surface, and the volume of water ejected is roughly 300 times as large as that of the leg itself (Fig. 2.19(f)). This water repellency is the origin of its high surface supporting force, and the force from just one leg is enough to support 15 times the total body weight on the water surface. This striking supporting force is the surest guarantee that the water strider can move freely on water surface without drowning even in violent storm and torrent. This discovery can be helpful in producing new waterproof materials. More importantly, it may shed light on the design of new aquatic devices.

2.4.2 *Superhydrophobic surface mimicking the legs of water striders*

Zhang's group[34] from Tsinghua University reported a method for the fabrication of floating superhydrophobic gold thread to mimic the leg of water striders by combining the layer-by-layer (LbL) self-assembly technique and electrochemical deposition of gold aggregates (Fig. 2.20). In their earlier paper, they had proposed a way for fabricating superhydrophobic surfaces.[35] First, an ITO electrode, which was modified with a polyelectrolyte multilayer by LbL assembly, was covered with dendritic gold clusters by electrochemical deposition. After further chemisorption of a self-assembled monolayer of *n*-dodecanethiol, a superhydrophobic film was formed. The contact angle on the as-prepared film surface varied obviously with the time of electrochemical deposition. The superhydrophobic surface prepared by this assembly method was stable. The process of electrochemical deposition was independent of the size and shape of substrate, so using this method to form superhydrophobic surface could be applied to the various materials, such as the gold threads above. This research may have new and promising applications for biomimetic drag-reducing and quick propulsion techniques.

Shi's group[36] from Tsinghua University reported a simple route for producing copper wire with superhydrophobic submicrofiber coating to mimic the legs of water striders. To compare with natural

(a)

(b)

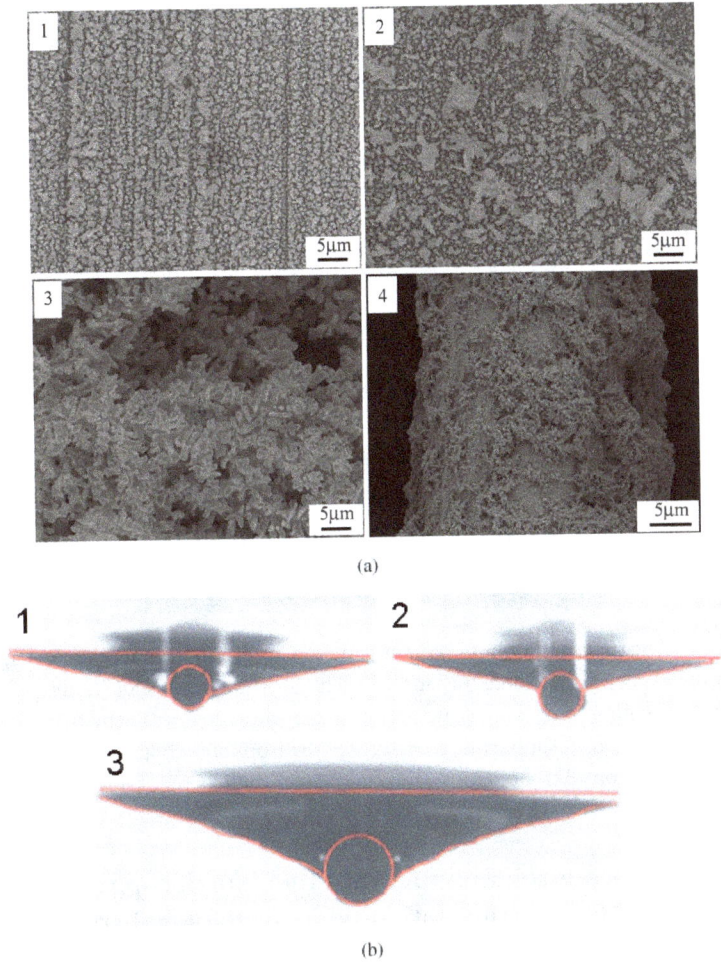

Fig. 2.20 Floating golden thread mimicking the structures of water strider's leg.[34] (a) SEM image of the as-prepared gold threads (magnified gradually from 1 to 4). (b) Gold threads floating on the water surface: (1) Superhydrophic modified gold thread with the diameter of 0.5 mm; (2) Unmodified one with the diameter of 0.5 mm; (3) Modified one with the diameter of 0.8 mm. (Copyright Wiley–VCH Verlag GmbH & Co. KGaA. Reproduced with permission.)

water strider's leg, they measured the maximal supporting force of the superhydrophobic copper wire on water surface (Fig. 2.21). To prepare the superhydrophobic coating, first, they coated a copper layer on arbitrary solid substrate, then the copper was anodized in a

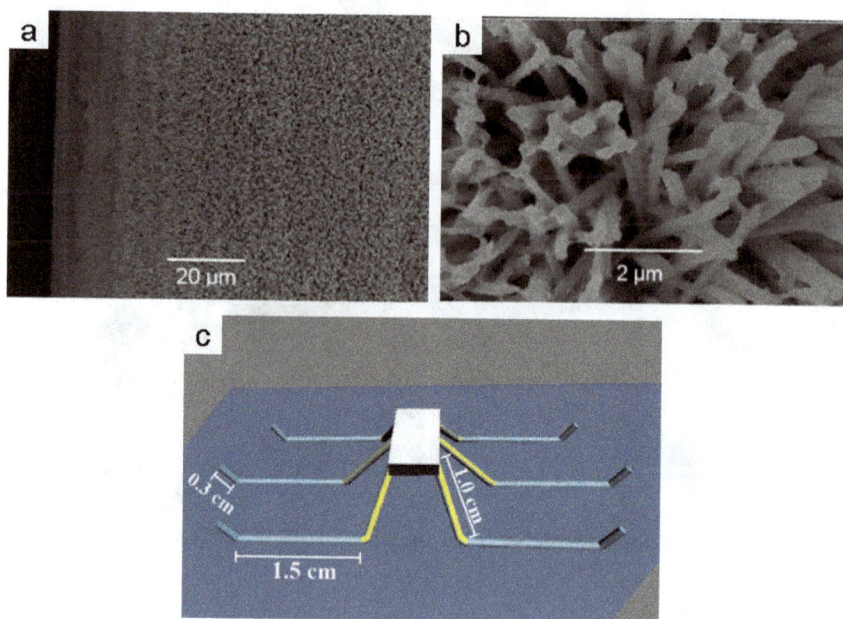

Fig. 2.21   The superhydrophobic coating on the Cu wire and the schematic diagrams of its maximal supporting force. (a) and (b): The SEM image of $Cu(SC_{12}H_{25})_2$ super-hydrophobic coating mimicking structures of water-strider legs. (c) Schematic diagram to estimate the maximal supporting force of superhydrophobic copper wire. (Reprinted with permission from Ref. 36. Copyright 2006 American Chemical Society.)

2 M KOH aqueous solution to form a thin film of copper hydroxide nanoneedles. After reacting with n-dodecanethiol, a thermally stable $Cu(SC_{12}H_{25})_2$ superhydrophobic coating formed. The contact angle of the modified nanoneedle surface was higher than 150°, and the tilt angle was lower than 2°. Furthermore, this film was thermally stable, and kept its superhydrophobic properties after heating at 160°C in air for over 42 hours.

2.5   Climbing Up the Wall — Gecko

2.5.1   *The mechanism of the high adhesive force of gecko's feet*

Geckos are well-known for their unique ability to climb rapidly up smooth vertical surfaces, even adhere to the ceilings upside down.

This phenomenon means that there is a special high adhesive force between gecko's feet and the surfaces. However, how does the force come from? The origin of the adhesive force has been a controversial question for a long time. In 2000, Prof. Full from University of California at Berkeley, USA, and co-workers published a report in *Nature* to reveal the mechanism of the high adhesive force of gecko's feet.[37] They considered that the special high adhesive force was due to the accumulation of van der Waals force between the numerous hairs in the feet of gecko and surface molecules of the solid materials. Van der Waals force is a weak attractive or repulsive electromagnetic force between neutral molecules in close contact. The van der Waals force is so weak that it is usually neglected when studying the mechanism of the high adhesion. Therefore, the relationship between this molecular scale force and gecko's adhesion seems unacceptable.

The foot of a Tokay gecko, a kind of gecko inhabiting Southeast Asia, was studied. Microscopy shows that there are nearly five-hundred-thousand keratinous hairs or setae on a gecko's foot. Each hair is about 100 µm long (twice as long as the diameter of a human hair), and contains 400–1000 fine terminal branches, called spatulae. For the hierarchical fine structures, the distance between the surface molecules and the setae is much closer, and the van der Waals force works. Although the force generated by each single hair is negligible, the accumulation of the force from thousands of setae becomes strong enough (Fig. 2.22). According to the calculation, a single foot of an adult gecko could produce about 10 atm (1 atm = 101 325 Pa) of adhesive force if all setae are simultaneously and maximally attached. Furthermore, the force produced by a single seta could support the weight of an ant, and even 196 N of adhesive force is produced by one million setae with the area smaller than a coin. If all the setae are attached, more than 1225 N of adhesive force can be produced by a gecko. In fact, a gecko could support its body with just a single toe.

On the other hand, gecko's feet have self-cleaning adhesive property with the antifouling ability.[38] Although gecko climbs up and down the vertical surface with its high adhesive feet, the pads of foot always keep clean. Scientists proposed that this self-cleaning property occurred on the arrays of the aligned setae. Contact mechanical

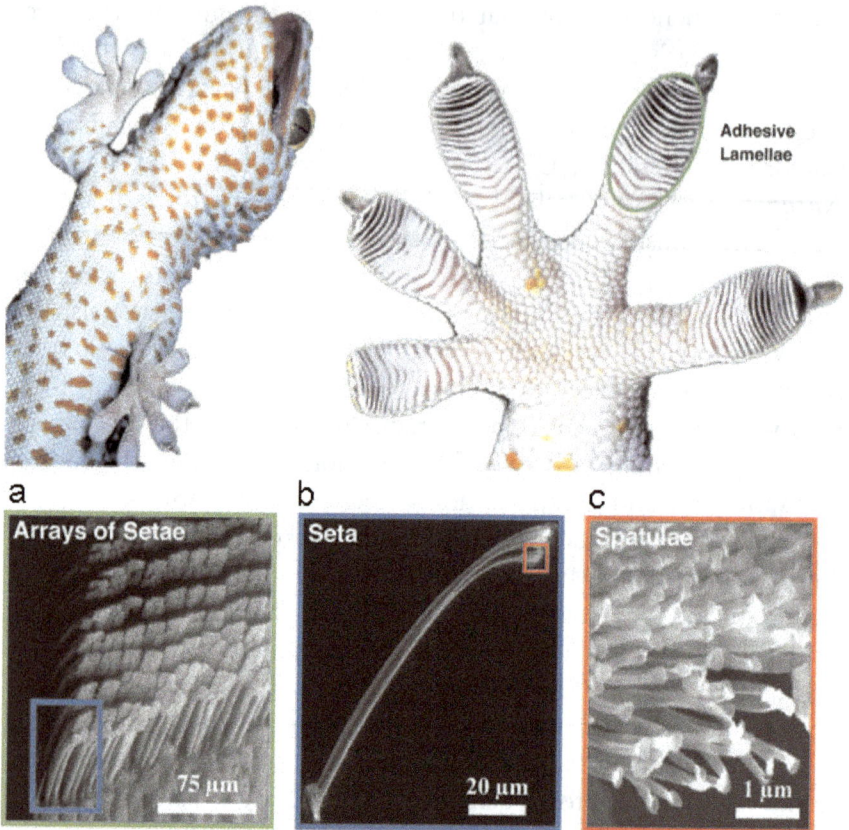

Fig. 2.22 The morphology of the setae on gecko's toe. (a) Aligned micrometer-scale setae. (b) A single seta. (c) The finest terminal branches of a seta, called spatulae. (From Ref. 38. Copyright 2005 National Academy of Sciences, U.S.A.)

models suggest that the energetic disequilibrium between the adhesive forces attracting a dirt particle to the substrate and that attracting the same particle to one or more spatulae lead to the self-cleaning phenomenon.

Similarly, Kesel[39] revealed that the spider's impressive ability to cling to and climb on overhanging smooth surfaces was based on the van der Waals force. After calculation, they found that the force could afford the spider to carry a load of 170 times the weight of its body. This force is little affected by the surrounding environment, so the spider could climb up wet or smooth surfaces. After observing the legs of spider *Evarcha arcuata* by SEM, as illustrated in Fig. 2.23(a), they found that a scopula on the leg was composed of many single scopula hairs (setae), which again were covered by an immense number of cuticular processes (setules) with the width of several hundred nanometers. Due to these setae and setules, the spider could cling to the surface of solid. Further AFM measurement revealed that each setule could generate a force of about 40 nN, which is quite high for a tiny spider weighed only 15 mg. The scientist pointed out that this adhesive force came from the van der Waals interaction among the intimately contacted setules, with the distance between each other being several nanometers. The accumulation of the tiny interactions on all the eight legs of a spider could lead to a very strong force macroscopically.

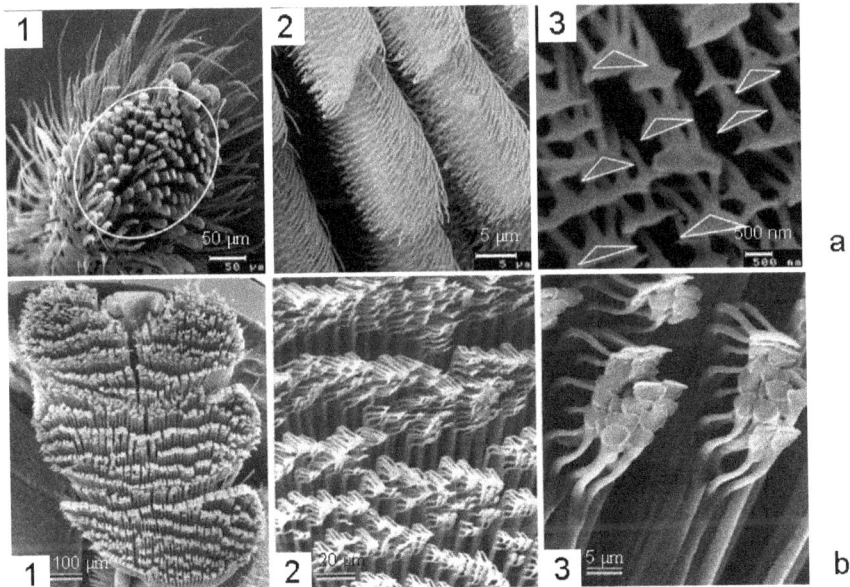

Fig. 2.23 High adhesive force generated by microstructures. (a) SEM images of spider legs[39] and (b) beetle ventral tarsi.[40] [Figure (a) — Reprinted with Permission from IDP Publishing Ltd. Figure (b) — Copyright 2000 National Academy of Sciences, U.S.A.]

Apart from gecko and spider, the high adhesion of beetle *Hemisphaerota cynea* with the substrates was proved to have something to do with their microscopic structures, as illustrated in Fig. 2.23(b).[40] SEM observation showed that the pads on the ventral tarsi of the beetle are stuck together in clusters, while the cluster arranged in rows.

Most animals and insects with high adhesive force on their feet obtain their ability of adhesion through the capillary force of water. However, geckos, spiders, flies and some other beetles can climb up the vertical or inverted surface without water. Through mimicking the microstructures of these animals, scientists considered that new type of strong adhesive tape, which could be stripped off easily without damage and used repeatedly, could be designed. Using this technique, a stamp may still be pasted on the envelope stably, even if the latter is wetted or polluted by oil. Furthermore, a special spacesuit based on the mechanism of van der Waals force may be produced, with which the spaceman could cling himself to the wall of spaceship, just like the gecko and spider on the ceiling.

2.5.2 *Design of biomimic "gecko tape"*

Inspired by the study above, Geim *et al.*[41] reported a method for manufacturing self-cleaning, re-attachable dry adhesives, named "gecko tape". Relying on the inter-molecullar attraction as two objects contacting each other, people with the help of the adhesives might become a real spiderman who could move on the smooth ceiling effortlessly. This special material was made of films with dense array of flexible polyimide pillars, prepared by electron-beam lithography and oxygen plasma etching (Fig. 2.24). While the fabricated pillars were thin and dense enough, strong adhesive force as that of the gecko setae could be obtained. The special adhesive materials have potential applications in many fields, for example, using as clips and sutures in surgery, safety devices for mountain climber, single-side nylon fastener tape, easy-getting-off bandages, adhesive gloves for goalkeeper in the soccer or cricket matches, and so on.

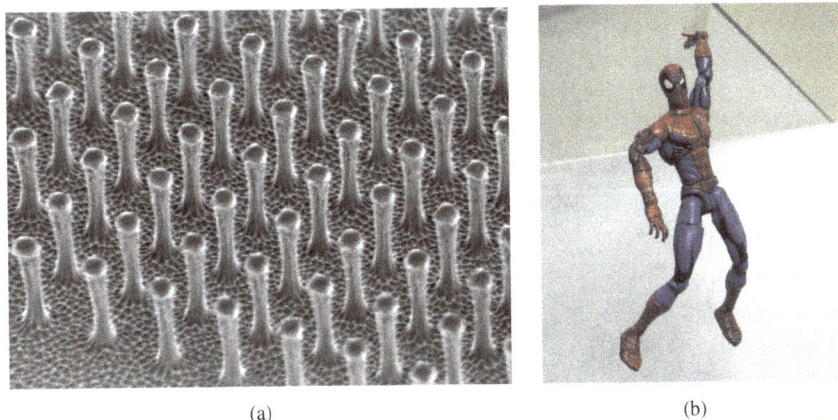

(a)   (b)

Fig. 2.24 SEM image of microfabricated polyimide hairs mimicking gecko setae and its applications.[41] (a) SEM image of microfabricated polyimide hairs. (b) A toy of spiderman with as-prepared polyimide hairs covered on its palm. (Reprinted by permission from Macmillan Publishers Ltd: *Nature Materials*, copyright 2003.)

For the next step, the researchers' major concern was the durability of such microfabricated adhesives. New type of materials must have pillars which are flexible enough and do not stick to each other or to the base surface, so as to ensure the adhesion to any surface. They even predicted that durability could probably be improved by trying other materials, e.g. keratin as in gecko hair, which would be sufficiently flexible as hydrophilic polyimide, but strongly hydrophobic.

2.5.3 *Superhydrophobic aligned polystyrene nanotube films mimicking the structure of gecko foot*

Recently, through mimicking the microstructures of gecko foot, the authors' group prepared an aligned polystyrene (PS) nanotube film via template-wetting method (Fig. 2.25(a)).[42] After wetting measurement, we found that the prepared PS film showed superhydrophobicity, while smooth PS film was weakly hydrophobic. Much surprisingly, the film showed strong adhesion to water, and water droplet would not roll

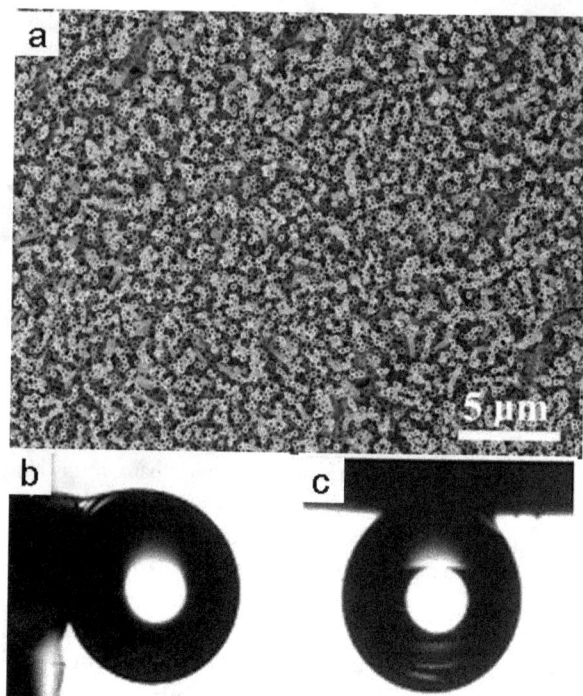

Fig. 2.25   Aligned PS nanotube film and its adhesive behavior.[42] (a) SEM image of aligned PS nanotube film mimicking gecko foot. (b) and (c): The adhesive behavior of a water droplet on the film. (Copyright Wiley–VCH Verlag GmbH & Co. KGaA. Reproduced with permission.)

off even when the PS film was tilted vertically or turned upside down, as illustrated in Fig. 2.25(b) and (c).

The adhesive force between the water droplet and the nanotube film was measured by a high-sensitivity microelectromechanical balance system (Fig. 2.26), and the result showed that the maximal adhesive force was about 60 µN. Figure 2.26 displayed the force-distance curves recorded before and after the water droplet came into contact with the nanotube film, and shapes of water droplet during measuring process were also shown in the insets. The experiment was carried out as follows. First, a 3-mg water droplet was suspended on a metal ring, and the shape of the droplet was kept spherical. Meanwhile, the PS nanotube film was placed on the plate of the

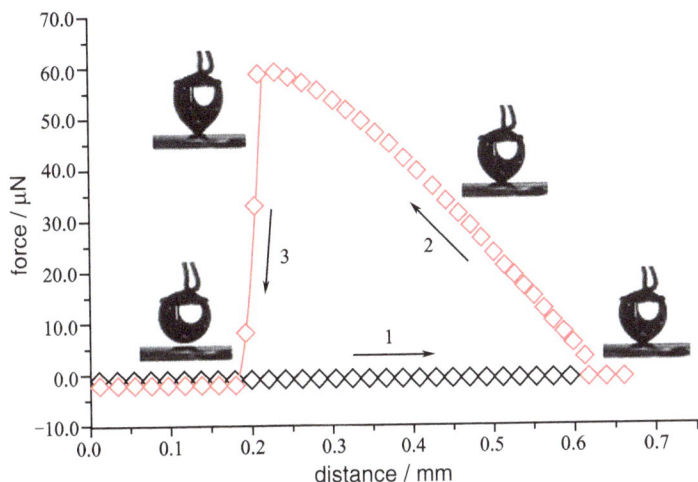

Fig. 2.26 Adhesive force measurement via a high-sensitivity microelectro-mechanical balance system.[42] (Copyright Wiley–VCH Verlag GmbH & Co. KGaA. Reproduced with permission.)

balance system. As the plate moved up and down, the variation of the weight on the balance could be recorded. In process 1, the PS nanotube film was brought into contact with the water droplet while maintaining the balance force at zero. Then, as the PS nanotube film left the water droplet immediately after contact, the recorded force increased gradually and reached its maximum of 60 µN. Finally, the balance force decreased to zero and the shape of water droplet came back to spherical from elliptical immediately when the PS nanotube film broke away from the water droplet in process 3. Under different preparing conditions, the magnitude of the adhesive force could be controlled by changing the diameter and density of the nanotube, and a decrease in the diameter might lead to an increase of the force. The as-prepared high adhesive PS nanotube film was similar to the gecko foot in structures and functions, and could potentially be used as a "mechanical hand" to transfer liquid without loss (Fig. 2.27).

Based on the research above, Wang[43] and Parkin[44] prepared similar superhydrophobic surfaces with high adhesive force via self assembly and chemical vapor deposition, respectively.

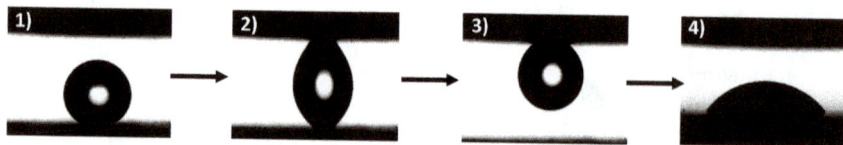

Fig. 2.27 Transfer process of a water droplet from a superhydrophobic surface to a hydrophilic one using high adhesive PS nanotube film as a "mechanical hand".[42] (Copyright Wiley–VCH Verlag GmbH & Co. KGaA. Reproduced with permission.)

2.6 A Desert Water-Collecting Insect — Desert Beetle

There is a family of beetle (*Stenocara*) living in Namib Desert in Africa. Each beetle is just as big as a thumb nail, and many "bumps", which are big or small, are distributed everywhere on its back elytra. Life cannot survive without water, how can these beetles find water in the boundless desert? Parker from the University of Oxford and Lawrence from Mechanical Sciences Sector, QinetiQ, have found the secret.[45] As reported in *Nature*, although rain is very rare while wind is frequent in the Namib Desert, dense fog is usual in the early morning, and the secret of how these beetles find water is hidden in the fog. The scientists discovered that, "bumps" are like peaks, and there is a valley between two adjacent bumps. "Bumps" are distributed in a near-random array, 0.5–1.5 mm apart, each about 0.5 mm in diameter (Fig. 2.28(a)). As seen under the electron microscope, the peaks of those "bumps" are smooth, with no covering, and hydrophilic (Fig. 2.28(b)), whereas the "valleys" are covered by a

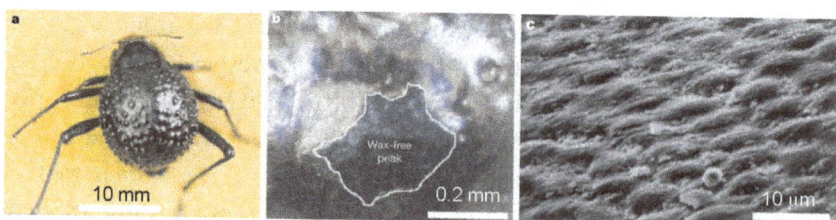

Fig. 2.28 Desert beetle and the microstructure on its elytra.[45] (a) The desert beetle. (b) The hydrophilic "peak" on elytra. (c) SEM image of the hydrophobic "valley" on elytra. (Reprinted by permission from Macmillan Publishers Ltd: *Nature*, copyright 2001.)

microstructure coated in wax. The microstructure consists of flattened hemispheres with the diameter of 10 µm and arranged in a regular hexagonal array, creating a superhydrophobic system reminiscent of the lotus leaf (Fig. 2.28(c)). When dense fog comes, the desert beetle tilts its body forwards into the wind and makes use of "bumps" on its top elytra. Micro-sized water drops in the fog condense on the "bumps", and then roll down through the hydrophobic "valley" to its mouthparts, pushed by wind.

Inspired by these beetles, the scientists then used glass spheres to mimic the structure. They made a water-collecting apparatus which had very small glass spheres embedded into the water-collecting surface coated with wax. Compared with common water-collecting surface made by flat glass slide or covered by wax, the new apparatus can collect more water. So here comes the conclusion: the key of collecting water from fog in desert lies in the combination of hydrophilic areas and hydrophobic areas in the desert beetle's elytra. This technology can be used in the future to clear up fog above the airport ground, collect water for agriculture, and gather drinking water from the rich fog in dry districts.

To mimic the character of water-harvesting wing-surface of the Namib Desert beetle, Zhai *et al.*[46] reported a method to create hydrophilic patterns on superhydrophobic surfaces with a polyelectrolyte solution, by selectively depositing multilayer films onto the hydrophilic patterns to introduce different wettability properties to different areas (Fig. 2.29). Potential applications of such water-harvesting surfaces include controlled drug release coatings, open-air microchannel devices and lab-on-chip devices.

2.7 Master of Hiding — Color-Changing Desert Beetle

As early as the year 1979, the *Science* magazine reported a desert beetle *Cryptogloassa verrucosa*, which could appear in different colors when subjected to different humidity (light blue in low humidity while jet black in high humidity).[47] The color variations are created by "wax filaments" covering the cuticle surface that spread from the tips of miniature tubercles. The meshwork that accumulates at low

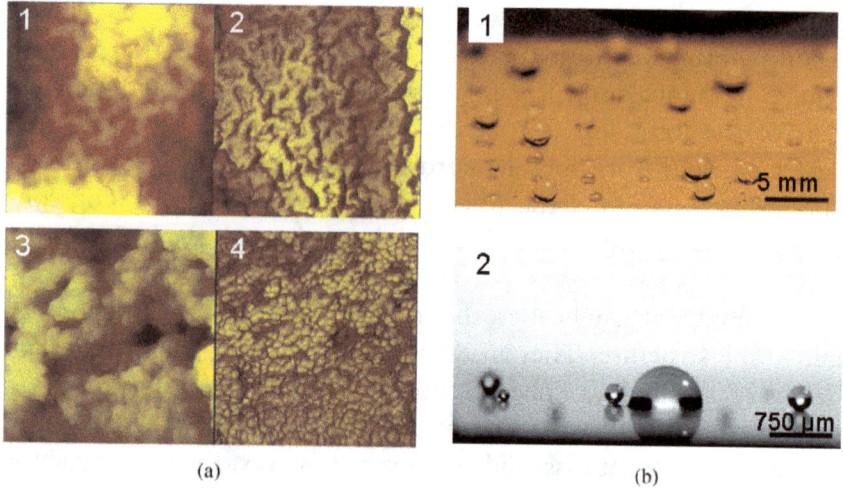

(a)   (b)

Fig. 2.29   Patterned surface mimicking the water-harvesting beetle wing surface.[46] (a) AFM images (1 μm × 1 μm): 1, 2 — Superhydrophobic surface, 3, 4 — Hydrophilic surface. (b) Small water droplet on the surface: 1 — Droplets sprayed on the superhydrophobic surface, 2 — Sprayed small water droplet accumulate on the patterned hydrophilic area. (Reprinted with permission from Ref. 46. Copyright 2006 American Chemical Society.)

humidity reduces transcuticular water loss and may lower the rate at which body temperature rises by increasing reflectance.

As shown in SEM images (Fig. 2.30), the elytra of this species are characterized by numerous miniature tubercles, which were distributed over the sides of each large tuber as well as the cuticle surface. When the beetle is in saturated humidity, the tip of each tubercle opens, amorphous material, similar in appearance to toothpaste gently squeezed from a tube, is exuded from the tip of each tubercle. This structure results in reflection of incident light, producing a dark or black coloration. When black beetles are transferred to low humidity, the amorphous secretion exuding from the tip of each tubercle organizes into numerous slender (0.14 μm) filaments, and these filaments from surrounding tubercles eventually spread and connect with each other to form a basketlike mesh (average thickness is 20 μm) that covers the entire surface of the beetle. This surface network now refracts incident light to produce the characteristic blue phase. Maximum

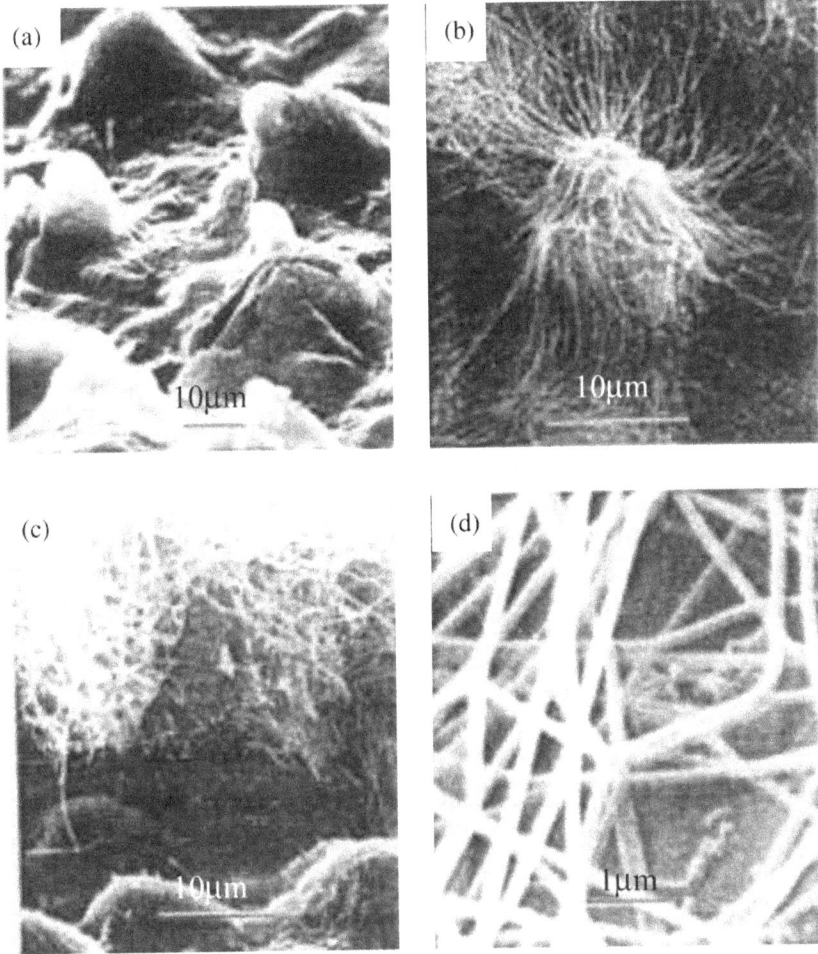

Fig. 2.30 SEM image of the desert beetle elytra. (a) and (b): Tubercles structure in micro-scale. (c) and (d): Filaments mesh-structure in submicro-scale. (From Ref. 47. Reprinted with permission from AAAS.)

development of this meshwork is simply to "dissolve" into the epicuticular wax, contributing to its thickness.

Chemical analysis of filaments from the surface of blue beetles indicated the presence of saturated hydrocarbons (mainly containing 20 to 36 carbon atoms, straight-chain), while cholesterol, alcohols, free fatty acids, and proteins were all detected. The tubercle exudate is the wax material that eventually covers the surface to waterproof the beetle; the type of protein present and the chemical nature of the lipid–protein interaction were not determined. The protein fraction probably plays a very important role in the morphological change observed at the various humidity extremes.

2.8 Structural Color in the Nature

According to our knowledge, colors in the nature are mainly produced by coloring matters, but some creatures (or minerals) have chosen photonic structures in their evolution. This means that they produce colors through the mutual reaction between natural light and micro-structures corresponding to the light-wave scale. Newton is the first man to point out that the colors of birds' wings, including the peafowl, and some insects, originate from thin-film interference. In recent years, reports about the photonic structure in biology have appeared in succession.[48]

2.8.1 *Colorful opal*

The precious stone opal, which is abundant in Australia, is a kind of mineral composed of close-packed aggregate of sub-micro SiO_2 particles. Its colorful appearance has nothing to do with pigments, but originates from its light band-gap structure which is formed through the periodic geometric structure. The color of reflected light varies with the wavelength of band-gap, thus beautiful colors shows up.[49]

Structures analogous to opal can be synthesized by stimulating the micro-structures of opal, and these are called synthetic or artificial opals.[50] They can be fabricated by spontaneous deposition through dilute solution of mono-disperse inorganic colloid particles (SiO_2, TiO_2, etc.), polymer colloid spheres (PS, PMMA, etc.), and many other compound colloid particles. The SiO_2, PS, and other opals can also be used as a template, and filled with materials with high refraction coefficient (usually the refraction coefficient is >2.8) or its precursor. After the filling process and materials mineralization in void, the SiO_2 or polymer template is removed by methods of

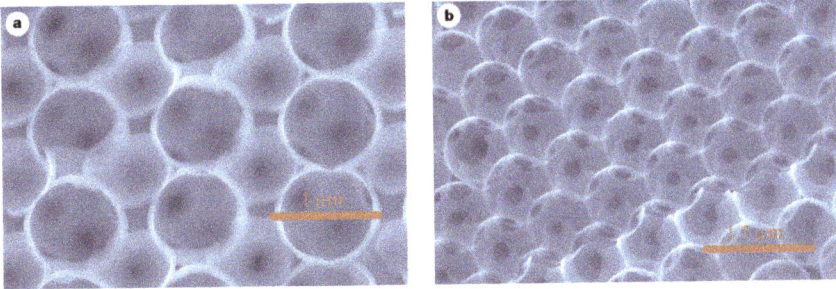

Fig. 2.31 SEM images of internal facets of silicon inverse opal.[51] (a) [110] facet. (b) [111] facet. (Reprinted by permission from Macmillan Publishers Ltd: *Nature*, copyright 2000.)

calcination, chemical corruption or dissolution, etc., to get arrays of global air holes, and the structure of inverse opal (Fig. 2.31).[51] Opal and inverse opal are so-called photonic crystals.[52–55] They not only exist in the nature, but also can be artificially designed and fabricated, through modern manipulate technology of high-tech materials. Photonic crystal has potentially comprehensive applications in the fields of biology, electronics, environment, and martials.

2.8.2 *The colorful butterfly wings*

As mentioned in Sec. 2.1.2, butterflies flying above flowers have remarkable colors in their wings, and these colors are produced by selectively reflecting sunlight using the orderly arrays of sub-micro-structures in the scales.[56] For example, P. Vukusic *et al.*[57,58] observed different microstructures in wing surfaces through their research of different varieties of butterflies. Taking *Ancyluris meliboeus* as an example, the lepidopteran wing is made of several layers of small scales, and each scale is only 3 to 4 μm thick. The scales overlap like tiny roof tiles, each having elaborate architecture. Such orderly assemblies can let light of only a certain wavelength to pass through, thereby we can see the unique color of butterflies (Fig. 2.32(a)). Modulated multilayering leads to dual color in butterfly *Papilio palinurus* (Fig. 2.32(b)).

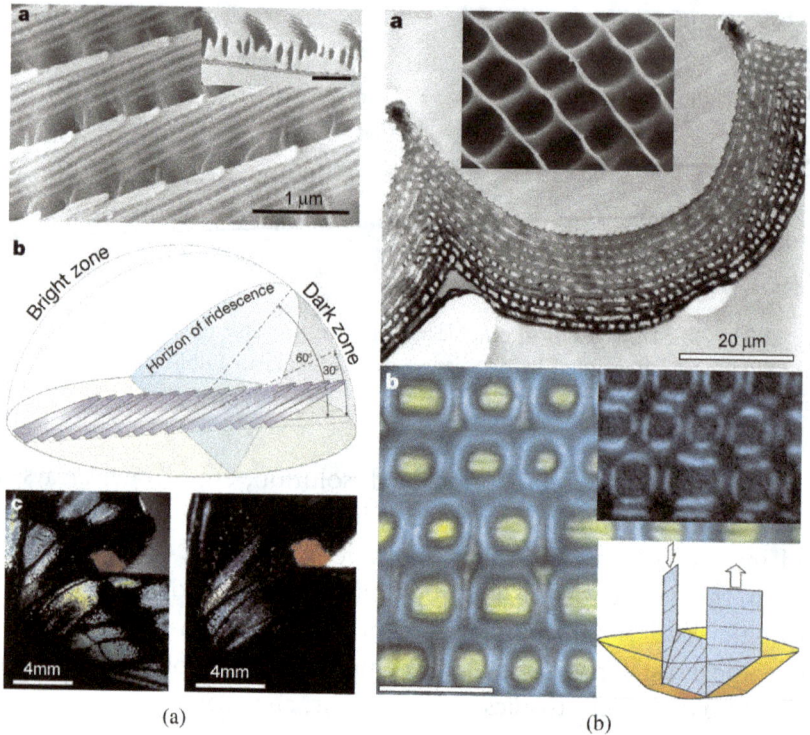

Fig. 2.32   Different microstructures on the surface of butterfly wings. (a) Butterfly (*A. meliboeus*) wings' photonic crystal structure and its color.[57] (b) Butterfly (*P. palinurus*) wings' hierarchical structure produces dual color.[58] (Reprinted by permission from Macmillan Publishers Ltd: *Nature*, (a) copyright 2001, (b) copyright 2000.)

Vukusic group[59] studied the wing structure and color of an African swallowtail butterfly (*Princeps nireus*) recently, and found that its wings were natural light-emitting diodes (LEDs). *P. nireus* has black wings with bright green and blue stripes. The wings are covered by tiny scales made up of nanostructures which absorb ultraviolet light and then emit the light back; those re-emitted light reacts with fluorescent pigment and produces bright blue and green color. Butterflies distant to each other use those remarkable colors to correspond. The principle, which is used by beautiful butterfly wings to deal with light, is strikingly similar to that used in digital displays. This discovery is important for human to study and design new types of light-emitting devices.

2.8.3 *Iridescent colors in peacock feather*

People have already known that there is periodic structure existing in the feather of many birds, including peacock, but the precise physical mechanism that produces the diversified colors in peacock tail feathers has not been established. In order to explain the coloration in male peacock's tail feathers, Prof. Zi's research group from Fudan University, China used optical microscope, scanning electron microscope and theory simulation to study the origin of peacock feather colors, and found the physical mechanism of its brilliant coloration.[60]

The researchers have discovered that, different from thin-film interference, peacock feather has a very ingenious coloration strategy. The cortex in differently colored barbules, which contains a two-dimensional (2D) photonic-crystal structure, is responsible for coloration (Fig. 2.33). The experiments and simulations reveal that the 2D periodic structure strongly reflects light of frequencies within a certain range along the direction parallel to the surface. There are two ways of modulation: one is controlling the length and the other is controlling the number of periods. Varying the lattice constant

Fig. 2.33 Structural images of peacock barbule under a microscope. (a) Side facet of green barbule. (b) and (c): Magnified images of the side-view of (b) green and (c) brown barbules. (d) Longitudinal cross-section of the green barbule with the surface keratin layer removed. (From Ref. 60. Copyright 2003 National Academy of Sciences, U.S.A.)

produces different colors, i.e. in the sequence of blue, green, yellow and brown, the lattice constant in colorful barbules increases accordingly. Based on the Fabry–Perot interference effect, brown barbules with the least periodic number can form additional blue color, so mixed coloration is formed and brown color is presented.

This foundational research illuminates the smart strategy used by nature to modulate color production, meanwhile inspires human that there are new ways to control coloration, e.g. via enhancing vision or bringing about light interference, with the possibility of exploring novel display technology in the future.

2.8.4 *Opal analogue discovered in beetles*

Beetles in dimly-lit tropical forests often display structural colors, but in direct sunlight only a part of the insect can be seen from any direction — it appears as a spot of light because multilayer reflectors on its rounded surface act like mirrors. Parker *et al.*[61] described a kind of beetle (*Pachyrhynchus argus*, found in forests of northeastern Queensland, Australia) which had a metallic coloration that was visible from any direction, even if the surrounding light was strongly directional. Such metallic color owes to a photonic crystal structure analogous to that of opal. To our knowledge, this is the first recorded example of an opal-type structure in an animal.

The beetle has scales that are about 0.1 mm in diameter (Fig. 2.34(a)) and occur in patches on the top and sides of the beetle's roughly hemispherical body. Individually, the scales are flat, lying parallel to the body, and consist of two parts: an outer shell and an inner structure. When the internal structure of sectioned scales are examined under scanning and transmission electron microscopes, the inner structure of the scales appears as an array of transparent solid spheres, each with a diameter of 250 nm. They are arranged in flat layers and have a precise, hexagonal close-packing order (Fig. 2.34(b)). As the lattice constant of the crystal approaches half of the wavelength of light, a single scale acts as a three-dimensional (3D) diffraction grating, forming an optical reflection domain. This is the most common nanostructure found in opal (although opal

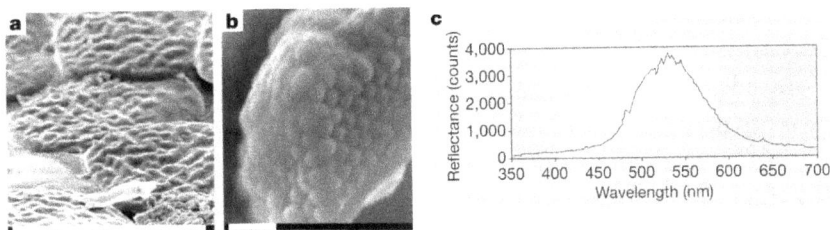

Fig. 2.34　Scales of the weevil and the reflection spectrum.[61] (a) and (b): Scale structure of weevil (*Pachyrhynchus argus*). (c) Analysis result of reflection spectrum. (Reprinted by permission from Macmillan Publishers Ltd: *Nature*, copyright 2003.)

microspheres occasionally form face-centered cubic packing), and allows for the reflection of just a narrow range of wavelengths over a wide range of incident angles. The inner structure is absent in some scales, causing them to appear transparent. Otherwise, in transmitted white light, the scales appear as the negative of the reflected color: yellow–green in reflected light and purple in transmitted light from most directions. When analyzed using an ultraviolet–visible spectrometer, with white light incident at 20° to the normal of the scale's surface, the peak reflectance occurs at a wavelength of 530 nm, at an angle of 20° to the other side of the normal (Fig. 2.34(c)). This opaltype structure is an example of a 3D "photonic crystal". The researchers have already reported the first photonic crystal found in animal sea mouse which has a 2D structure. In contrast, the 3D properties described here allow for a relatively omnidirectional optical effect, which explains why this weevil appears to be strongly colored from every direction *in situ*. This could be useful for interspecific color or pattern recognition.

Opal is difficult to manufacture in solid form. However, transmission electron micrographs of the weevil's opal-like structure reveal repeating patterns of light and dark areas in each microsphere, providing a clue to their molecular structure (which may be revealed by X-ray diffraction analysis) and production. These microspheres can only be constructed by molecular self-assembly, a process that could potentially be reproduced by the synthetic-opal industry. After all, the self-assembly technique of the abalone has been successfully copied in the manufacture of a nano-composite coating that is analogous to the nacre of its shell.[63]

Besides the above reports, there are other examples of structural color in nature. For example, pearls have intriguing colors mainly because of the reflection of its surface and intrinsic irradiance, and also due to the metal elements and pigments. According to some research results, white pearls contain more chromium, while golden and creamy pearls contain more cuprum and argentums, and pink pearls contain more calcium. The pearl color includes two parts: body color and fellow color. The former is the intrinsic color of the pearl itself, which is decided by the contained trace metal elements; the later is formed by a comprehensive interaction of light reflection and interference, from both the surface and inside of the pearl, plus upon the body color. Fellow color is usually cloudy.

After several billion years' evolution, organisms have already adapted to natural environment and what's more, they almost perfectly adapt to nature. Some ingenious functions in organisms are far smarter than human's designation, offering methods to settle difficult problems. Researching the intelligent surfaces in the nature provides an effective way of designing bio-mimic materials, and lays a bridge between biology and other scientific subjects. Intensive study on the structures and functions of natural biological materials in the view of material science, demonstration of their principles and reproduction of the unique functions, will finally help us find the approaches to solve problems.

References

1. Barthlott W, Neinhuis C. *Planta*, 1997, **202**: 1.
2. Neinhuis C, Barthlott W. *Annals of Botany*, 1997, **79**: 667.
3. Patankar N A. *Langmuir*, 2004, **20**: 8209.
4. Feng L, Li S H, Li Y S, Li H J, Zhang L J, Zhai J, Song Y L, Liu B Q, Jiang L, Zhu D B. *Adv. Mater.*, 2002, **14**: 1857.
5. Adamson A W, Gast A P. *Physical chemistry of surfaces*, Wiley, New York, 1997.
6. Mandelbrot B B. *The fractal geometry of nature*, Freeman, San Francisco, 1982.

7. Barthlott W, Neinhuis C, Cutler D, Ditsch F, Meusel I, Theisen I, Wilhelmi H. *Botanical Journal of the Linnean Society*, 1998, **126**: 237.

8. Li H, Wang X, Song Y, Liu Y, Li Q, Jiang L, Zhu D. *Angew. Chem. Int. Ed.*, 2001, **40**: 1743.

9. Patankar N A. *Langmuir*, 2003, **19**: 1249.

10. Onda T, Shibuichi S, Satoh N, Tsujii K. *Langmuir*, 1996, **12**: 2125.

11. Chen W, Fadeev A Y, Heieh M C, Öner D, Youngblood J, McCarthy T J. *Langmuir*, 1999, **15**: 3395.

12. Fürstner R, Barthlott W, Neinhuis C, Walzel P. *Langmuir*, 2005, **21**: 956.

13. Sun M H, Luo C X, Xu L P, Ji H, Ouyang Q, Yu D P, Chen Y. *Langmuir*, 2005, **21**: 8978.

14. Lee S M, Kwon T H. *Nanotechnology*, 2006, **17**: 3189.

15. Otten A, Herminghaus S. *Langmuir*, 2004, **20**: 2405.

16. Kennedy R J. *Nature*, 1970, **227**: 736.

17. Zheng Y M, Gao X F, Jiang L. *Soft Matter*, 2007, **3**: 178.

18. Sun T, Wang G, Liu H, Feng L, Jiang L, Zhu D. *J. Am. Chem. Soc.*, 2004, **125**: 14996.

19. Wagner T, Neinhuis C, Barthlott W. *Acta Zoologica*, 1996, **77**: 213.

20. Gorba S N, Keselb A, Berger J. *Arthr. Struc. Develop.*, 2002, **29**: 129.

21. Guo C, Feng L, Zhai J, Wang G, Song Y, Jiang L, Zhu D. *Chem. Phys. Chem.*, 2004, **5**: 750.

22. Lee W, Jin M, Yoo W, Lee J. *Langmuir*, 2004, **20**: 7665.

23. Chen G, Cong Q, Feng Y, Ren L. *Design and Nature II*, Collins M W, Brebbia C A. (Eds.) WIT Press, 2004, p. 245.

24. Gu Z, Uetsuka H, Takahashi K, Nakajima R, Onishi H, Fujishima A, Sato O. *Angew. Chem. Int. Ed.*, 2003, **42**: 894.

25. Watson G S, Watson J A. *Appl. Surf. Sci.*, 2004, **235**: 139.

26. Stavenga D G, Foletti S, Palasantzas G, Arikawa K. *Proc. R. Soc. B.*, 2006, **273**: 661.

27. Gao X F, Yan X, Yao X, Xu L, Zhang K, Zhang J H, Yang B, Jiang L. *Adv. Mater.*, 2007, **19**: 2213.

28. Keller J B. *Phys. Fluids*, 1998, **10**: 3009.

29. Sun S M, Keller J B. *Phys. Fluids*, 2001, **13**: 2146.

30. Hu D L, Chan B, Bush J W M. *Nature*, 2003, **424**: 663.

31. Lauder G V, Drucker E. *News Physiol. Sci.*, 2002, **17**: 235.

32. Dickinson M. *Nature*, 2003, **424**: 621.

33. Gao X, Jiang L. *Nature*, 2004, **432**: 36.

34. Shi F, Wang Z, Zhang X. *Adv. Mater.*, 2005, **17**: 1005.

35. Zhang X, Shi F, Yu X, Liu H, Fu Y, Wang Z, Jiang L, Li X. *J. Am. Chem. Soc.*, 2004, **126**: 3064.
36. Wu X, Shi G. *J. Phys. Chem. B*, 2006, **110**: 11247.
37. Autumn K, Liang Y A, Hsieh S T, Zesch W, Chan W P, Kenny T W, Fearing R, Full R J. *Nature*, 2000, **405**: 681.
38. Hansen W R, Autumn K. *PNAS*, 2005, **102**: 385.
39. Kesel A B, Martin A, Seidl T. *Smart Mater. Struct.*, 2004, **13**: 512.
40. Eisner T, Aneshansley D J. *PNAS*, 2000, **97**: 6568.
41. Geim A K, Dubonos S V, Grigorieva I V, Novoselov K S, Zhukov A A, Shapoval S Y. *Nat. Mater.*, 2003, **2**: 461.
42. Jin M, Feng X, Feng L, Zhai J, Sun T, Li T, Jiang L. *Adv. Mater.*, 2005, **17**: 1977.
43. Song X Y, Zhai J, Wang Y L, Jiang L. *J. Phys. Chem. B*, 2005, **109**: 4048.
44. Boscher N D, Carmalt C J, Parkin I P. *J. Mater. Chem.*, 2006, **16**: 122.
45. Parker A R, Lawrence C R. *Nature*, 2001, **414**: 33.
46. Zhai L, Berg M C, Cebeci F C, Kim Y, Milwid J M, Rubner M F, Cohen R E. *Nano Lett.*, 2006, **6**: 1213.
47. Hadley N F. *Science*, 1979, **203**: 367.
48. Vukusic P, Sambles J R. *Nature*, 2003, **424**: 852
49. Sanders J V. *Nature*, 1964, **204**: 1161.
50. Newton M R, Morey K A, Zhang Y, Snow R J, Diwekar M, Shi J, White H S. *Nano Lett.*, 2004, **4**: 875.
51. Blanco A, Chomski E, Grabtchak S, Ibisate M, John S, Leonard S W, Lopez C, Meseguer F, Miguez H, Mondia J P, Ozin G A, Toader O, van Driel H M. *Nature*, 2000, **405**: 437.
52. Yablonovitch E. *Phys. Rev. Lett.*, 1987, **58**: 2059.
53. John S. *Phys. Rev. Lett.*, 1987, **58**: 2486.
54. Zakhidov A A, Baughman R H, Iqbal Z, Cui C, Khayrullin I, Dantas S O, Marti J, Ralchenko V G. *Science*, 1998, **282**: 897.
55. Vlasov Y A, Bo X, Sturm J C, Norris D J. *Nature*, 2001, **414**: 289.
56. Srinivasarao M. *Chem. Rev.*, 1999, **99**: 1935.
57. Vukusic P, Sambles J R, Lawrence C R, Wootton R J. *Nature*, 2001, **410**: 36.
58. Vukusic P, Sambles J R, Lawrence C R. *Nature*, 2000, **404**: 457.
59. Vukusic P, Hooper I. *Science*, 2005, **310**: 1151.
60. Zi J, Yu X, Li Y, Hu X, Xu C, Wang X, Liu X, Fu R. *PNAS*, 2003, **100**: 12576.

61. Parker A, Welch Victoria L, Driver Dominique, Martini N. *Nature*, 2003, **426**: 786.
62. Parker A, McPhedran R C, McKenzie D R, Botten L C, Nicorovici N A P. *Nature*, 2001, **409**: 36.
63. Mayer G. *Science*, 2005, **310**: 1144.

Chapter 3

Wettability of the Solid Surface

The wetting of the solid surface by liquids is a common interface phenomenon. Wettability is one of the important properties of the solid surface. The most familiar example is the flashing water droplet on the lotus leaf surface in the early morning. While water spreads easily on the glass surface, mercury tends to be a ball. Wetting phenomenon not only affects the behavior of the creatures in nature, but also plays an important role in industry and our daily lives. The application of wettability covers broad fields including the flotation separation of minerals, oil extraction, water repellency, washing and so on. Moreover, the spreading of agriculture pesticide, the painting of photographic emulsion in the fabrication process of the film, the properties of the lubricant and painting are all concerned with wettability.

Water droplet exists like a ball on the lotus leaf and rolls off easily due to the gravity when the leaf is slightly tilted. As a result, we say that water do "not wet" the lotus leaf. Meanwhile, we say that our hands are wetted when we pull them out of water with a layer of residue sticking on them. These are the most familiar examples of wetting and unwetting. From the macroscopic view, wetting is a process that one liquid replaces the other on a solid surface. While on the microscopic view, after the original liquid on the solid surface is replaced by another liquid, the contact between the solid surface and the later liquid is on the molecular level and there is no molecules of the former liquid between them. The most common phenomenon is the replacement of the air by a liquid on the solid surface. For example, the replacement of air by water on a glass surface causes the water to spread.

There are two main factors that affect the wettability of a solid surface: a) surface free energy and b) surface microstructure. We will focus our discussion on these two factors and point out the meaning of the unique wetting property and its relationship with these two factors. Meanwhile, the dynamic behavior of the liquid on the solid surface will be introduced.

3.1 Basic Theory of Wettability

3.1.1 *Surface free energy*

For a solid surface, the larger the surface free energy (or surface tension) γ_{SV}, the easier it can be wetted by the liquids. For liquids, the surface tension is mostly under 100 mN/m except for liquid mercury. Accordingly, the solid surfaces can be divided into two types: a) High energy surfaces including common metals and their oxides, sulfides and inorganic salts. These surfaces have higher surface free enthalpy which is about hundreds to thousands mJ/m^2. As a result, they can be wetted by common liquids. b) Low energy surfaces such as organic solids and polymers. Their surface free enthalpy is about 25–100 mJ/m^2 which is comparable to the liquid. Their wetting property is heavily concerned with the component and property of the liquid/solid interface. For solid surfaces, they can be divided into hydrophilic and hydrophobic types according to the surface free energy.[1] Common hydrophilic surfaces include glass, metals, and so on. Hydrophobic/oleophilic surfaces include polyolefin, silica and so on. An example of hydrophobic/oleophobic surface is Teflon (Fig. 3.1).

3.1.1.1 *Wetting property of low energy surface*

Generally speaking, the measurement of γ_{SV} is very difficult. Zisman *et al.*[2] used to do a plenty of systematic work on low energy surfaces. They measured the contact angle θ of a variety of liquids on the same

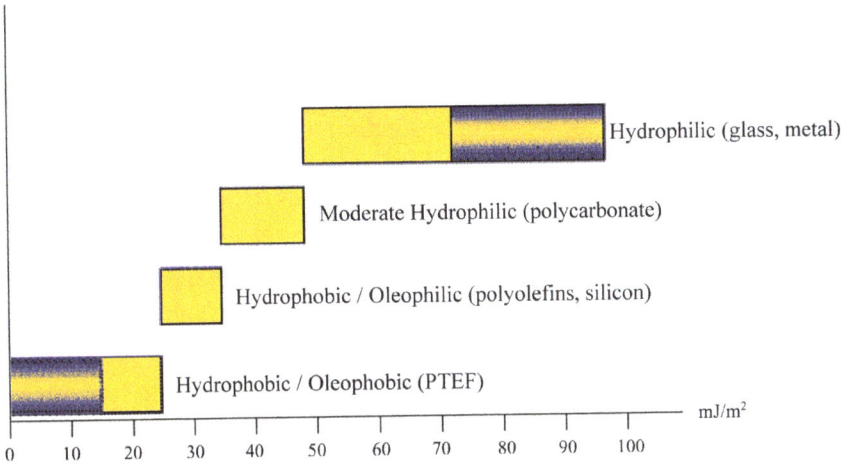

Fig. 3.1 Surface free energy and wetting behavior.

polymer surface which is smooth, clean, without plasticizer and approaching perfect. The results indicate that when $\cos\theta$ is plotted against the surface tension, the graph will be a straight line for the same series of liquids. Extend the data line to cross with the line $\cos\theta = 1$, then we call the surface tension value at the cross point to be the critical surface tension (γ_c) of the solid. For the liquids which are not in a series, we may get a narrow belt. If $\cos\theta$ is fluctuating due to γ_{LV} in the narrow belt region, we choose the surface tension value at the cross point with the lower critical line of this region to be the critical surface tension. γ_c is a key important experience constant in describing the wetting property of the low energy surfaces. Only the liquids which have a surface tension equal to or lower than the γ_c of a solid can spread on this solid surface. It is more difficult for a solid to be wetted when it has a lower γ_c value since the surface tension of the liquids needs to be lower in order to spread.

Researchers have found that the wetting property of polymers can be tuned by replacing the hydrogen atoms in their hydrocarbon groups or introducing other elements into them. Replacing the hydrogen atoms by fluorine atoms can reduce γ_c of polymers, and this trend can be strengthened by more replacements. On the other hand, the replacement of the hydrogen atoms by other halogen elements or introducing oxygen and nitrogen atoms into the hydrocarbon chain increases the γ_c of polymers. The order among several common elements in increasing the γ_c of the polymer solid is as follows:

$$N > O > I > Br > Cl > H > F.$$

Some solid surfaces have very low γ_c values such as poly(tetrafluoroethylene) and methyl silicone resin. The surface tension of water and most organic liquids are higher than their γ_c values. As a result, these solid surfaces cannot be wetted by the liquids mentioned above and these surfaces are called amphiphobic (hydrophobic and oleophobic) surfaces. These kinds of surfaces have very important applications in industry such as avoiding wax forming on the oil tube during the production process of the raw oil. When the raw oil flows out in a tube from the oil layer, as the temperature of the ground surface is lower than the oil layer, the temperature of the oil will decrease. As a result, the wax originally dissolving in the oil will precipitate out as crystals and cling on the tube wall. The wax will plug the oil tube and affect the production of raw oil if it is not cleaned in time. On the other hand, if we paint a layer of polymer coating which has a low γ_c value on the inner wall of the oil tube, neither water nor the raw oil can spread on the coating surface. By doing so, the wax crystal is hard to cling onto the tube wall under the impact of the oil flow. Preliminary experiments demonstrate that these kinds of coatings have good effect on preventing wax crystals from clinging to the tube wall.

3.1.1.2 *Wetting property of high energy surface*

From the view of surface free energy, common liquids seem to be able to spread on the surfaces of clean metals, metal oxides, and high-melting-point inorganic solids. However, a number of experiments have revealed that if the liquids are polar organics or contain polar organics, they cannot spread on the high energy surface. The experimental data from Zisman *et al.*[3] indicate that for some liquids, no matter what measures have been taken to purify them, they still have quite high contact angle on some high energy surfaces. They explain this phenomenon as follows: the polar organic liquids may turn the polar groups towards the high energy surface and nonpolar groups away from it. By forming a monolayer in this way, the surface becomes a low energy surface and the wetting property of the surface is only determined by that of the monolayer. If the surface tension of the liquid is higher than the critical surface tension of the aligned monolayer, it does not spread on its own monolayer surface. The liquid that has this property is called autophobic liquid. Conversely, a liquid will spread on its monolayer if its surface tension is lower than the critical surface tension of its monolayer. Methyl silicone oil is a good example. Its surface tension is 19–20 mN/m which is lower than the γ_c of the absorbed monolayer (usually larger than 22 mN/m), so it is able to spread on any high energy surface.

From the discussions above, we can figure out that the γ_c value of the polymer solid is concerned with the surface component. Introducing fluorine atoms decreases the γ_c value while introducing other atoms increases the value. From further research on the γ_c value of the polar organic monolayer, we discover that the γ_c value of the polymer solid or monolayer is mainly determined by the property of their outermost atoms or groups.

3.1.2 Surface roughness

3.1.2.1 Contact angle and Young's equation

Assume that a droplet is placed on a solid surface. Instead of spreading entirely, the droplet turns to have an angle to the solid surface which we call the contact angle θ (Fig. 3.2). The contact angle is defined as the *angle* at which a *liquid/vapor* interface meets the solid surface. The advantage of exploiting the contact angle to describe the wetting property of a liquid on a solid is intuitive. The disadvantage is the lack of representation on the energy variation during the wetting process.

The contact angle of a drop on a solid surface is a result of the equilibrium among the surface tensions of the solid/gas/liquid

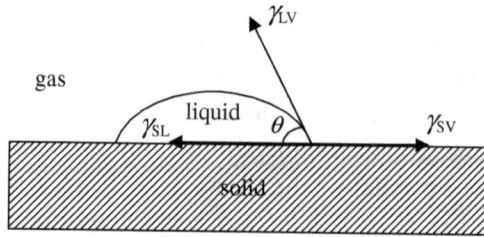

Fig. 3.2 Definition of the contact angle.

interfaces. The equilibrium makes the whole energy of the system the lowest and thus leads the droplet to be in the stable or metastable state. Generally speaking, the contact angle of a droplet on a smooth surface can be characterized by the Young's equation[4]:

$$\gamma_{SV} = \gamma_{SL} + \gamma_{LV} \cos\theta$$
$$\cos\theta = \frac{\gamma_{SV} - \gamma_{SL}}{\gamma_{LV}}, \tag{3.1}$$

in which γ_{SV}, γ_{SL} and γ_{LV} represent solid/gas, solid/liquid, liquid/gas interface tension, respectively. θ is called the balance contact angle or the intrinsic contact angle of the material and can be written like θ_e.

Young's equation is the foundation of research on the solid/liquid wetting interaction. The value of the contact angle θ is the criterion of the wetting property:

- $\theta = 0$ entirely wetting; liquid spreads on the solid surface
- $0 < \theta < 90°$ liquid wets the solid; the wetting property is better for smaller θ
- $90° < \theta < 180°$ liquid does not wet the solid
- $\theta = 180°$ completely unwetting; liquid turns to be a ball on the solid.

It should be pointed out that the application condition of the Young's equation is the ideal surface. In other words, the solid surface is smooth, isotropic, undeformed (under the impact of the vertical component of the liquid surface tension) and has a homogeneous component. Only on surfaces like this the liquid has a settled stable contact angle.

3.1.2.2 *Contact angle of the nonideal solid surface*

(1) Wenzel's model

Let us consider a droplet on a rough surface. It is hard to measure the genuine contact angle. The experimentally measurable one is the apparent contact angle (θ^*), as indicated in Fig. 3.3.

The relationship between the apparent contact angle and surface tension is not in accordance with Young's equation. However, we can get similar equations like Young's in the view of thermodynamics.

Assume that the liquid could always fill in the grooves on the rough surface as in Fig. 3.4(a). We call it "wet contact".

Under constant temperature and pressure, the free energy variation of the system caused by the tiny change of the interface can be written as:

$$dE = r(\gamma_{SL} - \gamma_{SV})dx + \gamma_{LV}\,dx\cos\theta^*, \qquad (3.2)$$

where dE is the energy needed when a contact line moves in an infinitesimal displacement. When the equilibrium is reached, $dE = 0$, we can figure out the relationship between the apparent contact angle θ^* and the intrinsic contact angle θ:

$$\cos\theta^* = r(\gamma_{SV} - \gamma_{SL})/\gamma_{LV}. \qquad (3.3)$$

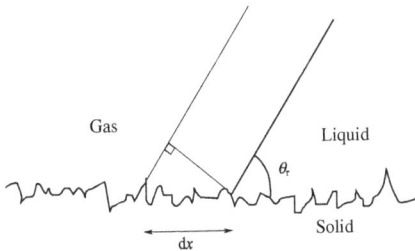

Fig. 3.3 The schematic diagram of contact edge between the droplet and the rough surface. (Reprinted from Ref. 10 with permission.)

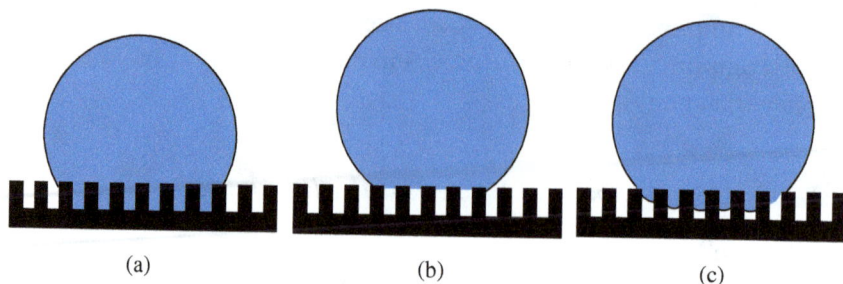

Fig. 3.4 (a) Wenzel's model. (b) Cassie's model. (c) Coexistence of Wenzel's and Cassie's model.

Comparing the above equation with the Young's equation, we get

$$\cos \theta^* = r \cos \theta. \tag{3.4}$$

This is the famous Wenzel's equation which is proposed by Wenzel in 1936.[5]

In the equation, r is the roughness factor which is defined as the ratio of the actual contact area of the solid/liquid interface to the geometric projected contact area ($r \geq 1$). θ^* is the contact angle of the rough surface under the Wenzel state.

Wenzel's equation indicates that the rough surface makes the actual solid/liquid contact area larger than that of the apparent visible one. This geometrically makes the surface hydrophobic or hydrophilic. In other words: (a) when $\theta < 90°$, θ^* decreases with the increase of the roughness and the surface becomes more hydrophilic; (b) when $\theta > 90°$, θ^* increase with the increase of the roughness and the surface becomes more hydrophobic. The enhancement of the rough structure on the surface wettability cannot be achieved by only varying the surface chemical components. Nakamae *et al.*[6] measured the free energy of the surface composed with the densely packed hexagonal –CF3 groups. This surface was considered to be the solid surface that has the lowest surface free energy. However, the experimental results show that even on the solid surface that has the lowest surface free energy, we can only get the contact angle as high as 119°. The corresponding surface free energy is 6.7 mJ/m².

It should be pointed out that Wenzel's equation is only suitable in the thermodynamic equilibrium state. However, due to the heterogeneity of the surface, the droplet needs to overcome the energy barriers caused by the surface microstructure fluctuation while it begins to spread. When the vibration energy is smaller than these energy barriers, the droplet is unable to reach the equilibrium state required by Wenzel's equation. Instead, the droplet chooses to be in a kind of metastable state.

(2) Cassie's model

Wenzel's equation reveals the relationship between the apparent contact angle and the intrinsic contact angle on the homogeneous rough surface. But the equation is no longer valid when the surface is composed of different kinds of chemical substances. Cassie and Baxter[7] extend the assumption made by Wenzel mentioned above and propose that the heterogeneous rough surface could be considered as a composite surface. In other words, they think that the contact of a droplet on the rough surface is a kind of composite contact. Assume that the solid surface is composed of substance 1 and substance 2, and these two different kinds of surfaces are well-distributed in the form of extremely tiny parts. The area of every tiny part is far smaller than the droplet dimension. Their intrinsic contact angles are θ_1 and θ_2 respectively. f_1 and f_2 are assumed to be their area fractions on a unit area ($f_1 + f_2 = 1$) and do not change during the spreading process. Additionally, consider θ^* to be the intrinsic contact angle of this surface. Then, under the state where the temperature and pressure remain constant, the energy variation caused by the tiny changes of the surface can be written as:

$$dE = f_1(\gamma_{SL} - \gamma_{SV})_1\, dx + f_2(\gamma_{SL} - \gamma_{SV})_2\, dx + \gamma_{LV}\, dx \cos \theta^* \quad (3.5)$$

When the equilibrium is reached, we get

$$f_1(\gamma_{SL} - \gamma_{SV})_1 + f_2(\gamma_{SL} - \gamma_{SV})_2 = \gamma_{LV} \cos \theta^*. \quad (3.6)$$

According to Young's equation, the equation above can be transferred as

$$\cos \theta^* = f_1 \cos \theta_1 + f_2 \cos \theta_2. \quad (3.7)$$

This equation is the so-called Cassie–Baxter equation.

In this equation, θ_1 and θ_2 are the apparent contact angles of the liquid on substance 1 and substance 2, respectively. f_1 and f_2 are the apparent area fractions of substance 1 and substance 2 ($f_1 + f_2 = 1$).

This equation can also be applied on the solid surfaces composed of solid substance and air, e.g. the surfaces with porous structures or other rough structures that can preserve air. When the hydrophobic property of the surface structure is high enough, Cassie and Baxter think that the droplet cannot fill in the grooves on the rough surface and leaves the air under it. In this way, the apparent contact area of the liquid/solid interface is actually composed of solid and air together (Fig. 3.4b).

Here, f_2 is the area fraction of the trapped air. Since the contact angle between air and water θ_2 is 180°, the equation can be rewritten as

$$\cos \theta^* = f_1 \cos \theta_1 - f_2. \tag{3.8}$$

It must be pointed out that the equations above are obtained from the empirical results after modeling. As a matter of fact, the solid surfaces may not be the same with the ones mentioned above since they are concerned with the surface morphology. For example, for surfaces which have parallel grooves and concave structures, although they have the same roughness, they present completely different properties. So, the roughness may not be characterized by the roughness factor r if we are not completely aware of the morphology of a composite surface.

3.1.2.3 *Research on the wettability of the rough surface*[8]

(1) Hydrophilic surface ($\theta < 90°$)

To some extent, the increasing contact angle means the increase of the surface free energy. In the Wenzel's model, the extra interface

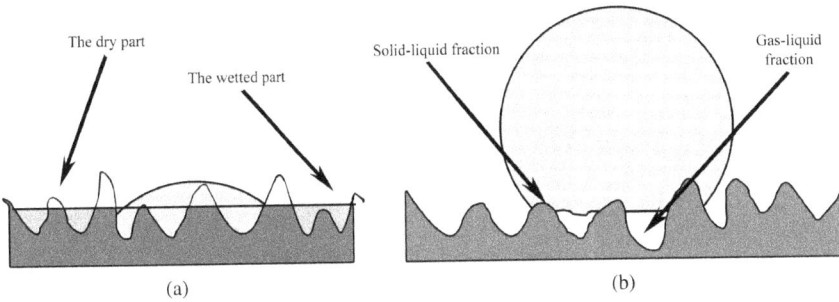

Fig. 3.5 (a) The schematic diagram of the apparent contact angle on the rough surface under the condition of capillary effect. (b) The schematic diagram of the apparent contact angle on the rough surface under the condition of air trapping.

free energy is provided by the extra solid/liquid interface, while in the Cassie's model it is provided by the extra solid/gas interface.

Wenzel's equation reveals that the surface roughness can enhance its wettability. As a result, the contact angle for a hydrophilic surface increases when its roughness increases. On the hydrophilic rough surface, the capillary effect occurs and sucks water into the grooves on the surface. Now, the apparent contact angle will be determined together by the wet part (ϕ_s) and dry part ($1 - \phi_s$) which are the area fractions of the liquid/liquid interface and solid/liquid interface, respectively, as in Fig. 3.5(a). Then, the apparent contact angle equation can be written as

$$\cos \theta^* = \phi s \cos \theta + (1 - \phi_s). \tag{3.9}$$

ϕ_s and $1 - \phi_s$ are the area fractions of the solid and liquid phase under the droplet, respectively.

Under the condition of constant temperature and pressure, the free energy variation of the system caused by the tiny change of the interface dx is described as below:

$$dE = (r - \phi_s)(\gamma_{SL} - \gamma_{SV})dx + (1 - \phi_s)\gamma_{LV} dx. \tag{3.10}$$

When this variation is preferable in energy, we get $\cos\theta > \dfrac{1-\phi_s}{r-\phi_s}$.

Here, we define $\cos\theta_c = \dfrac{1-\phi_s}{r-\phi_s}$, θ_c is the critical contact angle $(0 < \theta_c < 90°)$.

When $0 < \theta < \theta_c$, the droplet fills in the rough structure and can be superhydrophilic with a contact angle nearly 0°. The Wenzel's equation is valid here.

When $\theta_c < \theta < 90°$, the solid surface under the droplet remains dry. The Wenzel's equation is also valid here.

(2) Hydrophobic surface

For the hydrophobic surface, it will trap some air because of the roughness and make the surface itself heterogeneous. At this point, the contact angle of the rough surface can be written as below:

$$\cos\theta^* = -1 + \phi_s(\cos\theta + 1). \qquad (3.11)$$

Here, ϕ_s and $(1 - \phi_s)$ are the area fractions of the solid/liquid and gas/liquid interface, respectively.

Similarly, there is a critical contact angle (θ_c) for the hydrophobic surface as well:

$$\cos\theta_c = \dfrac{\phi_s - 1}{r - \phi_s}.$$

When $90° < \theta < \theta_c$, the system sits in the metastable state which gradually changes between the Wenzel's and Cassie's states.

When $\theta > \theta_c$, air is easily trapped in the structure and makes the contact heterogeneous. Cassie's equation can be applied here and the system is in the stable state. When the area fraction of the gas/liquid interface is big enough, the surface can be superhydrophobic with a contact angle as high as 150° as in Fig. 3.5(b).[9]

(3) The relationship between Wenzel state and Cassie state

The relationship between the intrinsic contact angle (θ) and the apparent contact angle (θ^*) of different kinds of rough surfaces can be

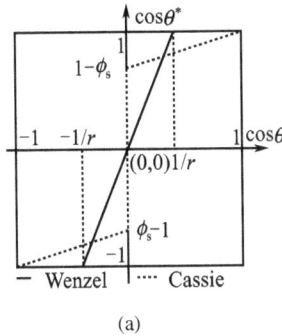

(a)

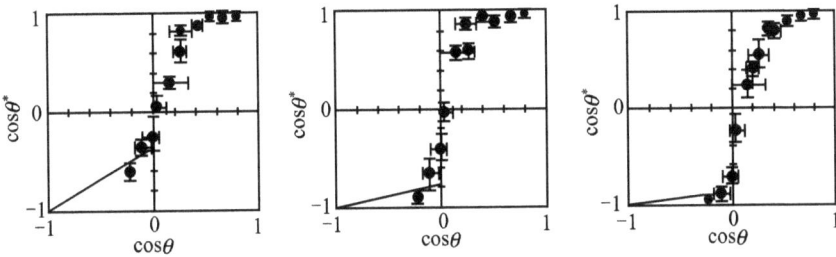

(b)

Fig. 3.6 (a) The cosine relationship between the intrinsic contact angle θ and the apparent contact angle θ^* of the rough surface.[10] (b) The experimental results of the relationship between the intrinsic contact angle θ and the apparent contact angle θ^* of the rough surface.[11] [Figure (a) — Reprinted with permission from EDP Sciences. Figure (b) — Copyright 1996, American Chemical Society.]

united as in Fig. 3.6(a).[10] The figure also shows the linear relationships between θ and θ^* which are valid both in Wenzel state and Cassie state. These relationships are in accordance with the experimental results reported by Shibuichi *et al.*[11] Dettre and Johnson[12] discovered the critical value of the roughness factor by modeling the rough surface on the bases of Wenzel's and Cassie's equations. According to their report, exceeding this critical value will make the solid surface wettability change from the case where Wenzel's equation is suitable to the case where Cassie's equation is better. With the surface roughness increasing, the energy barrier between the Cassie state and Wenzel state is higher and this makes the Cassie state more stable.

3.1.2.4 *Energy analysis*

Let us use a dispensing tube under the bottom of the substrate to dispense liquid as illustrated in Fig. 3.7 so as to form a droplet on the substrate. The radius of the dispensing tube is far smaller than the droplet and thus the initial surface free energy of the droplet can be negligible.[13]

Assume that the efficient area of the liquid/solid contact interface (dry substrate) is $(\gamma_{SL}-\gamma_{SV})^{\text{eff}}$, if the substrate is rough and the contact below the droplet is heterogeneous, then we get

$$
\begin{aligned}
(\gamma_{SL} - \gamma_{SV})^{\text{eff}}/\gamma_{LV} &= -\phi_s [(\gamma_{SV} - \gamma_{SL})/\gamma_{LV}] + (1 - \phi_s) \\
&= -\phi_s \cos\theta_e + (1 - \phi_s).
\end{aligned} \tag{3.12}
$$

For wetting contact, we get

$$
(\gamma_{SL} - \gamma_{SV})^{\text{eff}}/\gamma_{LV} = r[(\gamma_{SL} - \gamma_{SV})/\gamma_{LV}] = -r\cos\theta_e. \tag{3.13}
$$

From the state of no droplet to the state where a droplet forms ultimately, the energy variation can be written as

$$
G = S\gamma_{LV} - \cos\theta_r A,
$$
$$
\cos\theta_r = (\gamma_{SV} - \gamma_{SL})^{\text{eff}}/\gamma_{LV},
$$

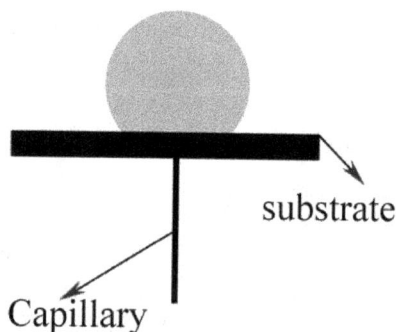

substrate

Capillary

Fig. 3.7 Form a droplet on the substrate by dispensing liquid from a tube under the substrate. (Reprinted with permission from Ref. 13. Copyright 2003, American Chemical Society.)

where S is the contact surface area between the droplet and air, A is the apparent contact surface area between the liquid and the substrate (neglecting gravity).

Let ψ be the contact angle of the droplet on the substrate. When $\psi = \theta_r$, G reaches its minimum value and can be written as:

$$\frac{G}{\gamma_{LV}} = 2\pi a^2 (1 - \cos\psi) - \pi a^2 \sin^2 \Psi \cos\theta_r. \qquad (3.14)$$

Considering the restriction of the volume, then

$$g = V - \frac{\pi a^3}{3} (1 - \cos\Psi)^2 (2 + \cos\Psi) = 0,$$

where a is the radius of the drop when the contact angle is ψ. V is the volume of the drop. $A = 2\pi a^2 (1 - \cos\psi)$, $S = \pi a^2 \sin^2\psi$.

So, the minimum value of the energy can be rewritten as:

$$E = \frac{G}{\gamma_{LV}} + \lambda g,$$

where λ is Lagrange coefficient. When we set $g = 0$, $\partial E / \partial a = 0$, $\partial E / \partial \psi = 0$, we can get the equations containing a, λ and ψ:

$$\frac{\partial E}{\partial a} = 0 \rightarrow a\lambda + \frac{2\cos\psi\cos\theta_r + 2\cos\theta_r - 4}{(1 - \cos\psi)(2 + \cos\psi)} = 0 \qquad (3.15)$$

$$\frac{\partial E}{\partial \psi} = 0 \rightarrow a\lambda + \frac{2(\cos\psi\cos\theta_r - 1)}{a(1 - \cos\psi)(1 + \cos\psi)} = 0. \qquad (3.16)$$

The contact angle on a rough surface can be written as two forms. One is θ_r^c (composite surface contact angle which is in accordance with Cassie's equation and $\psi = \theta_r^c$) and the other is θ_r^w (wetting surface contact angle which is in accordance with Wenzel's equation and $\psi = \theta_r^w$). The kind of contact angle it belongs to is determined by the contact type between the droplet and the substrate. Furthermore, according to the discussions above, no matter which kind of contact it is, it represents the minimum energy state. So, the energy equation of the droplet on the substrate which is in the equilibrium state can be written as:

$$\frac{G}{\sqrt[3]{9\pi V^{2/3}}\gamma_{LV}} = (1 - \cos\theta_r)^{2/3}(2 + \cos\theta_r)^{1/3}. \qquad (3.17)$$

Obviously, the right side of the equation is monotonic increasing. G is increasing with the increase of θ_r. As a result, if the surface structures are the same, the state which has the lowest apparent contact angle θ_r will have the lowest energy.

3.1.3 *The new limit between hydrophilic and hydrophobic surfaces — ~65°*

There are still a lot of arguments on the concepts of hydrophilicity and hydrophobicity. At all times, the prevailing definition of the limit is 90°. In other words, solid surfaces which have contact angles $\theta <$ 90° are considered to be hydrophilic and the surfaces which have contact angles $\theta > 90°$ are hydrophobic. However, the latest researches show that the limit of hydrophilicity and hydrophobicity should be about 65°.[14] According to this new limit, the range of hydrophobic surfaces has been broadened.

As we know, the wettability of a solid surface is usually characterized by its contact angle. According to Young's equation, the contact angle is determined by the interactions across the three interfaces on the smooth surface. Here, water droplet is considered as a mathematical entity whose surface and bulk phase are consistent. As a matter of fact, the structures and activities of water at the surface or in the bulk are different and vary with the change of the contact face. Bearing this in mind, we can understand surface wettability profoundly. The interaction among the liquid molecules is a physical process and is controlled by the surface tension and intersolubility. As for water, the interaction force among molecules is the strong three-dimensional net-like hydrogen bond. The research results from R. H. Yoon *et al.*[15] show that the local changes of the chemical potential of water (for example, caused by the contact of water and solid surface) can be detected by the surface tension apparatus and other assistant technology even though these changes are at the scale of tens of nanometers. In accordance with the predictions of DLVO theory, these measurements indicate the repulsive and attractive forces between the two oppositely placed surfaces when these surfaces are immersed into water. These forces depend on the variations of water density which is induced by the change of the structure and arrangement of water due to the wettability difference. Since the water on the surface and the water in the bulk are different in terms of self crossing, there are at least two types of water that have different structures and interactions. One has relatively small density and forms the hydrophobic surface through open hydrogen network. The other has relatively high density and forms hydrophilic surface through collapsed hydrogen network.

Figure 3.8 shows the data measured by improved atomic force microscope indicating the relationship between the adhesive force and the surface tension of water. Here, the adhesive force of pure water τ^0 is used to define the hydrophobic surface. $\tau^0 = \gamma^0 \cos\theta < 30$ dyn/cm, where γ^0 is the surface tension of water and $\gamma^0 = 72.8$ dyn/cm.

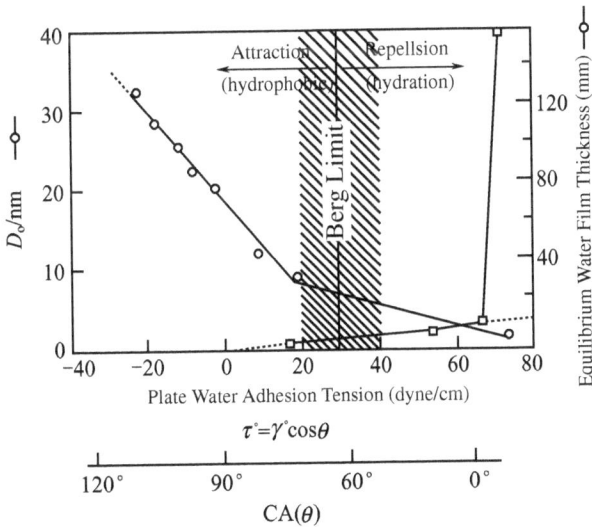

Fig. 3.8   The relationship between the adhesive force and the thickness of water film. (Reprinted from Ref. 14. Copyright 1998, with permission from Elsevier.)

Moreover, D_0 (-0.554 ± 0.027) $\tau^0 + 18.12 \pm 0.41$. It can be figured out that when $-40 < \tau^0 < 30$ dyn/cm, D_0 decreases linearly. When $20 < \tau^0 < 40$ dyn/cm, the attractive force (hydrophobic force) is changed to the repulsive force (hydrophilic force). Extend this straight line to $D_0 = 0$. It shows that above $\tau^0 = 33.7$ dyn/cm, i.e. $\theta = 62.4°$, the hydrophobic force no longer supports the surface wetting. This result is close to the limit of 65° proposed by J. M. Berg *et al.*[16] In other words, when the surface contact angle θ is below 65°, the surface shows repulsive force (hydrophilic surface, $\tau^0 > 30$ dyn/cm).

Now, let us focus on the experimental results reported by J. M. Berg *et al.* two different series of mixed Langmuir–Blodgett (LB) films with a controllable degree of polarity, deposited on mica, have been studied by wetting and surface-force techniques. Both series contain 50% eicosylamine (EA). Films of one series consist of EA, arachidic acid, and docosandioic acid, while those of the other consist of EA, 1-eicosanol, and 1,22-docosandiol. Carboxylic acid groups give lower contact angles than hydroxyl groups. Concerning the stability of the LB films in aqueous solutions, repeated exposure to a three-phase line and high-salt solutions is found to cause breakdown. Surface force measurements on carboxylic acid-containing films show that films with a 0% (contact angle = 113°) or 25% (contact angle $\approx$ 90°) content of diacid interact with a long-range (hydrophobic) attraction across water. No similar long-range attraction is observed for the 50% case (contact angle $\approx$ 65°).

Figure 3.9(a) shows advancing contact angles against water as a function of DDA ratio for some deposited layers in the carboxylic series. The contact angle decreases with increasing ratio of bipolar substance, and since the contact angles, for a given content of bipolar substance, are higher for hydroxyl system monolayers than for carboxylic system monolayers, the COOH group is apparently more hydrophilic than OH.

Figure 3.9(b) shows the advancing (θ_A) and receding (θ_R) contact angles as a function of pH for carboxylic system monolayers with DDA ratio of 0%, 25%, and 50%. The pH was adjusted by adding

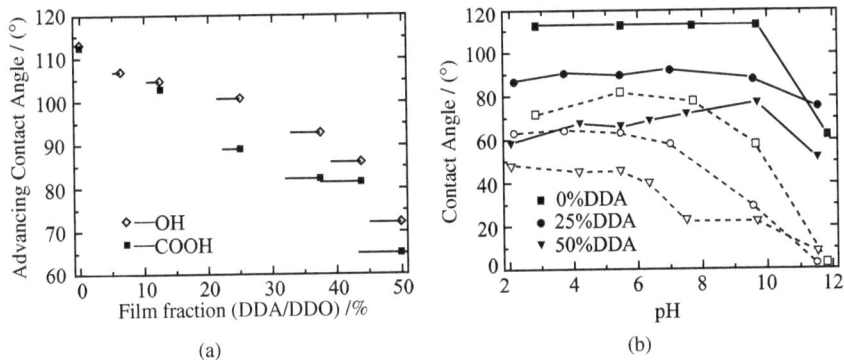

Fig. 3.9 (a) Advancing contact angles for water on deposited carboxylic (diamonds) and hydroxyl (filled squares) system monolayers, as a function of bipolar substance (DDA or DDO) ratio. (b) Advancing (filled symbols, solid lines) and receding (open symbols, broken lines) contact angles for aqueous solutions containing HCl or NaOH on deposited carboxylic system monolayers, at 0%, 25%, and 50% DDA ratio, as a function of pH. (Reprinted with permission from Ref. 16. Copyright 1994, American Chemical Society.)

hydrochloric acid or sodium hydroxide. Up to a pH value of about 10, the advancing contact angles for 0% and 25% DDA ratio are nearly constant, while the advancing contact angle for 50% DDA ratio actually increases moderately with increasing pH. Above pH 10, the advancing contact angles decrease significantly for all three surfaces, as well as for the 0% DDA monolayer, the outer surface of which should expose no ionized groups. This indicates that a breakdown of the LB film is taking place. The receding contact angles, which tend to be less reproducible than the advancing angles, start to decrease at lower pH values than the advancing angles. Note that while θ_A increases somewhat for 50% DDA surface in the pH 7–10 range, θ_R actually decreases. The breakdown of the deposited film at pH values above 10, indicated by the advancing angle for 0% DDA films, results in very low receding contact angles, which are below about 5° when the bare mica surface is exposed.

Our group[17] acquired the rough structure by sanding polymer surfaces. The structure enhances the wettability of the surface, making

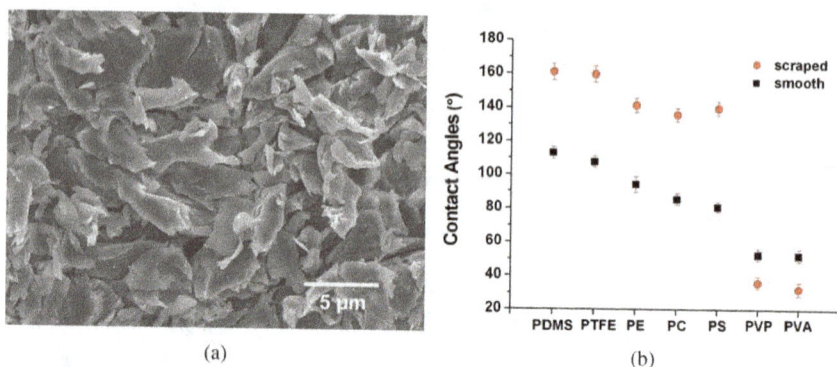

Fig. 3.10 (a) The microstructure of polytetrafluoroethylene surface after sanding. (b) The relationship between the intrinsic and the apparent contact angles of different kinds of polymer surfaces.[17]

the hydrophilic surface more hydrophilic and the hydrophobic surface more hydrophobic.

By measuring the intrinsic and apparent contact angles of different kinds of polymer surfaces, we can get an equation as below:

$$\cos\theta_{rf} = 4.751\cos^3\theta + 1.1254\cos^2\theta + 0.1999\cos\theta - 0.7894.$$

According to this equation, we can predict the limit of hydrophilic and hydrophobic surfaces to be 62.7°.

3.2 Surfaces with Special Wettability

From previous studies, we know that surface roughness can enhance the surface wettability, consequently leading to two extreme cases of special wettability: superwettable (the static CA inclines to 0°) and superrepellent to a liquid. Strictly speaking, for a solid substrate, when the contact angle of a liquid droplet on it is smaller than 10°, it is called superwettable; when the contact angle of a liquid droplet on it is larger than 150°, it is called superrepellent. Special wettability includes super-hydrophilicity, super-hydrophobicity, super-oleophilicity, and super-oleophobicity. Next we will discuss the surfaces with special wettability in detail.

3.2.1 *Structure effect of surfaces with special wettability*

3.2.1.1 *Theoretical simulation of the micro/nano structure effect*

Since our group discovered that the micro/nano hierarchical structure induces the lotus leaves self-cleaning effect, the lotus effect has attracted much attention. Patankar *et al.*[18] have theoretically simulated the lotus structure effect. According to the Wenzzle's or Cassie's formula, they depict a model of fine scale roughness made of square pillars arranged in a regular array. The first generation roughness is the square pillars with the size of $a \times a$, height H, with the periodic spacing of regular array being b. The second generation roughness is the micro-pillars growing on the square pillars. The double-scale structures were assumed periodic in a comparatively large area (Fig. 3.11).

Based on the theoretical model, the apparent contact angles are given by Eq. (3.18) through extending the Wenzzle's formula, and the contact angles are related to the height of the pillar. The apparent contact angles do not have relationship with the pillar height in Eq. (3.19), which is derived from Cassie's formula. The two equations could be transferred from describing the first generation

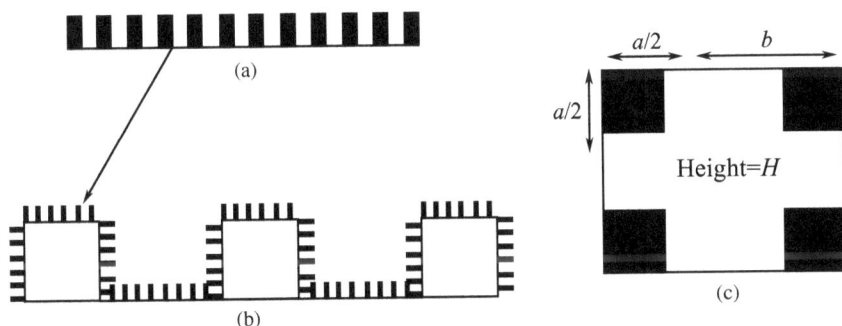

Fig. 3.11 A model roughness geometry for theoretical simulation of the lotus structures. (Reprinted with permission from Ref. 18. Copyright 2004, American Chemical Society.)

structure to the second generation structure by replacing subscript "1" by "2".

$$\cos\theta_r^c = A_1(1+\cos\theta_e) - 1 \qquad (3.18)$$

$$\cos\theta_r^w = \left(1 + \frac{4A_1}{(a_1/H_1)}\right)\cos\theta_e, \qquad (3.19)$$

where $A_1 = \dfrac{1}{((b_1/a_1)+1)^2}$, in which a_1 is the length of the pillars, b_1 is the periodic spacing of the regular array, H_1 is the height.

To mimic the lotus effect, two conditions are needed:

(1) A composite drop formed on the two-level scales roughness. This condition could ensure that the drop has the minimum sliding angles, because the wetted drop formed according to the Wenzzle's formula exhibits much more hysteresis than the composite drop. The sliding angle of the same-sized drop is ten times compared to a composite drop.

(2) The apparent contact angle of the composite drop needs to be as high as possible. The composite drops are formed on the lotus leaf type. It is assumed that the composite state should represent the minimum in energy for self-cleaning surfaces. The geometric parameters of the rough surface must ensure that the composite drop has lower energy than the wetted drop. This indicates that the apparent angle of the wetted drop must be larger than that of the composite drop. In such cases, even though the composite drop has a smaller apparent angle, it should be chosen because it leads to lower hysteresis.

Based on the work mentioned above, the lotus structure effect could be theoretically analyzed:

(1) *Designing the first generation of roughness structured surface*
     As indicated by Eqs. (3.18) and (3.19), the wetted drop and the composite drop coexist. If the surface is very rough, the cosine of an angle in Wenzel equations might be less than −1.

(2) *Designing a surface without the need for double structure*

From Eqs. (3.18) and (3.19), it could be calculated that when (b_1/a_1) is 2.89 and $(H_1/a_1) > 37$, the theoretical contact angle of the first generation could reach 160°. This indicates that the slender or high aspect ratio pillars have excellent water repellent effect.

(3) *Designing a kind of two-generation roughness structured surface*

a. designing the first generation of a two-generation structure

In this case when the surface has the double scales structure, even if a wetted contact is formed with the first generation, it is still possible that a composite drop is formed with the second generation. It is because the contact area of the drop with the first generation roughness will still be small if a composite drop is formed on the second generation (Fig. 3.12).

b. designing the second generation of a two-generation structure

With such roughness, the cosine in Wenzel formula might be less than −1. Because of the presence of the second generation of roughness, the surface would exhibit superhydrophobic effect when $(H_1/a_1) = 5$, which is not possible when there is only the first generation of roughness (Fig. 3.13).

The results above indicate that the second generation of roughness greatly decreases the critical value of the

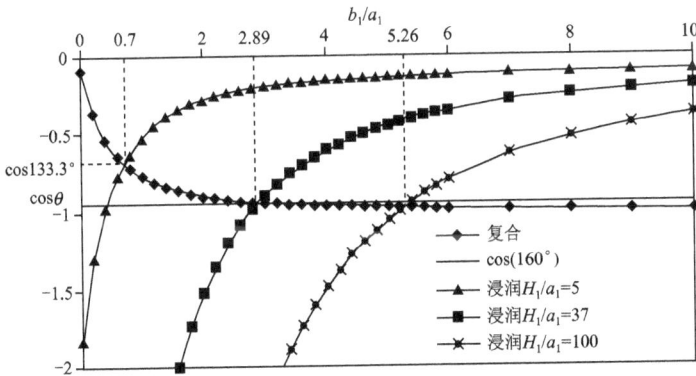

Fig. 3.12 Plot of the apparent contact angles for wetted and composite drops as a function of roughness geometry of the first generation. (Reprinted with permission from Ref. 18. Copyright 2004, American Chemical Society.)

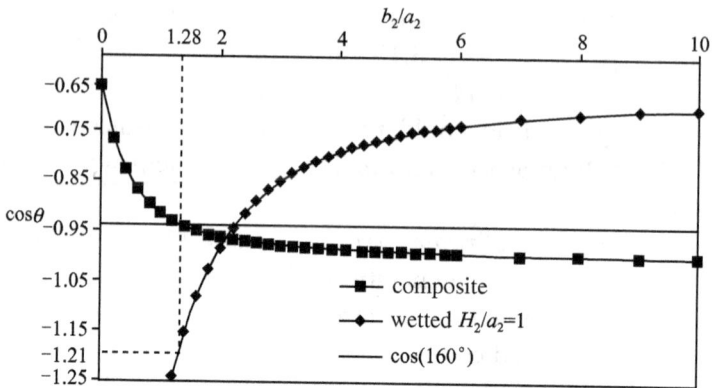

Fig. 3.13 Plot of the apparent contact angles for wetted and composite drops as a function of roughness geometry of the second generation. (Reprinted with permission from Ref. 18. Copyright 2004, American Chemical Society.)

(H_1/a_1), making it easier to realize the superhydrophobicity. Meanwhile, the composite drop ensures that the water drop could slide easily from the rough surface. This is a most important property for the self-cleaning surface. This result is consistent with the findings of the authors' group[19] that the self-cleaning property of the lotus leaf comes from the micro/nano structure of the surface. Besides, B. Bhushan *et al.*[20] have studied the hydrophobicity and friction of the plant leaves. They also conclude that the nano-structure has more important function than the micro-structure.

Herminghaus proposed that micro/nano structured surfaces modified with any materials will become non-wetting.[21] In other words, creating micro/nano hierarchical structures on the hydrophilic-materials modified surface will also produce the hydrophobic surface. Here the liquid drop will be pinned on the micro-structure of the surface. This is a metastable state between the Wenzel and Cassie state. If the energy barrier between the Wenzel and Cassie state could be conquered, the hydrophobic, even superhydrophobic surfaces could be created from traditional hydrophilic, materials. This theory is of great importance to broaden the methods of fabricating super-hydrophobic materials.

3.2.1.2 The contact angle of the fractal surfaces

Onda et al.[11,22] modified the contact angle equation of a rough surface and provided the contact angle equation of a fractal surface according to the fractal theory.

The contact angle θ between a flat solid surface and a liquid droplet is given by Young's equation:

$$\cos\theta = \frac{\gamma_{SV} - \gamma_{SL}}{\gamma_{LV}}. \tag{3.1}$$

For a rough solid surface, the apparent contact angle of the liquid droplet is given as:

$$\cos\theta^* = r\cos\theta. \tag{3.4}$$

A fractal surface is a kind of rough surface, so that the wettability of the fractal surface could be described by Eq. (3.4). However, the coefficient r of the fractal surface is very large and can be infinite for a mathematically ideal fractal surface. Therefore, the wettability of the fractal surface could be greatly enhanced due to the surface roughness; that is, the fractal surface will be superrepellent (superwettable) to a liquid when θ is greater (less) than 90°.

However, the application of Eq. (3.4) is limited. In fact, when the absolute value of its right-hand side exceeds unity, Eq. (3.4) could not give the contact angle value. Therefore, the correct expression of the contact angle equation could be given simply by introducing the effective interfacial tension of the solid–liquid fractal interface γ_{fSL}, and that of the solid-gas fractal interface γ_{fSV}. Then the contact angle θ_f is given by

$$\cos\theta_f = \frac{\gamma_{fSV} - \gamma_{fSL}}{\gamma_{LV}}. \tag{3.20}$$

Since γ_{f1j} ($j = 2, 3$) could be considered as the total interfacial energy of the fractal surface per unit projected area, γ_{f1j} could be approximated to $\gamma_{f1j} = (L/l)^{D-2}\theta_{1j}$. Herein, D ($2 \leq D < 3$) is the fractal dimension of the surface; L and l are the upper and the lower limit lengths of the fractal behavior respectively. Substitute the γ_{f1j} into Eq. (3.20) in which $(L/l)^{D-2}$ is substituted for r.

In order to estimate θ_{f1j} more precisely, they take account of the adsorption of a gas on the solid–liquid fractal interface and that of a liquid on the solid–gas fractal interface. Then θ_{f1j} can be expressed by:

$$\theta_{fSL} = \min_{x}\left\{\gamma_{SV}\left(\frac{L}{l}\right)^{D-2} x + \gamma_{SL}\left(\frac{L}{l}\right)^{D-2}(1-x) + \gamma_{LV}S_{LV}(x)\right\} \quad (3.21)$$

$$\theta_{fSL} = \min_{y}\left\{\gamma_{SV}\left(\frac{L}{l}\right)^{D-2}(1-y) + \gamma_{SL}\left(\frac{L}{l}\right)^{D-2} y + \gamma_{LV}S_{LV}(y)\right\} \quad (3.22)$$

where x ($0 \leq x \leq 1$) is the fraction of the area of the fractal surface covered with the adsorbed gas, y ($0 \leq y \leq 1$) is that covered with the adsorbed liquid, and S_{LV} is the area of the gas–liquid interface per unit projected area. The symbol "min" is the operation of minimizing the succeeding quantities with respect to x or y. This operation corresponds to the fact that adsorption so occurs as to minimize the total interfacial energy. Typically, $S_{LV}(z)$ ($z = x, y$) could be approximated by a cubic function of z, given by:

$$S_{LV}(z) = z + \left\{\left(\frac{L}{l}\right)^{D-2} - 1\right\}z(1-z)\left\{\frac{(1-\xi)(1-3\xi)}{1-2\xi} - z\right\}, \quad (3.23)$$

where ξ is the contacting ratio of the fractal surface (the ratio of the area touching a flat plate placed on the fractal surface). θ_f determined by Eqs. (3.20)–(3.23) is shown schematically in Fig. 3.14, in which $\cos\theta_f$ is plotted as a function of $\cos\theta = (\gamma_{SV} - \gamma_{LS})/\gamma_{LV}$.

The curve in Fig. 3.14 coincides with the line of $\cos\theta_f = (L/l)^{D-2}\cos\theta$ in the vicinity of $\theta = 90°$; however, the curve deviates from it when θ approaches $0°$ or $180°$ owing to the adsorption. From this we could find that the two endpoints are respectively representative of the two extreme cases of special wettability: superrepellent ($\theta_f \approx 180°$) and superwettable ($\theta_f \approx 0°$) to a liquid. According to the above

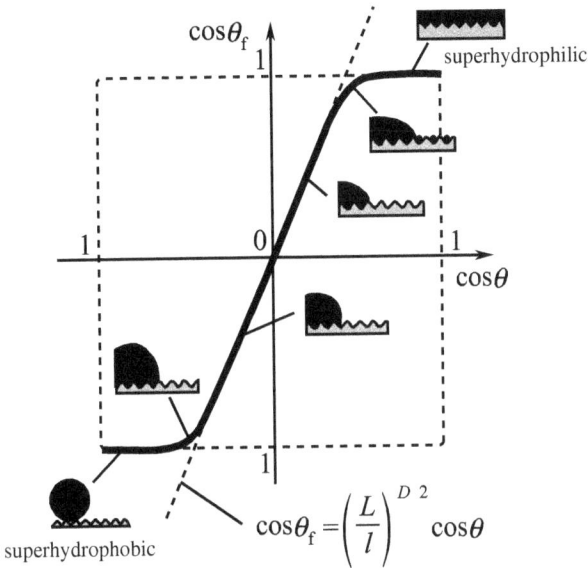

Fig. 3.14 Schematic illustration for $\cos\theta_f$ vs $\cos\theta$ theoretically predicted. (Reprinted with permission from Ref. 22. Copyright 1996, American Chemical Society.)

theoretical calculations, the contact angle θ_f of a fractal surface could be given by Eq. (3.24):

$$\cos\theta_f = \left(\frac{L}{l}\right)^{D-2} \cos\theta. \tag{3.24}$$

3.2.2 The length scales effect of surfaces with special wettability

As mentioned above, the special micro-structures of the solid surface will induce special wettabilities, such as superhydrophobicity and superhydrophilicity. Herein, we will mainly introduce the theory how the mesoscopic length scales influence the superhydrophobicity which was proposed by D. Chandler *et al.*,[23] and particularly discuss the critical separation distance between the two micro-structures that will produce the superhydrophobicity (the value is 100 nm as indicated by calculation).

Hydrophobic interactions are the effective attractions between apolar groups in water, and play a central role in the stability of mesoscopic assembly and biological structure in aqueous environment. However, a quantitative understanding of this role has always been elusive. One obstacle to this understanding is the multifaceted nature of hydrophobic interactions. D. Chandler *et al.* have studied this problem. They particularly show how hydrophobic interactions between small apolar groups at low concentrations in water are very different from those between large assemblies or those at relatively high concentrations of hydrophobic groups in water. The former is the aqueous solvation such as a butane or butanol molecule, and the latter is relevant to the solvation of macromolecules such as proteins.

Figure 3.15 displays hydrophobicity on small and large length scales. Figure 3.15(a) presents that hydrophobic units do not form hydrogen

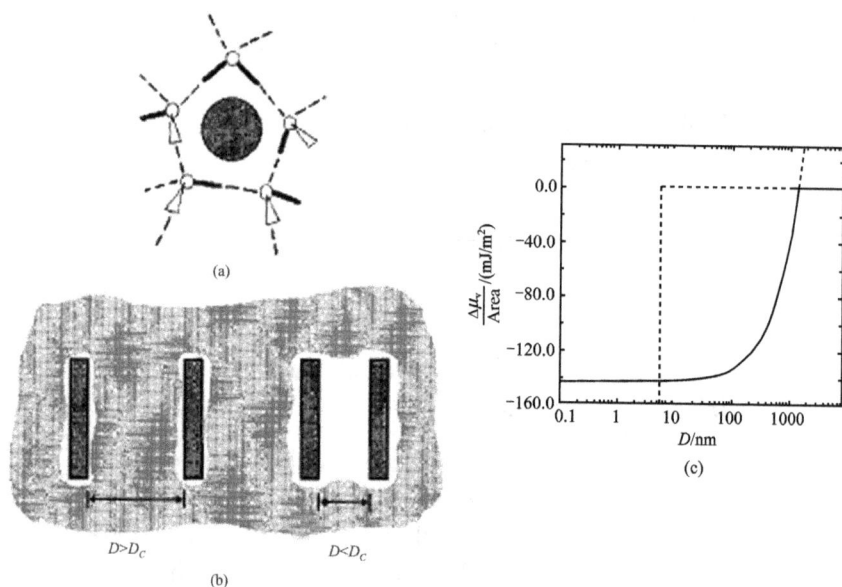

Fig. 3.15   (a) and (b): Schematic view of water structure near a small and a large hydrophobic sphere, respectively. (c) Free energy per unit surface area for water confined between two parallel hard surfaces separated with a distance *D*. (Reprinted with permission from Ref. 23. Copyright 1999, American Chemical Society.)

bonds to water, and certain volume regions are produced due to the vanishing density of water molecules. The sphere in the figure stands for the hydrophobic unit. When the units are small enough, water can reorganize near them without breaking the hydrogen bonds. The entropic cost of this structural change leads to low solubility for small apolar species in water. The cost and corresponding solubility are readily understood and computed according to the properties of homogeneous bulk water. Meanwhile, there is no strong inducement for small numbers of small hydrophobic groups to associate in water. The water is likely to separate such species rather than drive them together.

On the other hand, at a region close to a large hydrophobic unit, several apolar units may assemble, possibly interspersed with a few hydrophilic units; however the persistence of a hydrogen-bond network is geometrically impossible. Drying is the result of the energetic effect. The drying would also induce strong attractions between large hydrophobic objects. This phenomenon has been proved by the surface force measurement. For example, thin vapor layers would be produced because the loss of hydrogen bonds near the two extended hydrophobic surfaces (depicted in Fig. 3.15(b)) causes water to move away from those surfaces. Both the fluctuation in the interfaces formed in this way and the remaining liquid contained between these surfaces can be destabilized and expelled. The surface will attract each other due to the resulting pressure imbalance. If the liquid is close to coexistence with the vapor phase, just as the case for water at ambient conditions, this phenomenon occurs with widely separated surfaces.

For the geometry depicted in Fig. 3.15(b), when D is small enough for this destabilization to occur, macroscopic considerations provide an estimate of when the intersurface separation will happen. When the surface tension γ is the constant of proportionality, the bounded liquid has an unfavorable surface energy proportional to the net surface area. This energy is counteracted by the favorable bulk free energy proportional to the average number of molecules in the bounded liquid, where the proportionality constant is the difference between liquid and gas chemical potential, $\mu_l - \mu_g$. When D is large enough, provided $\mu_l - \mu_g < 0$, the bulk energy dominates over the surface energy, resulting in stable bounded liquid. On the other hand, when D is less than the critical separation, surface energy dominates and the bounded liquid is destabilized with respect to the vapor.

$$D \approx \frac{2\gamma}{n_l |\mu_l - \mu_g|}. \qquad (3.25)$$

In summary, when two parallel plates are put into water, if the distance (D) between them is larger than the critical value, $D > Dc$, there is no thin vapor layer produced; if $D < Dc$, the hydrogen bond between the two parallel plates dramatically reduces, which means water is moved away from the surfaces, producing the vapor layers. Therefore it can be found that the separation distance between the two parallel plates is very important in this theory model.

In Eq. (3.25), n_l is the molecular density of the bulk liquid (the average number of molecules per unit volume). Therefore, for water at room temperature and atmospheric pressure, $Dc \approx 100$ nm. At such large length scale, the evaporation of water will lead to dry surfaces, and it seems to be a generic and well-understood phenomenon. However, its pertinence to hydrophobicity at large length scales has not gone without attention. Since drying is a nonlinear phenomenon, its manifestation at large length scales is affected by small length scale structure.

Figure 3.15(c) depicts the theory proposed by D. Chandler *et al.*, from which the free energy per unit surface area for water confined between two parallel hard surfaces separated a distance D is deduced. When the $D \to \infty$ limiting value, the free energy is taken as zero. The solid line in the figure stands for the steady branches, while the dashed line stands for the metastable branches, and the dotted line shows the small D limit of stability for the liquid density phase. When the distance of the two hydrophobic materials is smaller than D, the water between the two plates will be moved away.

More importantly, only when local concentrations of hydrophobic units are large enough or extended enough to induce drying, hydrophobic effect that separates oil group from aqueous solution will appear. Even in the presence of hydrophilic surfaces, hydrophobic forces of assembly are so powerful that they can arise from drying transitions. Since powerful hydrophobic interactions require the beginning of a phase transition, the dynamics of assemblies stabilized by this interaction will depend on the dynamics of the phase transition.

3.2.3 *Examples of creation of superhydrophobic surfaces from hydrophilic materials (65° < CA < 90°)*

The authors' group reported the creation of a superhydropobic surface consisting of nanofibers fabricated from hydrophilic poly(vinyl alcohol) (PVA).[24] The PVA nanofibers were prepared using the template-based extrusion method. We also developed a simple method for preparing large-area surfaces of well-aligned polycarbonate (PC) nanopillar arrays using porous anodic alumina film as a template.[25] In the two examples, the contact angles of water drops on the smooth PVA and PC surfaces are 72.5° and 85.7° respectively. Therefore, we have succeeded in realizing the creation of superhydrophobic surfaces from traditional hydrophilic materials. Xu's group[26] also have done important work on fabricating superhydrophobic polymer films with self-cleaning lotus-like microstructures. For example, they first reported creating a superhydrophobic surface with a simple casting process under ambient atmosphere. The polymer surface possesses natural lotus-like micro- and nano-hierarchical structure.

Zhou *et al.* fabricated a perpendicular nanopin superhydrophobic film using a bottom-up process of high-crystalline materials such as metal oxide and hydroxide based on chemical bath deposition (CBD).[27] The results showed that the contact angles of smooth surface and roughness surface are 75.2° and 178° respectively. This is another typical example of creation of superhydrophobic surfaces from hydrophilic materials. Based on the electrospinning technology, M. Zhu *et al.* reported a method to create superhydrophobic surfaces using PHBV (Poly(hydroxybutyrate-co-hydroxyvalerate)) — a kind of hydrophilic materials. The studies showed that by varying the concentration of polymer solutions, the surface morphology and wetting property could be controlled. The polymer surface is characterized by microscale valley-and-hill structure, which was formed by aggregating electrospun beads. With the changing polymer concentrations, the bead structure changed from smooth to porous and popcorn-like. The water contact angles range from 110.7° to 158.1° on these surfaces. However, it is only 75.9° on the smooth PHBV surface.[28] The authors' group reported a facile method for the fabrication of colloidal crystal films with tunable wettability by simply modifying temperature.[29] The film was assembled from amphiphilic materials polystyrene-block-poly(methyl methacrylate)-block-poly(acrylic acid) (P(St-MMA-AA)). The superhydrophobic film (CA = 150.2°) could be fabricated at high assembly temperature.

3.2.4 *The wettability of the solid surface influenced by the air phase*

We know that the wettability of the solid surface is determined by the three phases: solid, liquid, and air. People often pay attention to the solid or liquid phase but ignore the effect of air phase. In recent years, we have found that the air composition and pressure also play key effects on the property of interface. There is a mesostate (energy barrier) between the Wenzel model and Cassie model. Pressure needs to be applied on the drop to compress the air. Quéré *et al.*[30] indicated that by changing the pressure of the liquid, the energy barrier could be conquered (Fig. 3.16).

L. Xu *et al.* investigated the corona splash due to the impact of a liquid drop on a smooth dry substrate.[31] They found a striking phenomenon that splashing can be completely suppressed by decreasing the pressure of the surrounding gas. Suppressing the splashing has relationship with the gas pressure, gas composition and the intrinsic property of the liquid drop. The threshold pressure where a splash first occurs is measured as a function of the impact velocity. This value is also found to scale with the molecular weight of gas and the viscosity of the liquid. In liquid solid impacts, the compressible effects in the gas are responsible for splashing.

Cheng *et al.* observed water condensation and evaporation on lotus leaf surfaces *in situ* inside an environmental scanning electron microscope (ESEM).[32] This real-time observation is propitious to illustrate how water drops grow to large contact angles and the contact angle hysterisis phenomenon, especially when the size of water drop is compared to the roughness scale. The results show that water drops grow to large contact angles during water condensation, and

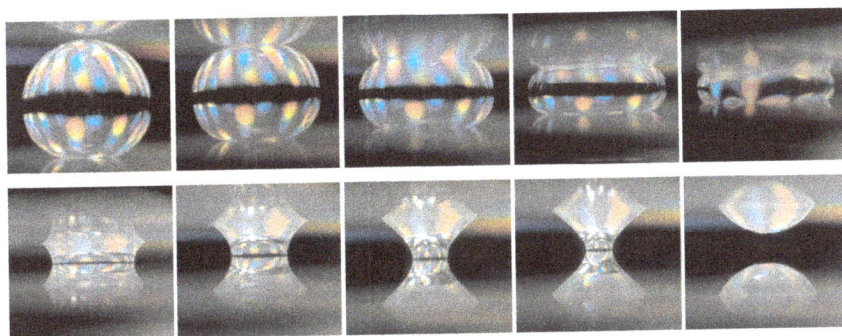

(a)

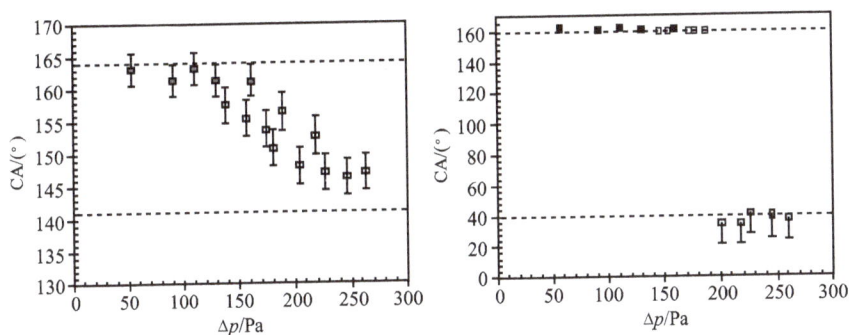

(b)

Fig. 3.16 (a) Compressing and separating the water drop between two identical microstructured hydrophobic surfaces. (b) Relationship between the apparent contact angle and applied pressure.[30] (Reprinted by permission from Macmillan Publishers Ltd: *Nature Materials*, copyright 2003.)

decrease in size and contact angle during the evaporation phase. Meanwhile, they have found that the condensed water forms "sticky" drops on the lotus leaf surface, and they remain on the surface at any tilt angle despite having a large contact angle. Strong adhesion exists between water drops and the bumps on lotus leaves after the lotus surface is exposed to water vapor. Even when drop size decreases, the contact angle hysteresis is still obvious.

ESEM was used to investigate the condensation and evaporation of water on an initially dry lotus leaf. With the increase of water vapor pressure, microscale droplets began to form in areas adjacent to bumps.

Fig. 3.17   ESEM images captured when water droplets are on a lotus leaf. (a) and (b): The droplets grow during condensation. (c) and (d): The droplets shrink during evaporation. (Reprinted with permission from Ref. 32. Copyright 2005, American Institute of Physics.)

The apparent contact angle of these droplets were less than 90°; when the droplets growed in size, the apparent contact angle increased to greater than 90° and the droplets appeared to be lifted up to rest atop the bumps (Fig. 3.17(a) and (b)). Smaller drops were absorbed by larger ones during the coalescence because of the Laplace pressure difference. With the water drops volume increasing, the larger drop was observed to undergo two processes: the water drops' base was pinned by the bumps, and its apparent contact angle continued to increase; then the drop expanded its base by jumping to contact the neighboring bumps. These processes continuously repeated, and larger drops were obtained. All drops were stationary. Even when the leaf was tilted 45°, there was still no sign of rolling or sliding. Reducing the water vapor pressure to cause evaporation during the real-time observation in

ESEM, the drop volume began to decrease. With the droplet volume shrinking, the contact angle began to decrease because the drop base was pinned by the bumps. During the evaporation process, there was not enough water to sustain the base area, and the contact area kept reducing. This process continuously repeated causing the droplet to shrink until the entire droplet disappeared (Fig. 3.17(c) and (d)).

In order to explain the phenomenon reasonably, Cheng *et al.* assumed a geometric model for forming a high contact angle on a rough hydrophilic surface when the droplet size is comparable to the toughness size. As is shown in Fig. 3.18, θ_c is the intrinsic contact angle of the smooth surface; θ_a is the apparent contact angle of the roughness surface when the drop size is comparable with a characteristic length scale of the roughness, such as the microscale water drop on the lotus leaf surface. This geometric model is different from the conventional view of liquid in contact with rough surfaces, including the Wenzel and Csssie–Baxter model because these models assume that the roughness scale is much smaller than the size of the drop. Meanwhile, this model also illustrates how the high apparent contact angle (CA > 90°) and hysterisis form on the hydrophilic surface (CA < 90°). This result and that of superhydrophobic surfaces created with hydrophilic materials supplement each other.

In practical applications, the sizes of droplets are always different. For example, the sizes of water droplets are dramatically different between the fog and rain. A surface is superhydrophobic for the

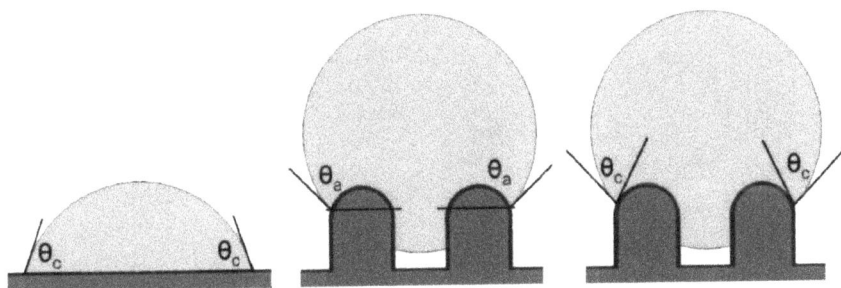

Fig. 3.18 A geometric model for forming a high contact angle droplet on a rough hydrophilic surface when the droplet size is comparable to the roughness size. (Reprinted with permission from Ref. 32. Copyright 2005, American Institute of Physics.)

raindrop size water droplets, but for the fog-drop and even smaller droplets it is not. Lots of micro-sized droplets will stick on the surface because the roughness scale is not small enough. These micro-sized droplets will influence the air trapped under Cassie state, further breaking its superhydrophobic property.

3.3 Contact Angle Hysteresis

Generally, it means more hydrophobic for the surface that has a higher water contact angle. In Fig. 3.19, for example, the hydrophobicity of the surfaces decreases successively from left to right, and the left one is more hydrophobic, because it has a much bigger static contact angle. However, if all the surfaces are tilted at a small angle from the horizontal, as presented in the bottom part of the picture, we can see that the water droplet on the right can slide off easily, while the one on the left will be pinned on the substrate. With these facts we can find the intrinsic difference between the static contact angle and the dynamic contact angle. To study the fabrication of hydrophobic surfaces and the practical application of the self-cleaning surfaces, the

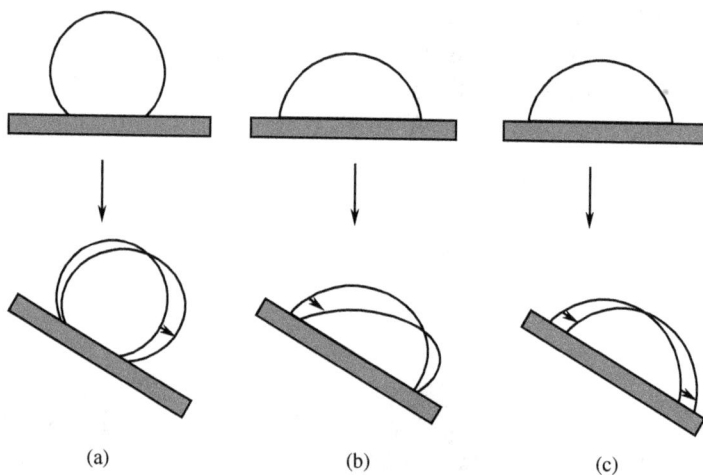

Fig. 3.19 States comparison of liquid droplet on three different solid surfaces. (Reprinted with permission from Ref. 33. Copyright 1999, American Chemical Society.)

dynamic behavior of the liquid droplet at minor force interaction must be considered. So the research on dynamic wettability and contact angle hysteresis is very important.

Generally, there are three kinds of dynamic behaviors of liquid droplet on solid surface: tilting, sliding, and pinning. Liquid droplets in these states have different interactions with the different solid surfaces. For example, an ideal self-cleaning surface needs an extremely small contact angle hysteresis, that is, liquid droplet can easily roll off on the surface.[34] Taking the lotus leaf as an example, the contact area between the water droplet and the leaf surface takes just 2–3% of the whole surface area, and there is a lot of air trapping on the leaf surface. The size of contaminating particles (such as dust) is generally larger than that of the surface microtextures, so the contaminating particles can just contact with the tips of the microtextures, and the actual contact area is very small. Because the adhesive force between the contaminating particles and leaf surface is much lower than that between water droplet and leaf surface, the contaminating particles on the leaf surface can be "trapped" by the rolling water droplet which slides off the surface easily to achieve self-cleaning effect, when the leaf surface tilts (Fig. 3.20(a)). Nonetheless, on hydrophobic

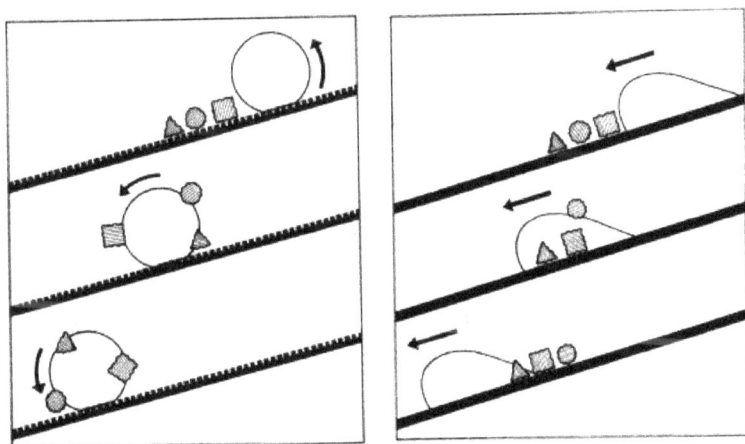

Fig. 3.20 Interactions of contaminating particles with two kinds of hydrophobic surfaces. (Reprinted with permission from Ref. 34. Copyright 1997, Springer Berlin/Heidelberg.)

smooth surface, the contact area between contaminating particles and the smooth surface is large enough, and the water droplet can only leave the surface in the tilting form. In this condition, water droplet cannot take away the contaminating particles, and the particles will just re-distribute in the liquid interaction. So the smooth surface does not exhibit self-cleaning property (Fig. 3.20(a)). Meanwhile, the sliding liquid droplet possesses much more powerful ability to bring contaminating particles away than the tilting droplet.

Recently, researches on superhydrophobic surfaces mainly focus on the fabrication of surfaces with high water contact angles and low sliding angles. Water droplets on these surfaces can easily roll off and take off the contaminating particles, showing good self-cleaning effect. However, little has focused on the field that liquid droplet has high adhesion with solid superhydrophobic surfaces. In fact, the superhydrophobic surfaces of high water adhesion represent a sort of interfacial materials that have valuable importance in theory and application. Compared with lotus leaf, on the one hand, the structures of gecko's feet and lotus leaf are both composed of similar micro–nano binary structures, so they can exhibit superhydrophobic effect. On the other hand, due to the different micro-structures of the lotus leaf and gecko's feet, the state of the trapping air will be different when water droplet contacts the surfaces. The structure of lotus leaf is constructed with micro-papilla and nano-protrusions on it. This kind of nano-protrusions form patterns randomly, and the trapping air pockets are continuous and connected to atmosphere; this is called lotus state. Nonetheless, the structure of gecko's feet is constructed with allied seta, and this kind of structure will cause two kinds of trapped air: There are sealed pockets of air trapped in the seta (act as micro capillaries), as well as air pockets continuous with the atmosphere trapped between seta; this is called Gecko state. It can be estimated that besides the two states, there is another state in which all the trapped air is sealed. However, the dynamic behavior of liquid droplet in this state still needs further investigation.

On the research of high water-adhesive solid surface, recently the authors' group have fabricated a novel superhydrophobic film with high water-adhesive property by mimicking the microstructure of gecko's feet.[35] This kind of tube allied film surface can produce high Van der Waals force with water droplet so that it will not roll off even the surface is inversed. Different from the self-cleaning effect of the lotus leaf, this kind of dynamic wetting behavior cannot be explained by traditional theory. The significance of its application is that it can be used as "mechanical hand" in the field of no-loss liquid transport.

3.3.1 *Advancing and receding contact angle*

Considering a liquid droplet with an equilibrium contact angle on a horizontal smooth surface, if the surface is ideally smooth and homogeneous, the edge of the droplet will extend or retreat when some liquid is added or withdrawn, but the contact angle will keep constant during the processes. On the contrary, if the surface is rough or heterogeneous, the added liquid will just make the droplet higher but the edge stands still. Thus the contact angle will increase and this is the so-called advancing contact angle, θ_A (Fig. 3.21(a)). If adequate liquid is added, the edge of the droplet will extend suddenly, and the contact angle at the just extending moment is called maximal advancing contact angle, $\theta_{A, max}$. When some liquid is drawn off from the droplet, the standing edge of the droplet will make the droplet flatter, and the droplet contact angle will decrease, and this is called receding contact angle, θ_R (Fig. 3.21(b)). If adequate liquid

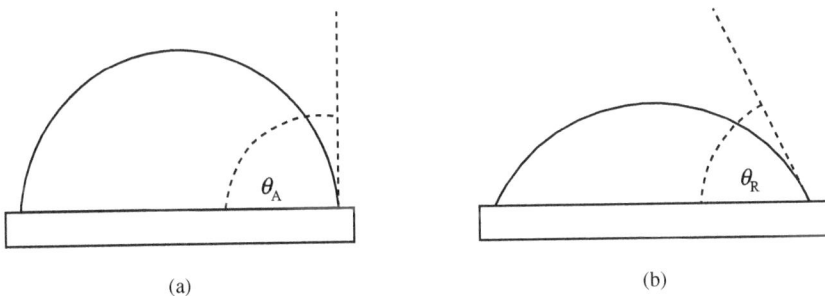

(a)   (b)

Fig. 3.21 Morphology of tiny liquid droplet on solid surface. (a) Increasing liquid, contact line extending. The advancing contact angle (θ_A) can be exhibited at the ending of liquid movement. (b) Drawing off liquid, contact line retreating. The contact angle decreases and exhibits receding contact angle (θ_R).

is drawn off, the edge of the droplet will suddenly retreat, and the contact angle at this moment is called minimal receding contact angle, $\theta_{R,min}$.

Nonetheless, the advancing and receding contact angles can simultaneously coexist on an inclined plane. If there is no contact angle hysteresis, the liquid droplet will roll off just when the plane tilts a little. Contact angle hysteresis can make liquid droplet stably sit on the inclined plane. This phenomenon demonstrates that there is an energy barrier existing in the liquid edge movement.

3.3.2 *Sliding angle*

There is a difference between the measured contact angle after the solid–liquid interface extension (advancing contact angle) and that before the solid–liquid interface retreat (receding contact angle). Generally, advancing contact angles are larger than receding contact angles, that is, $\theta_A > \theta_R$ (Fig. 3.22), and the difference of them is called sliding angle. The values of sliding angles represent the contact angle hysteresis of a solid surface. The advancing contact angle and receding contact angle can also be described as follows: the advancing contact angle is the contact angle at the moment just before triple-phase contact line moves when adding to the droplet volume; it can be considered as the tilted angle that the droplet will not slide until the inclined plane researches. Receding angle is the

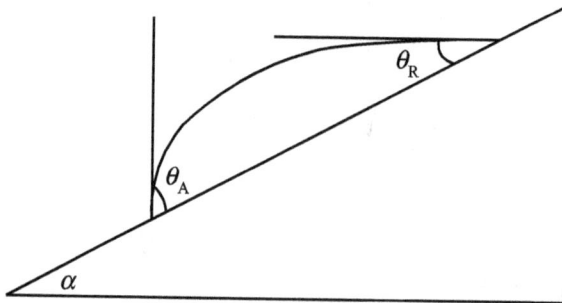

Fig. 3.22 Droplet on an inclined plane, and its corresponding advancing contact angle (θ_A) and receding contact angle (θ_R).

contact angle at the moment just before triple-phase contact line moves when drawing off from the droplet volume; it can be considered as the tilted angle that the droplet will not slide until the downhill side researches the value. Surfaces in Fig. 3.19(a) and (b) exhibit water contact angles of $\theta_A/\theta_R = 120°/180°$, and the surface in case (c) exhibits $\theta_A/\theta_R = 70°/70°$. In surface (a), the downhill side of the droplet can advance, but the uphill side stays pinned until the receding angle is reached. In surface (b), it is easier to slide, because the contact angle of its uphill side is smaller than (a). While on surface (c), the droplet can slide simultaneously at a tiny tilted angle.

3.3.3 *Factors influencing contact angle hysteresis*

3.3.3.1 *Roughness induced hysteresis*

In the case of a rough surface, the real surface area is different from the apparent surface area. Defining the ratio of real surface area (A) and apparent surface area (A') to be surface roughness:

$$r = A/A'$$

The contact angle of a given rough surface can be expressed by Wenzel equation:

$$\cos \theta_r = r \cos \theta.$$

It reveals that the absolute value of $\cos\theta_r$ on a rough surface is always larger than that of $\cos\theta$ on a smooth surface. In other words, creating rough structures on a hydrophobic surface $(\theta > 90°)$ will result in the increase of contact angles. While on a hydrophilic surface $(\theta < 90°)$, it will result in the decrease of contact angles. It is reported that contact angle hysteresis is related to surface roughness. For example, on a rough surface of wax, the water advancing contact angle is $110°$, receding angle is $89°$, and the hysteresis is $21°$. While on the smooth wax surface, the water advancing contact angle is $97°$, receding angle is $89°$, and the hysteresis is only $8°$. Meanwhile, for the surface with $\theta < 90°$, the advancing contact angle θ_A increases with increasing r, and this is in contradiction to the Wenzel equation. Obviously, this is in relation with the hysteresis induced by surface roughness. It is considered that except for the lowest surface free energy point, there is still metastable state existing on the surface. The liquid wetting and dewetting will rest on different metastable states, which will induce hysteresis.

3.3.3.2 Hysteresis from surface heterogeneity and multiphase

The heterogeneity and multiphase of solid surface is also the reason of contact angle hysteresis. There is a energy barrier at the phase junction, and the frontier of the liquid generally rests on the phase junctions. Advancing contact angles generally reflect the area with low surface energy, or the solid surface that shows weak liquid affinity. While receding contact angles generally reflect the area with high surface energy, or the solid surface that shows strong liquid affinity.

The apparent contact angle on a composite surface with two components of different wettability can be expressed by the Cassie equation:

$$\cos\theta_r = f_1\cos\theta_1 + f_2\cos\theta_2.$$

Due to the energy barrier at the S_1/S_2 junction, liquid frontier is inclined to rest on the phase junction. Advancing contact angles tend to reflect the low surface energy area (with larger intrinsic contact angle), while the receding contact angles tend to reveal high surface energy area (with smaller intrinsic contact angle). The relationship of θ_r, θ_A and θ_R can be expressed approximately as follows:

$$\cos\theta_r = 1/2(\cos\theta_A + \cos\theta_R). \tag{3.26}$$

3.3.3.3 Surface contamination

Surface contamination is the most common reason of contact angle hysteresis, and it can cause hysteresis on both liquid and solid surfaces. For example, water will completely spread on clean glass surface with contact angle being zero. However, when water contacts the smeary glass, most oil contamination will spread on the water surface, thus there is higher γ_{SV} on the solid surface in the process of testing receding contact angles than that of testing advancing contact angles, which will make the receding contact angle smaller than advancing contact angle. It is reported that in the experiments of grapheme and talc, the absolute cleaning surface of liquid and solid can eliminate hysteresis.[36]

Surface contamination mostly comes from liquid and solid surface adsorption, which results in remarkable change of contact angles. It is reported that even the change of the air composition can influence water contact angle on gold surface. Thus, it can be speculated what will happen when the experimental condition is unclean or contacting samples directly with hands.

3.3.4 *Triple-phase contact line*

When a liquid droplet is deposited on a solid surface, the contacting part of solid/liquid/air triple phases is a dimensional line, which is called contact line. The shape of contact line is attributed to the microstructure of the solid surface, which can be simply divided into two cases, continuous and discontinuous (Fig. 3.23). Generally, liquid droplet with continuous contact line is hard to roll off, and it will be pinned on the solid surface with a bigger sliding angle; on the

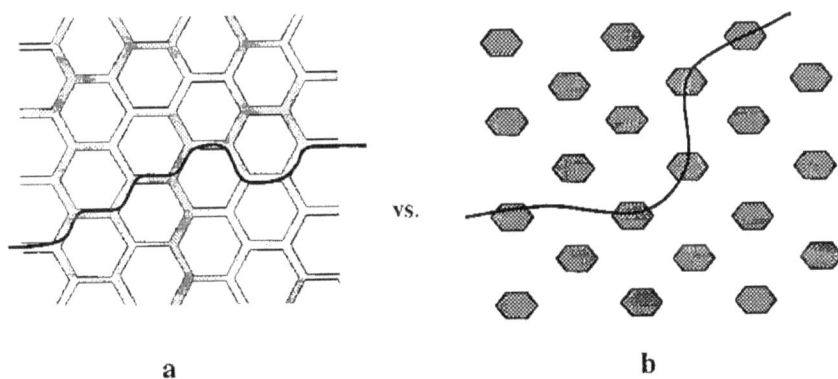

vs.

a   b

Fig. 3.23 Triple-phase contact line: (a) continuous; (b) discontinuous. (Reprinted with permission from Ref. 33. Copyright 1999, American Chemical Society.)

contrary, liquid droplet is easier to roll off in the case of discontinuous contact line, with a smaller sliding angle.

Quéré and his colleagues have done detailed researches in the field of dynamic behavior of liquid droplets on superhydrophobic surfaces.[37] The importance of one-dimensional triple-phase contact lines is revealed. It is well known that liquid–solid contact surface will decrease when contact angle increases, if the contact line does not easily collapse, the contact angle hysteresis will be small and it is in favor of liquid droplet moving; if the contact line is stable, larger external force or tilted angle is needed to make the droplet moving and sliding, or the droplet can just leave the surface by evaporation.

3.3.5 *Examples of dynamic behavior of water droplets*

Quéré's group have fabricated a kind of liquid marble, which is like the solid elastic ball, by covering water droplet with a non-stick film (Fig. 3.24).[38] The secret of this liquid marble attributes to the covered hydrophobic powder, which can suppress wetting with water. If the powders are dispersed into a water droplet, they spontaneously migrate to the water-air interface and form a film to protect the water. It is very easy to control the movement of the liquid marble using gravity field, electric field or magnetic field.

Fig. 3.24 A "liquid marble" covered by hydrophobic powder.[38] (Reprinted by permission from Macmillan Publishers Ltd: *Nature*, copyright 2001.)

This kind of droplet with diameter of 1 mm can exhibit nearly perfect spherical shape when sitting on a glass plate, which is different from the common lentoid shape of a water droplet caused by the affinity between water molecules and glass substrate. A tiny force such as electromagnetic force can make it move, due to the extremely small viscous friction. Moreover, the liquid marble can stably sit on water surface like a lightsome insect. It is believed that all these features are of potential benefit not only in microfluidic applications but also in the medicine and environmental inspection. For example, it can improve the technology that position tiny fluid onto the silicon substrate for chemical or biological analysis.

Furthermore, the researches monitored the quick singularity process of how water droplet rebounded on superhydrophobic surface using high speed camera (Fig. 3.25). Water droplets with different weights and impact velocities were used to strike the lotus-like waterproof surface. The whole striking process was monitored by a high-speed camera at 1000 frame/s, and the video was played

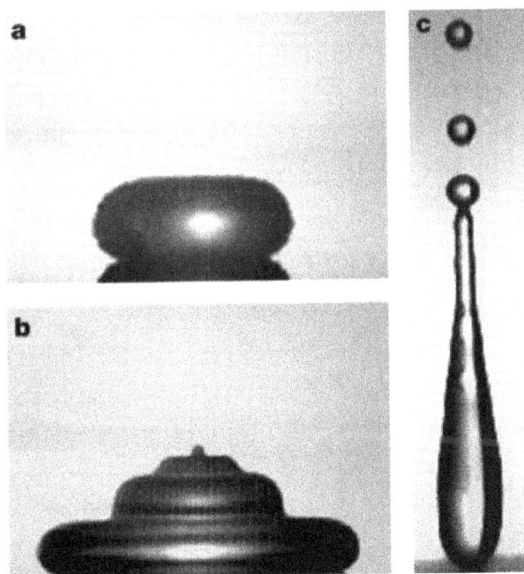

Fig. 3.25 Dynamic behavior of water droplet on superhydrophobic surface.[3] (Reprinted by permission from Macmillan Publishers Ltd: *Nature*, copyright 2002.)

back at a much slower speed to investigate the droplet moving characteristics. In comparison, the record speed of the common movie is 24 frame/s.

Through investigation, it was revealed that the water droplet would quickly deform to adsorb the striking energy before it bounced back. The deformation mode was decided by impact velocity. At a very small velocity, water droplet exhibited a tiny deformation, the whole process seemed like a bouncing ball. When the velocity was a little higher, it would be compressed like a chessman. As the velocity kept on increasing, it would turn to a ladder-like pyramid, then was compressed and congregated to form an elongated tear-like shape, and left the solid surface totally. While at much higher velocity, the droplet would turn to a flat shape, congregate and recover then bounce back like a reverse baseball bat.

Besides, researchers found that the contact time of the bouncing droplet had no relation with the impact velocity, but was in relation to the droplet size. The larger volume corresponds to a longer contact time. The contact time of a water droplet with radius about 0.1 mm is less than 1 ms, but that of a water droplet with radius about 4 mm reaches 0.1 s. Understanding the contact kinetic behavior of liquid droplet and solid surface is of benefit to designing waterproof materials and fog-spraying products, e.g. to improve the print quantity of ink jet printer.

3.3.6 *Theoretical study on contact angle hysteresis*

The importance of contact angle hysteresis in hydrophobicity research was reported 40 years ago by Furmidge *et al.*[39] However, for a long time, the subsequent research of hydrophobic materials only focused on the value of contact angle but ignored the consideration of θ_A, θ_R, contact angle hysteresis and the triple contact line. Just in recent researches, scientists began to focus on the influence of sliding angles and dynamic contact angles. This is because it might not be adequate to describe the hydrophobicity of a surface using static contact angle alone, and the contact angle hysteresis might be more important than the maximum contact angle. In 1962, Furmidge *et al.* reported a equation about the relationship of contact angle hysteresis and sliding angle:

$$F = (mg \sin \alpha)/w = \gamma_{LV}(\cos \theta_R - \cos \theta_A), \quad (3.27)$$

where F is the force that makes the droplet move on solid surface, and it is the linear critical force per unit length of droplet circle. α is the sliding angle, m is the weight of water droplet, w is the width of droplet, γ_{LV} is the free energy of liquid at liquid–air interface, θ_A and θ_R corresponds to advancing and receding contact angle, respectively. From this equation, we can see that two surfaces with the same hysteresis might not have the same sliding angles. This is because the value of m/w will change with different contact angles. So, it is better to use the sliding angles to compare the sliding characteristics of the surfaces with different contact angles.

Wolfram et al.[40] have reported the equation describing the sliding angle of liquid droplet on different smooth surfaces:

$$\sin \alpha = k\frac{2r\pi}{mg}, \quad (3.28)$$

where α is the sliding angle, r is the radius of contact cirque, m is the weight of droplet, g is gravity acceleration, and k is a constant.

Murase et al.[41] have converted the equation to describe the relationship between sliding angle and contact angle:

$$r = R \sin \theta, \quad (3.29)$$

$$\tfrac{3}{4}\pi R'^3 \rho g = mg, \quad (3.30)$$

$$R' = \left\{ \frac{1}{4}(2 - 3\cos \theta + \cos^3 \theta) \right\}^{\frac{1}{3}} R, \quad (3.31)$$

where ρ is the water density, R' and R are the droplet radii before and after sitting on the surface, respectively. Combing Eqs. (3.28)–(3.31) yields:

$$k = \left\{ \frac{9m_2(2 - 3\cos \theta + \cos^3 \theta)}{\pi^2} \right\}^{\frac{1}{3}} \frac{\sin \alpha (g\rho^{\frac{1}{3}})}{6\sin \theta}. \quad (3.32)$$

Using this equation, with the calculated value of α, θ and m, we can calculate the constant k on any smooth surface, which is in relation to the solid–liquid interaction energy.

Hashimoto et al.[42] have reported the expression of sliding angle α and contact angel θ' on rough surface based on research above. To simplify the research, they firstly gave some assumptions:

(1) The shapes of the tips of the protrusions are constant for all of the surfaces; that is r has a constant value. The contact angle varies only with the fraction value f.
(2) The interaction energy between water and substrate is proportional to the true contact area, which is rf times as large as the apparent contact area. Therefore, the constant related to the interaction energy k in Eq. (3.32) is assumed to be rf times as great as that for the flat surface.
(3) Experimental contact angles are equal to equilibrium contact angles.

Using assumption (2) and Eq. (3.32), we can obtain expression of rough surface:

$$\sin \alpha = \frac{2rfk \sin \theta'}{g}\left\{\frac{3\pi^3}{m^2\rho(2 - 3\cos\theta' + \cos^3\theta')}\right\}^{1/3}. \quad (3.33)$$

Combining Cassie equation and Eq. (3.33), we can obtain Eq. (3.34), and this is the equation expressing the relationship of sliding angle α and contact angle θ' on a rough surface:

$$\sin \alpha = \frac{2rfk \sin \theta'(\cos\theta')}{g(r\cos\theta + 1)}\left\{\frac{3\pi^3}{m^2\rho(2 - 3\cos\theta' + \cos^3\theta')}\right\}^{1/3}. \quad (3.34)$$

Fort's group[43] gave further explanation based on the studies of Hashimoto. They gave conceptive clarification to the assumptions of Hashimoto, and expanded them to more solid surfaces.

They started from the equation of sliding angle and droplet size provided by Furmidge:

$$\sin \alpha_c = k \frac{2r\pi}{mg}. \qquad (3.28)$$

Here, α_c is the sliding angle or equilibrium critical contact angle.

Figure 3.26 illustrates the equilibrium state of a tilted droplet. We can see that the downward component of weight can be balanced only if the back contact angle (θ_b) is smaller than the front one (θ_f). This equilibrium state can be expressed by Furmidge's another equation,

$$mg \sin \alpha \approx \gamma_{LV}(\cos \theta_b - \cos \theta_f)w.$$

When a droplet is tilted until it begins to slide down ($\alpha = \alpha_c$), one may assume that the back and front contact angles are close to their minimum and maximum possible values (θ_- and θ_+, respectively). Then,

$$mg \sin \alpha_c \approx \gamma_{LV}(\cos \theta_- - \cos \theta_+)w. \qquad (3.35)$$

We can conclude from Eqs. (3.28) and (3.35) that the sliding constant k is related to hysteresis (that $\theta_- \neq \theta_+$). Despite this evidence, Murase[41] assumed that k is "related to the interaction energy between solid and liquid", and Hashimoto *et al.*[42] assumed that "k is proportional to the product of roughness, r, and the fraction of wetted area, f" (assumption (2)).

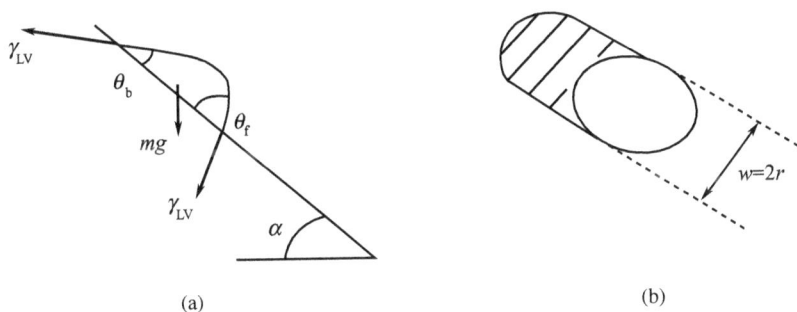

(a)   (b)

Fig. 3.26 Profile of a tilted droplet: (a) side view; (b) top view. (Reprinted with permission from Ref. 43. Copyright 2002, American Chemical Society.)

In fact, Eq. (3.35) is an equation of energy balance. When the solid surface is progressively inclined, the droplet will slide down just at the angle $\alpha = \alpha_c$ such that gravity can furnish the energy necessary to develop the back wetted surface. As long as Eq. (3.35) remains approximately valid, the energy used to create this surface will be:

$$\gamma_{LV}(\cos\theta_- - \cos\theta_+). \tag{3.36}$$

So the interpretation of constant k as being related to the solid–liquid energy of interaction seems possible.

A reversible process is an evolution through equilibrium states. This means that both the advancing and receding contact angles must be equilibrium contact angles and obey Young's equation:

$$\frac{\gamma_{SV\pm}^{R} - \gamma_{SL}^{R}}{\gamma_{LV}} = \cos\theta_{\pm}, \tag{3.37}$$

where γ_{LV} is the liquid–gas surface tension, and the superscript R is used to denote rough surfaces i.e. γ_{SV}^{R} and γ_{SL}^{R} correspond to the surface energy per unit apparent area of the solid–gas or solid–liquid interface, respectively.

Fort and co-workers have set up a physical model of droplet over partially wetted rough surfaces (Fig. 3.27). Using this model, it is easy to calculate the value of θ_- and θ_+.

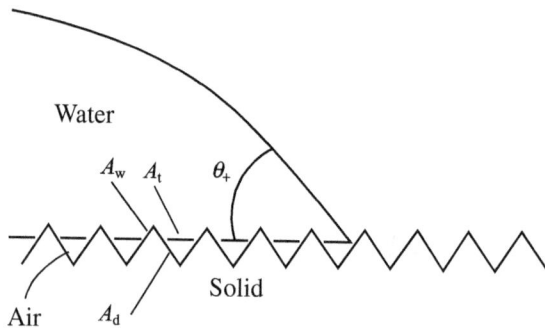

Fig. 3.27   Droplet over a partially wetted rough surface.[43] (Reprinted with permission from Ref. 43. Copyright 2002, American Chemical Society.)

Firstly, they gave some assumptions on the model:

(a) Defining the roughness r such that the real (microscopic) surface area of the solid surface is r times its apparent (macroscopic) area, $A_r = rA_a$ (with $r \geq 1$ and $A_r = A_w + A_d$); the subscripts w and d stand for wetted and dry surfaces respectively, and r refers to roughness.

(b) Roughness does not modify the surface energies per unit area of microscopic surface (γ_{SV} and γ_{SL}); that is, their values are identical to those of a flat surface.

(c) Defining the wetted fraction as $f \equiv A_w/A_r$. Thus, $A_w = frA_a$, and $0 \leq f \leq 1$.

(d) The liquid–vapor interface, resulting from the trapped air, occupies a fraction $(1 - f)$ of the apparent surface area below the drop; that is, $A_t/A_a = (1-f) = (A_r - A_w)/A_r = A_d/A_r$. Note that this yields $A_t = 0$ if $A_d = 0$, and $A_t = A_a$ if $A_d = A_r$, also as expected from Fig. 3.27.

(e) Once the droplet slides down, the solid surface will remain wetted at the same regions that were in contact with water when the droplet was above them. That is, a fraction fr of the apparent surface will remain wetted behind the drop.

With this model, it is easy to calculate the surface tensions of the rough surface (γ_{ij}^R) as a function of the corresponding values for an ideal flat surface (γ_{ij}) by simply adding the contributions to the surface energy at the microscopic level.

The surface energy of the interface below the droplet will be:

$$A_a \gamma_{SL}^R = A_w \gamma_{SL} + A_d \gamma_{SV} + A_t \gamma_{LV}$$
$$= frA_a \gamma_{SL} + (1-f)rA_a \gamma_{SV} + (1-f)A_a \gamma_{LV}.$$

That of the dry solid surface at the front of the drop is:

$$A_a \gamma_{SV+}^R = A_r \gamma_{SV} = rA_a \gamma_{SV}.$$

According to assumption (e) above, at the back of the drop we have:

$$A_a \gamma_{SV-}^R = A_w(\gamma_{SL} + \gamma_{LV}) + A_d \gamma_{SV}$$
$$= frA_a(\gamma_{SL} + \gamma_{LV}) + (1-f)rA_a \gamma_{SV}.$$

From these values and Eq. (3.37), the expression of advancing and receding contact angles can be easily obtained, as Eqs. (3.38) and (3.39) shown below in:

$$\cos\theta_+ = fr\cos\theta_s + f - 1 \qquad (3.38)$$

where θ_s is the horizontal contact angle for a smooth surface:

$$\frac{\gamma_{SV} - \gamma_{SL}}{\gamma_{LV}} = \cos\theta_s.$$

Equation (3.38) is in accordance with Hashimoto's result, and it can deduce to Wenzel ($f = 1$) and Cassie ($r = 1$) equations. The equation for receding contact angle is

$$\cos\theta_- = fr + f - 1. \qquad (3.39)$$

At this point, it must be noted that Eqs. (3.38) and (3.39) are exact under the hypotheses of the model in Fig. 3.27. The agreement of Eq. (3.38) with Hashimoto's result is due to the fact that for such a surface shape, the definition of f is the same. In fact, for a surface of arbitrary shape, it is not possible to derive Eqs. (3.38) and (3.39) as exact. However, in such a general case, f and r can be taken as phenomenological parameters.

Once θ_- and θ_+ are known, the sliding constant k in Eq. (3.35) can be calculated by introducing them into Eq. (3.28). So the interpretation of constant k as related to the solid–liquid energy of interaction seems possible. This procedure can be done for different solid surfaces. Instead of that, it will be more useful to calculate the critical droplet size $(V/w)_c$, defined as the maximum drop size that can stand in equilibrium on vertical surface ($\sin\alpha_c = 1$). According to this definition, Eqs. (3.28) and (3.35) yield:

$$\left(\frac{V}{w}\right)_c = \frac{k\pi}{\rho g} = \frac{\gamma_{LV}(\cos\theta_- - \cos\theta_+)}{\rho g}, \qquad (3.40)$$

where V is the drop volume and ρ is its density.

Substituting Eqs. (3.38) and (3.39) into Eq. (3.40), we can obtain:

$$\left(\frac{V}{w}\right)_c = fr\frac{\gamma_{LV}}{\rho g}(1 - \cos\theta_s). \qquad (3.41)$$

For rough surface, the factor $\dfrac{\gamma_{LV}}{\rho g}(1 - \cos\theta_s)$ is constant. So the variation of $(V/w)_c$ is interpreted as due to the product fr. The calculated results have good agreement with the experimental results of Hashimoto.

3.3.7 *Anisotropic rolling*

It is reported that the surface morphology not only can influence the apparent contact angles but also can influence the sliding behavior. The directional difference of microstructure on the same solid surface will contribute a difference in sliding behavior of a liquid droplet, which is called anisotropic rolling. Watanabe and colleagues[44] have systematically studied the contact angles and sliding behavior of water droplets on the hydrophobic surfaces with one dimensional groove structure and pillar structure. It was revealed that the water contact angles on the groove structure were 135° and 117° from the parallel and the orthogonal directions of the groove, respectively, while that of the pillar structure was 139°. Furthermore, the sliding behaviors of water droplets on the surfaces were quite different. It is well known that the sliding behavior of water droplet depends on the length and continuity of the solid/liquid/gas triple-phase contact line on the surface. A continuous short triple-phase contact line is preferable for a surface with an excellent water-rolling property. Figure 3.28 reveals the shape and length of triple-phase contact line for the groove and pillar structure. As shown in the figure, the triple-phase contact line toward the sliding direction is continuous for the parallel direction in the groove structures (Fig. 3.28(a)). For the pillar structure and orthogonal direction in the groove structure, the triple-phase contact line toward the sliding direction is discontinuous, and the length of the triple-phase contact line is longer toward the orthogonal direction

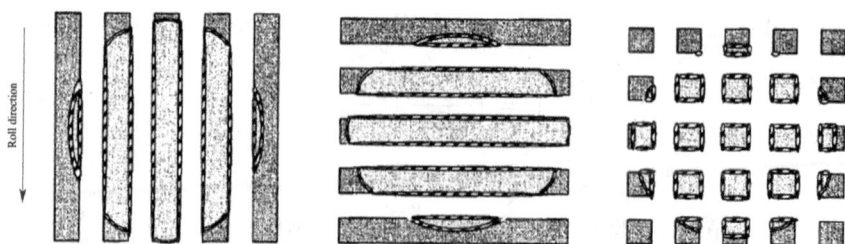

Fig. 3.28   Schematic drawing of the shape and length of triple-phase contact line for the groove and pillar structure. (Reprinted with permission from Ref. 44. Copyright 2002, American Chemical Society.)

in the groove structure (Fig. 3.28(b) and (c)). Thus, corresponding to the energy gap for the movement of the triple-phase contact line in the slidng direction, the order of the sliding angle is: parallel direction < pillar structure < orthogonal direction. The surface with groove structure, on which the water contact angle was lower than that on the pillar structure, showed better water-sliding behavior in the parallel direction than the pillar structure, due to the low energy barrier for the movement of the triple-phase contact line in the sliding direction.

Takahara *et al.*[45] have demonstrated that anisotropy is attributed to the difference in the energy barrier of wetting. Figure 3.29 illustrates the possible mechanism of the wetting anisotropy. The arrows show the wettability and sliding direction. When a contact line situated parallel to the line advances in the orthogonal direction (x-axis direction), the liquid-phobic region functions as the energy barrier of wetting. For the parallel direction (y-axis direction), there is no energy barrier of wetting. Therefore, the contact angles in the x-axis direction are higher than those in the y-axis direction, because the contact line must jump over the energy barrier of wetting for a droplet to advance in the x-axis direction. Based on the above analysis, we can conclude that it is more efficient to design the shape and length of triple-phase contact line than to simply decrease the solid–liquid contact area to increase the contact angles when producing surfaces with low sliding angles.

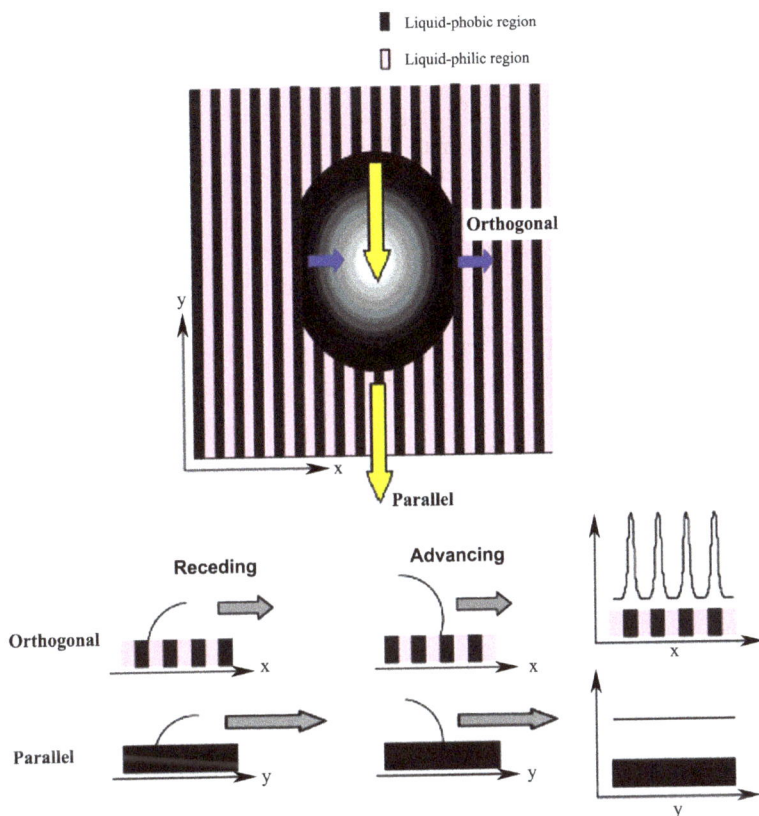

Fig. 3.29 Schema of mechanism of wetting anisotropy.[45] (Reprinted with permission from Ref. 45. Copyright 2005, American Chemical Society.)

The surface is the area that materials connect and interact with the environment. Surface property is one of the important physical-chemical characteristics of materials. Many processes such as adsorption, catalysis, lubrication and friction are all in close relation to the surface properties. Surface wettability has aroused much attention due to its great importance not only in basic research but also in practical applications. On the basis of material composition and structure, designing and fabricating interfacial materials with unique wettability deepens our understanding of the surface phenomenon, and expands the application fields of materials.

References

1. Gu T R, Zhu B Y, Li W L, Ma J M, Dai L R, Cheng H M. *Surface Chemistry*. Science Press, Beijing, 2001.
2. Ellison A H, Fox H W, Zisman W A. *J. Phys. Chem.*, 1953, **57**: 622.
3. Hare E F, Zisman W A. *J. Phys. Chem.*, 1955, **59**: 335.
4. Young T. *Trans. Roy. Soc. London*, 1805, **95**: 65.
5. Wenzel R N. *Ind. Eng. Chem.*, 1936, **28**: 988.
6. Nishino T, Meguro M, Nakamae K, Matsushita M, Ueda Y. *Langmuir*, 1999, **15**: 4321.
7. Cassie A B D, Baxter S. *Trans. Faraday Soc.*, 1944, **40**: 546.
8. Quere D. *Capillarity and Wetting Phenomena*. Springer-Verlag, New York, 2004.
9. Cathilde M, Quere D. *Soft Matter.*, 2005, **1**: 55.
10. Bico J, Marzolin, Quere D. *Europhys. Lett.*, 1999, **47**: 220.
11. Shibuichi S, Onda T, Satoh N, Tsujii K. *J. Phys. Chem.*, 1996, **100**: 19512.
12. Johnson Jr R E, Dettre R H. *Adv. Chem. Ser.*, 1963, **43**: 112.
13. Patankar N A. *Langmuir*, 2003, **19**: 1249.
14. Vogler E A. *Adv. Colloid Interface Sci.*, 1998, **74**: 69.
15. Yooh R H, Flinn D H, Rabinovich Y I. *J. Colloid Interface Sci.*, 1997, **185**: 363.
16. Berg J M, Tomas Eriksson L G, Claesson P M, Barvet K G N. *Langmuir*, 1994, **10**: 1225.
17. Guo C W. PhD thesis 2005, Institute of Chemistry, Chinese Academy of Sciences.
18. Patankar N A. *Langmuir*, 2004, **20**: 7097.
19. Feng L, Li S H, Li Y S, Li H J, Zhang L J, Zhai J, Song Y L, Liu B Q, Jiang L, Zhu D B. *Adv. Mater.*, 2002, **14**: 1857.
20. Bhushan B, Jung Y C. *Nanotechnology*, 2006, **17**: 2758.
21. Herminghaus S. *Europhys. Lett.*, 2000, **52**: 165.
22. Onda T, Shibuichi S, Satoh N. *Langmuir*, 1996, **12**: 2125.
23. Lum K, Chandler D, Weeks J D. *J. Phys. Chem. B*, 1999, **103**: 4750.
24. Feng L, Song Y L, Zhai J, Liu B Q, Xu J, Jiang L, Zhu D B. *Angew. Chem. Int. Ed.*, 2003, **42**: 800.
25. Guo C W, Feng L, Zhai J, Wang G J, Song Y L, Jiang L, Zhu D B. *Chem. Phys. Chem.*, 2004, **5**: 750.
26. Xie Q D, Fan G Q, Zhao N, Guo X L, Xu J, Dong J Y, Zhang L Y, Zhang Y J, Han C C. *Adv. Mater.*, 2004, **16**: 1830.

27. Hosono E, Fujihara S, Honma I, Zhou H. *J. Am. Chem. Soc.*, 2005, **127**: 13458.

28. Zhu M F, Zou W W, Yu H, Yang W, Chen Y M. *J. Mater. Sci.*, 2006, **41**: 3793.

29. Wang J X, Wen Y Q, Feng X J, Song Y L, Jiang L. *Macromol Rapid Commun.*, 2006, **27**: 188.

30. Lafuma A, Quéré D. *Nat. Mater.*, 2003, **2**: 457.

31. Xu L, Zhang W W, Nagel S R. *Phys. Rev. Lett.*, 2005, **94**: 184505.

32. Cheng Y T, Rodak D E, Angelopoulos A, Gacek T. *Appl. Phys. Lett.*, 2005, **87**: 194112.

33. Chen W, Fadeev A Y, Hsieh M C, Öner D, Youngblood J, McCarthy T J. *Langmuir*, 1999, **15**: 3395.

34. Barthlott W. *Planta*, 1997, **202**: 1.

35. Jin M H, Feng X J, Feng L, Zhai J, Sun T L, Li T J, Jiang L. *Adv. Mater.*, 2005, **17**: 1977.

36. Hu F Z, Chen G Y, Du Y J. *Interface of Materials*. East China University of Science & Technology Press, Shanghai, 2001.

37. Richard D, Clanet C, Quéré D. *Nature*, 2002, **417**: 811.

38. Aussillous P, Quéré D. *Nature*, 2001, **411**: 924.

39. Furmidge C G L. *J. Colloid Sci.*, 1962, **17**: 309.

40. Wolfram E, Faust R. Chapter 10, *Wetting, Spreading, and Adhesion*, Padday, J. F. (Ed.), Academic Press, London, 1978.

41. Murase H. *Proceedings of the Fifth Interface Meeting of the Science Council of Japan*. Tokyo, 1998 (in Japanese).

42. Miwa M, Nakajima A, Fujishima A, Hashimoto K, Watanabe T. *Langmuir*, 2000, **16**: 5754.

43. Roura P, Fort J. *Langmuir*, 2002, **18**: 566.

44. Yoshimitsu Z, Nakajima A, Watanabe T, Hashimoto K. *Langmuir*, 2002, **18**: 5818.

45. Morita M, Koga T, Otsuka H, Takahara A. *Langmuir*, 2005, **21**: 911.

Chapter 4

Biomimic Superhydrophobic Surface

Inspired by research of smart surfaces in nature, scientists learn that perfect combination of structure and property can be realized by mimicking the surface structures of organisms. From the perspective of fundamental science, we know that the wettability of solid surfaces is governed by the following two factors: (1) the chemical composition, and (2) the roughness of the surface. According to the lotus effect and the research on the wettability of rough surfaces, superhydrophobic surfaces can be fabricated in two ways: one is to create a rough surface with hydrophobic materials (CA > 90°), and the other is to modify the rough surface with materials of low surface free energy. Obviously, the main point of constructing superhydrophobic surfaces is the effective creation of rough surface structures and surface chemical modification. With in-depth study of superhydrophobic surfaces, many fabricating methods have emerged; here, we will review the methods of preparing superhydrophobic surfaces and the multifunctional surfaces developed with these methods.

4.1 Methods of Preparing Superhydrophobic Surfaces

4.1.1 *Heterogeneous nucleation method*

To obtain a superhydophobic surface with fractal structure, Onda et al.[1,2] firstly synthesized waxy alkylketene dimmer (AKD), then solidified it after melting on a glass plate. The prepared surface showed a contact angle of 174° for water droplets (Fig. 4.1).

(a)

(b)

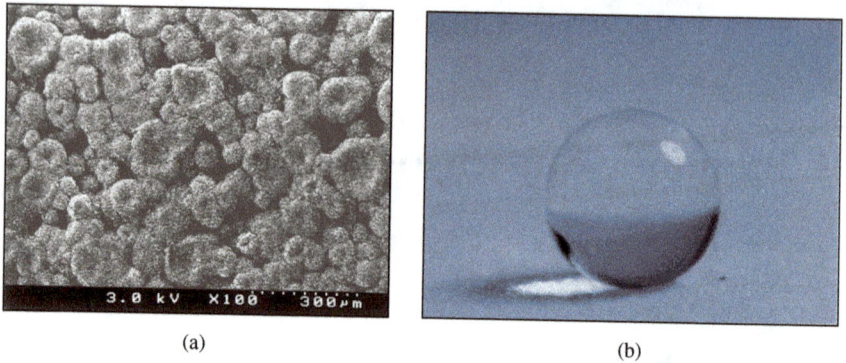

Fig. 4.1 (a) SEM images of the fractal AKD surface. (b) Water droplet on fractal AKD surface. (Reprinted with permission from Refs. 1 and 2. Copyright 1996, American Chemical Society.)

Experiments revealed that dialkylketone (DAK), which was produced simultaneously by hydrolysis of AKD, had no effect on the surface's superhydrophobicity. However, it served as crystalline nuclei in the process of forming rough surfaces of AKD. Water droplets rolled easily without any trace on the surface by tilting the substrate slightly. After studying the relationship of the contact angles to the fractal and smooth AKD surfaces by changing the surface tension of the probe liquid, they concluded that $(L/l)^{D-2} = (34/0.2)^{2.29-2} \approx 4.43$, which was consistent with the expected results of the contact angle equation of fractal surfaces as described in Chapter 3.

4.1.2 *Plasma treatment method*

Plasma treatment on the surface is an effective way to obtain rough structure, which has been widely used in the preparation of superhydrophobic surface. For example, McCarthy et al.[3] have produced superhydrophobic surface by using the plasma polymerization of 2,2,3,3,4,4,4-heptafluorobutyl acrylate on smooth poly(ethylene terephthalate) (PET) surface, with water contact angles of $\theta_A/\theta_R = 174°/173°$. They also prepared rough surfaces by etching of polypropylene (PP) using inductively coupled radio-frequency argon plasma in the presence of poly(tetrafluoroethylene) (PTFE).[7]

Surfaces become rougher with increasing etching time (Fig. 4.2). Contact angles on these surfaces are as high as $\theta_A/\theta_R = 172°/169°$. Schreiber and others[5] reported a hexamethyldisiloxane film with water contact angle up to 180°, which was deposited by low temperature

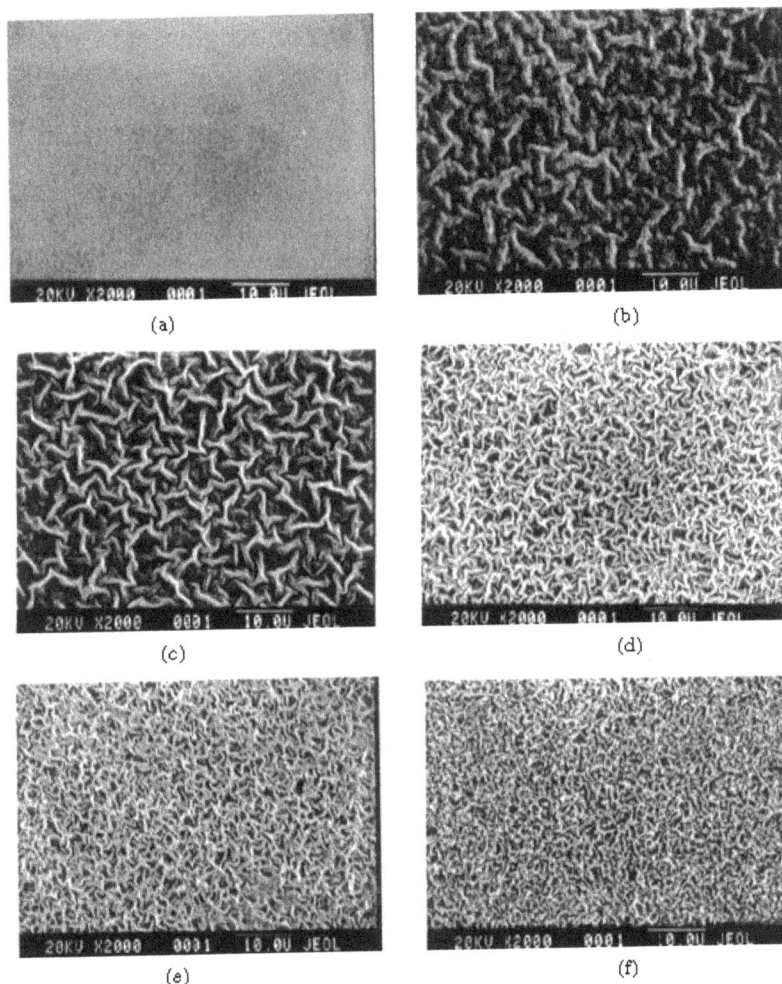

Fig. 4.2 Scanning electron microscope (SEM) images of polypropylene as a function of argon/PTFE plasma reaction time. (a) 0 min; (b) 30 min; (c) 60 min; (d) 90 min; (e) 120 min; and (f) 180 min. (Reprinted with permission from Ref. 7. Copyright 1999, American Chemical Society.)

plasma polymerization; Washo[6] has produced a plasma polymerized tetrafluoroethylene film with water contact angles of 165°–170°; Youngblood and McCarthy[7] have prepared oxygen-plasma treated PTFE film, in which the water contact angles are 170°/160°, and water droplet rolled easily on these surfaces. Matsumoto *et al.*[8] obtained a superhydrophobic plasma-polymerized fluorocarbon film using a mixture of CH_4/C_4F_8 gas. Zhang *et al.*[9] prepared a superhydrophobic film with water contact angle of 174°/135°, by using plasma polymerization of allypentafluorobenzene (APFB) on the plasma-pretreated polyimide (PI) films. F. Poncin-Epaillard *et al.*[10] have reported a stable and transparent superhydrophobic surface which was treated by plasma. There are two ways to achieve this: one is direct CF_4 plasma modification of low-density polyethylene (LDPE); and the other is CF_4 treatment on LDPE after O_2 plasma modification. Study has found that roughness of the samples' surfaces could be modulated by the plasma parameters. O_2 plasma allows production of a variable roughness, while CF_4 plasma enhances this roughness and creates an apolar layer at the same time. Researchers have also found that the superhydrophobic surface could be obtained even with a very low roughness (around 20 nm).

Recently, Badyal and co-workers[11] have prepared superhydrophobic nanospheres by continuous wave plasma polymerization; in addition, they also obtained superhydrophobic surfaces by using plasma fluorination of polybutadiene films.[12] The reasons for the superhydrophobicity are the combination of fluorine atoms and unsaturated alkene groups on the polymer surface, as well as surface roughening by the electrical discharge. Favia *et al.*[13] obtained ribbon-like superhydrophobic films by fluorocarbons deposition on silicon substrates with RF glow discharges plasma (Fig. 4.3).

The superhydrophobic surfaces can be obtained directly by using plasma treatment with low surface-energy materials that contain fluorine. In addition, they can also be prepared by roughening the ordinary materials' surfaces with plasma in combination with other methods of surface chemical modification. For example, Teshima *et al.*[14] obtained rough structure on poly(ethylene terephthalate) (PET) substrates by oxygen plasma etching and introducing hydrophilic

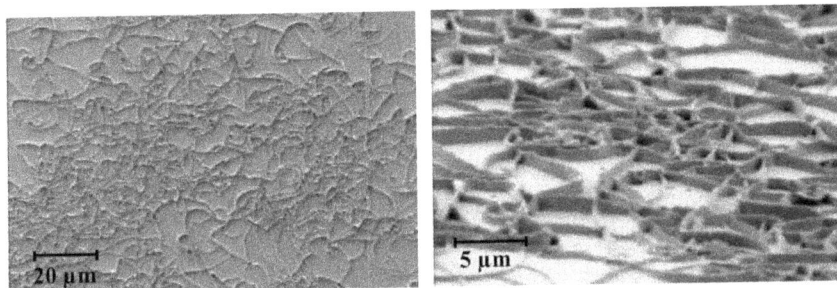

Fig. 4.3 SEM pictures of film deposited with a modulation DC of 5%, at two different magnifications. (Reprinted from Ref. 13. Copyright 2003, with permission from Elsevier.)

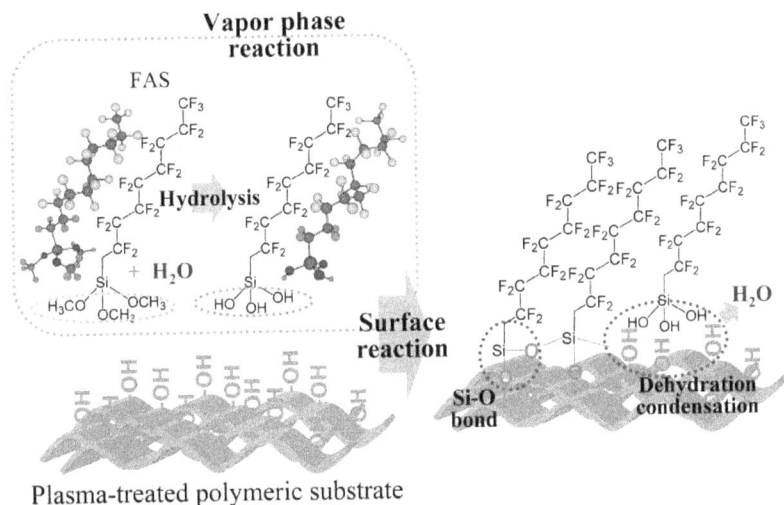

Fig. 4.4 Formation models of an FAS hydrophobic layer on the nanotextured and hydrophilic polymer surface by low-temperature CVD. (Reprinted with permission from Ref. 14. Copyright 2003, American Chemical Society.)

functional groups on these substrates at the same time, and then grafting various organosilane molecules on the hydrophilic groups by means of chemical vapor deposition (CVD) to form a hydrophobic layer; finally, a transparent and surface-modified superhydrophobic PET substrate was obtained (Fig. 4.4).

4.1.3 *Etching method*

McCarthy and co-workers[4] have reported a series of different-sizes-and-shapes arrayed silicon surfaces prepared by photolithography (Fig. 4.5). After treatment by silanization reagents, superhydrophobic surfaces were obtained. The study revealed that when the square posts showed X-Y dimensions of 2 and 32 µm and similar distances between the posts, the surfaces exhibited superhydrophobic behavior with high advancing and receding water contact angles. Water droplets rolled easily on these surfaces when the substrates were tilted slightly. Contact angles had nothing to do with the post height (20–140 µm) and the surface chemistry (siloxane-, hydrocarbon-, and fluorocarbon-modified surfaces were prepared). When the square posts showed X-Y dimensions of 64 and 128 µm and similar distances between the posts, the surfaces were not superhydrophobic; the water droplets were pinned on the surfaces and obviously intruded between the posts. Increasing the distance between the posts would increase the receding contact angles up to the water intrusion between the posts; the contact length of the three-phase contact line became shorter. When the shape of the posts changed from square to staggered rhombus, star, or indented square,

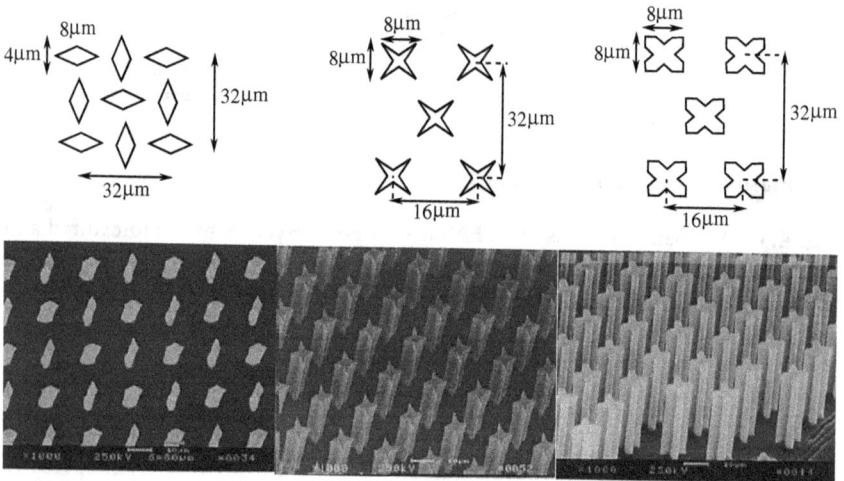

Fig. 4.5 Two-dimensional (X-Y) representations of surfaces containing different geometry posts (above: models; below: SEM images). (Reprinted with permission from Ref. 4. Copyright 2000, American Chemical Society.)

the receding contact angles increased due to the more contorted contact line. The maximum length scale of roughness is about 32 μm to achieve superhydrophobicity of the square posts' surfaces.

Chen *et al.*[15] have fabricated well-ordered single-layer polystyrene nano-beads arrays by nanosphere lithography firstly, then obtained rough surface by oxygen plasma treatment which reduced the size of the nano-beads. To increase the surface hydrophobicity, the arrays were then coated with a 20-nm–thick gold film and modified with octadecanethiol (ODT). A superhydrophobic surface was prepared by oxygen plasma treatment of double-layer PS nano-beads (400 nm), as shown in Fig. 4.6. The right inset shows that the water contact angle on the ODT modified surface is 170°; the left inset shows the change of surface morphology before and after the oxygen treatment.

It is also reported that Yoshimitsu *et al.*[16] have produced ordered micro-roughness patterned silicon surface by mechanical etching, then they obtained a superhydrophobic surface with water contact angle of about 150° after treatment by fluoroalkylsilane. Bartell and Shepard[17] have fabricated tetrahedral rough olefin surface by mechanical etching, and the water contact angles on the surface were about 160°/125°. Bico *et al.*[18] have prepared spikes-like, shallow cavities-like and stripes-like micro-roughness patterned surface on silicon wafer by using mask etching method; the contact angle of water was

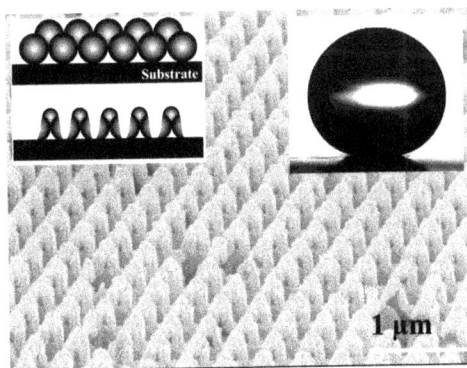

Fig. 4.6 SEM image of a 440-nm–diameter double-layer polystyrene surface after 120 s of oxygen plasma treatment. (Reprinted with permission from Ref. 15. Copyright 2004, American Chemical Society.)

as high as 167°, 131°, and 151° respectively after modification by fluoride. M. Natali *et al.*[19] also reported superhydrophobic surfaces prepared by nanoimprint lithography and wet chemical etching on silicon wafers. The specific process is described below: the glass molds were used to firstly imprint a positive photoresist layer, and then UV-ozone was used to irradiate the layer to remove the residues. The pattern was transferred to a SiO_2 layer after etching by buffered hydrofluoric, then to Si layer by isotropic hydrofluoric-nitric-acetic acid (HNA) or by anisotropic KOH etching. Lastly, the superhydrophobic surface was obtained after coating the samples with a self-assembled monolayer of octadecyltrichlorosilane. E. Mazur *et al.*[20] reported that superhydrophobic silicon surface could be prepared by femtosecond laser technology. They used femtosecond laser to irradiate the silicon surface, and created a micro/nanostructure on the surface. Then the surface was modified by fluoroalkylsilane. The surface morphology and wettability of the silicon can be modulated by changing the laser fluence.

The authors' group[21] have reported a polydimethylsiloxane (PDMS) superhydrophobic film prepared by laser etching method. The surface was composed of a regular array of micro-size convexes and cauliflower-like nanostructures on the microconvexes. The PDMS surface with micro-, submicro- and nano-composite structures had a very high water contact angle and a very low sliding angle compared to those with a single nano- or micro-scale structures. Moreover, the contact angles and sliding angles could be tuned by changing the sizes of convexes (Fig. 4.7).

4.1.4 *Sol-gel method*

Minani *et al.*[22-25] prepared alumina gel films on glass plates by sol-gel method. After immersing these films in boiling water, the flower-like porous alumina thin films could be obtained within 30 s. Then a superhydrophobic and transparent film with water contact angle of 165° was created, followed by fluoroalkylsilane modification. The roughness of the films increased when the immersing time increased. Shirtcliffe *et al.*[26] produced superhydrophobic organo-silica foam by

(1)

(2)

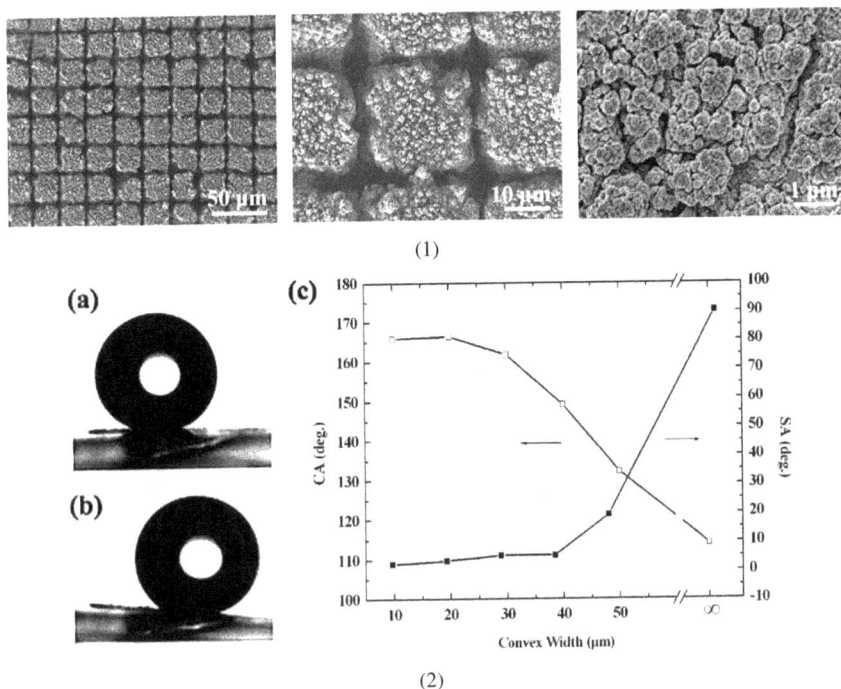

Fig. 4.7 (1) Typical SEM images of the laser-etched PDMS surface. (2) (a) Photograph of a water droplet on the rough PDMS surface; (b) Sliding behavior of a water droplet on the same rough PDMS surface shown in Fig. 1(a); (c) The relationship between the widths of microconvexes and the CA (left)/CA (right), respectively.[21] (Copyright Wiley–VCH Verlag GmbH & Co. KGaA. Reproduced with permission.)

sol-gel phase-separation process. A transition between hydrophobicity and hydrophilicity appeared when the material was heated to 400°C or above, and the structure was unchanged. K. Nakano *et al.*[27] have reported superhydrophobic surfaces which were created by fibrous aggregation of perfluoroalkyl chain–containing organogelators on glass plates. Low molecular-weight organogelators could form gels by heating them in proper solvent. The evaporation of the solvent from the gels resulted in xerogels, which have three-dimentional network structures of fibrous aggregates. The aggregation provided nanoscale rough surfaces, which was the main reason for forming superhydrophobic surfaces.

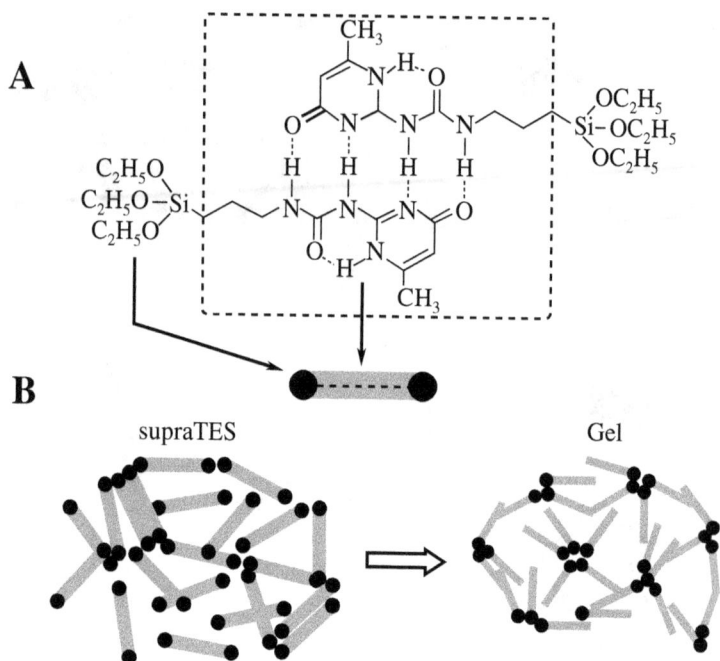

Fig. 4.8 (A) Possible structure of a synthesized supramolecular organosilane (supraTES) in solution. (B) Schematic of the inner structure of supraTES in the interconnected granular structure during sol-gel process. (Reprinted with permission from Ref. 28. Copyright 2004, American Chemical Society.)

Cho *et al.*[28] obtained superhydrophobic surfaces by sol-gel process using a supramolecular organosilane with quadruple hydrogen bonds (Fig. 4.8). They also reported that the addition of a small amount of low molecular-weight poly(dimethyl siloxane) (PDMS) in the process of experiment could reduce the contact area. The PDMS layer played a similar role as the waxy layer on the lotus leaves.

Shang *et al.*[29] created rough-structured films on glass plates by using sol-gel process with different components of SiO_2 sols acting as precursors (TEOS: $Si(OC_2H_5)_4$; MPS: $H_2CC(CH_3)CO_2(CH_2)_3$ $Si(OCH_3)_3$; MTES: $CH_3Si(OC_2H_5)_3$). The microstructures of the film surfaces could be tuned by controlling hydrolysis and condensation reactions. Modifying the film surface with two kinds of self-assembly

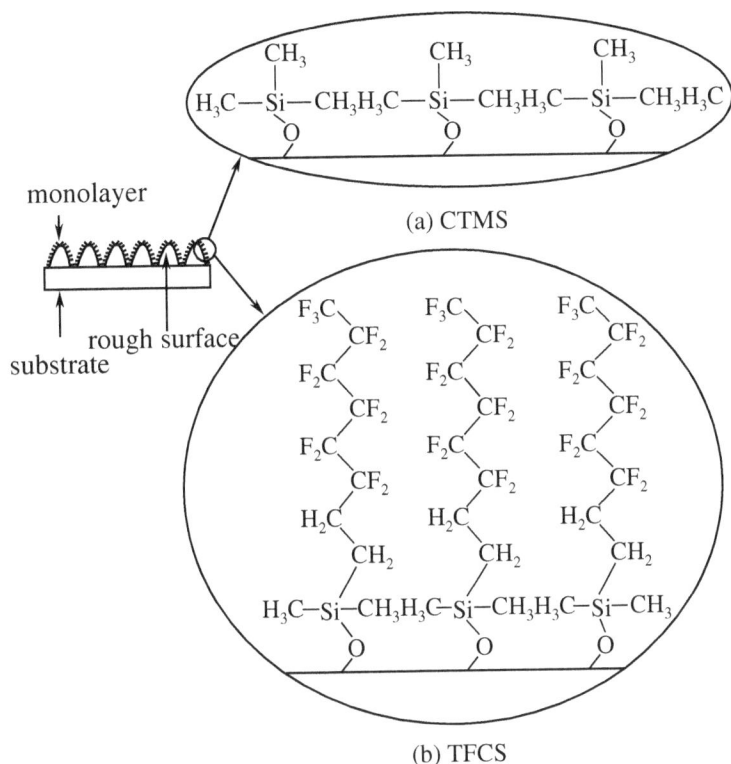

Fig. 4.9 Schematic drawing of surface chemistry after self-assembly with (a) CTMS and (b) TFCS. (Reprinted from Ref. 29. Copyright 2005, with permission from Elsevier.)

monolayer (one is chlorotrimethylsilane (CTMS, $(CH_3)_3SiCl$), which contains $-CH_3$ groups, and the other is tridecafluoro-1,1,2,2-tetrahydrooctyldimethylchlorosilane (TFCS, $CF_3(CF_2)_5(CH_2)_2(CH_3)_2SiCl$) with $-CF_3$ groups.) could be used to obtain a variety of transparent films, and the water contact angles on these surfaces could reach up to 165°/115° (Fig. 4.9).

Rao *et al.*[30] produced superhydrophobic silica aerogels by using methyltrimethoxysilane (MTMS) precursor. It was found that the molar ratios of precursor, concentrations of catalyst, water, and solvent played important roles in the process of structure growing of aerogels. The concentration of aerogels decreases when the molar ratios of $NH_4OH/MTMS$ (N), $H_2O/MTMS$ (H), and $MeOH/MTMS$ (M) increase, the volume shrinkage of the aerogels decreases with increase in N and H values, and decrease in M values; the gelation time increases with decrease in N and H values and it decreases with decrease in M values. Generally, MTMS sols do not form gels in the case of lower concentration of catalyst (<1 M); less number of Si-OH groups could be used for condensation which caused slow reaction with each other. However, the gel could form at higher catalyst concentration (≥ 2 M) because of the increase in the rate of condensation reactions. The three methoxy groups of MTMS monomer hydrolysis and condensation reactions are shown below:

Hydrolysis:

$$CH_3Si(OCH_3)_3 + 3H_2O$$
$$\rightarrow CH_3-Si(OH)_3 + 3CH_3OH. \tag{1}$$

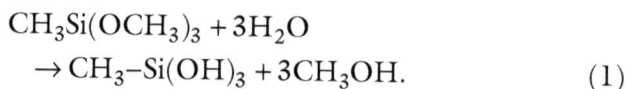

Water condensation:

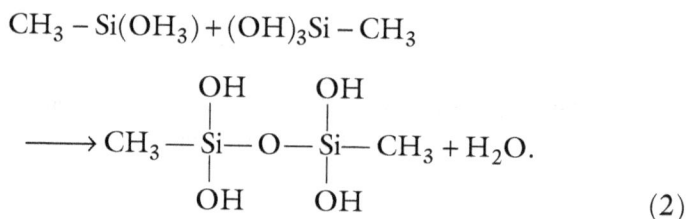

$$CH_3 - Si(OH_3) + (OH)_3Si - CH_3$$

$$\longrightarrow CH_3 - \overset{\overset{\textstyle OH}{|}}{\underset{\underset{\textstyle OH}{|}}{Si}} - O - \overset{\overset{\textstyle OH}{|}}{\underset{\underset{\textstyle OH}{|}}{Si}} - CH_3 + H_2O. \tag{2}$$

Alcohol condensation:

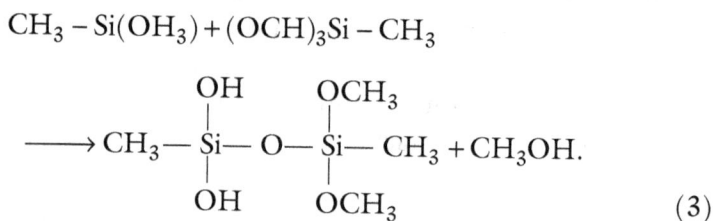

$$CH_3 - Si(OH_3) + (OCH)_3Si - CH_3$$

$$\longrightarrow CH_3 - \overset{\overset{\textstyle OH}{|}}{\underset{\underset{\textstyle OH}{|}}{Si}} - O - \overset{\overset{\textstyle OCH_3}{|}}{\underset{\underset{\textstyle OCH_3}{|}}{Si}} - CH_3 + CH_3OH. \tag{3}$$

The contact angle of aerogels varies with catalyst concentration, that is, the contact angle increases from 159° to 167° with N values increasing. In addition, the contact angle increases from 162° and 160° to 173° and 167°, respectively, and then decreases to 160° and 159° with H and M values increasing. The aerogels could also keep their hydrophobicity even when heated to 480°C.

4.1.5 *Chemical vapor deposition (CVD) method*

By making use of chemical vapor deposition (CVD) method, the authors' group have prepared various kinds of patterns on quartz-glass plate substrates,[31,32] such as honeycomb-, post-, and island-like aligned carbon nanotube films (Fig. 4.10). It turns out that these films have contact angles higher than 160° and sliding angles lower than 5°, which should be attributed to the arrays of nanostructure and microstructure on the surface.

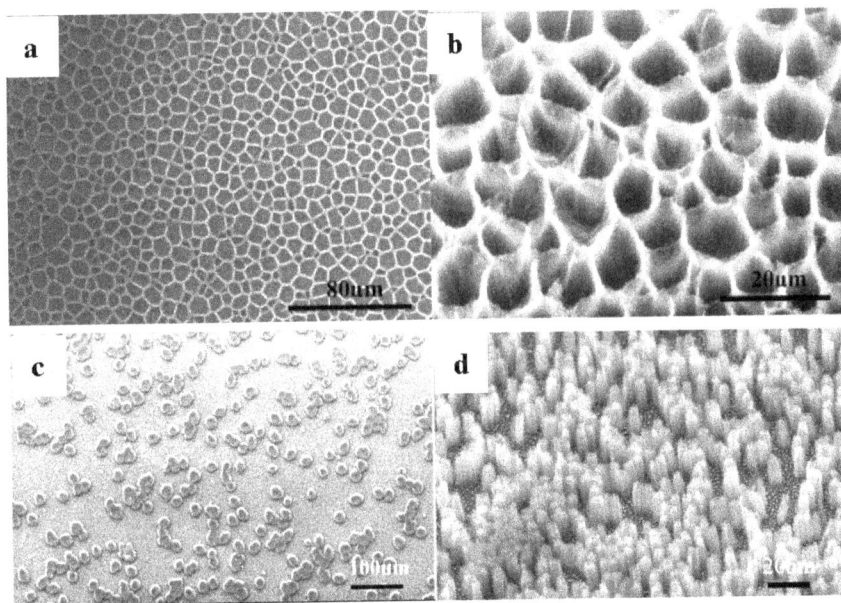

Fig. 4.10 SEM images of the aligned carbon nanotubes (ACNTs) films prepared by CVD method. (a) and (b): Honeycomb-like patterned ACNTs (different magnification). (c) Island-like ACNTs. (d) Post-like ACNTs. (Reprinted with permission from Ref. 31. Copyright 2002, American Chemical Society.)

Y. Wang *et al.* reported the preparation of amino-functionalized surfaces by depositing aminopropyltrimethoxysilane (APTMS) on silicon surfaces.[33] The wettability of the self-assembled films can be adjusted by altering the chain length of alkanoic acids, which are used for modification of the surface. Furthermore, by cooperating the self-assembled films with surface roughening, superhydrophobic surfaces with contact angles larger than 153° were obtained.

Also, aligned carbon nanotube forests can be obtained by utilizing a plasma-enhanced chemical vapor deposition (PECVD) technique. After coating a thin PTFE on the surface by a hot-filament chemical vapor deposition (HFCVD) process, researchers finally realized stable superhydrophobic properties (Fig. 4.11).[34]

Hozumi and Takai have prepared ultra water-repellent films using the mixture of tetramethylsilane (TMS) and fluoro-alkyl silane (FAS) by microwave plasma-enhanced CVD (MWPECVD),[35] whose equipment scheme is demonstrated in Fig. 4.12. The contact angles for water droplets increase with the total pressure of reactants, and the maximum contact angle is about 160°.

Besides, by employing trimethylmethoxysilane (TMMOS, $(CH_3)_3Si(OCH_3)$) as a raw material, transparent and ultra water-repellent thin films with nanoscale texture were prepared by MWPECVD on glass, silicon, and organic glass (PMMA) substrates.[36-38]

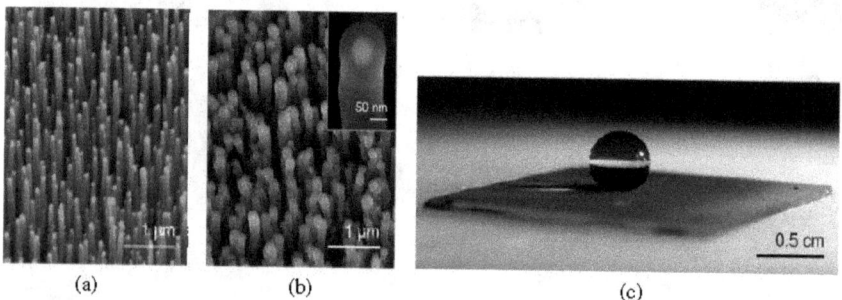

(a)   (b)   (c)

Fig. 4.11   (a) SEM images of carbon nanotube forests prepared by plasma-enhanced chemical vapor deposition (PECVD) technique. (b) PTFE-modified carbon nanotube forests. (c) A spherical water droplet on the PTFE-coated forest. (Reprinted with permission from Ref. 34. Copyright 2003, American Chemical Society.)

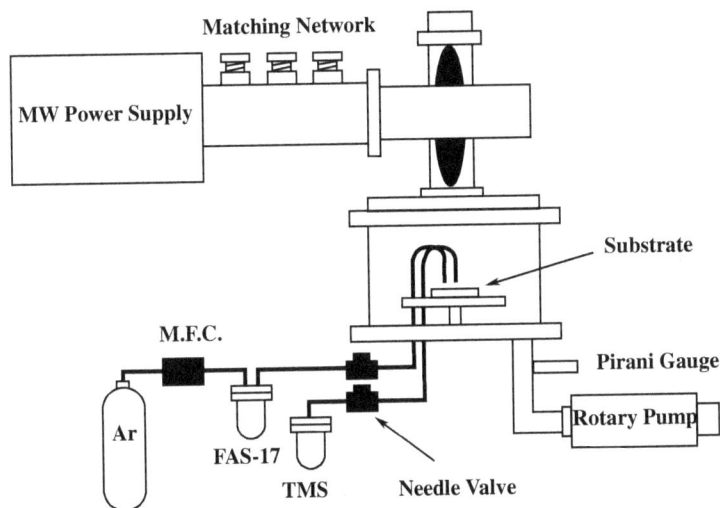

Fig. 4.12 Schematic diagram of experimental apparatus of microwave plasma-enhanced CVD (MWPECVD). (Reprinted from Ref. 35. Copyright 1997, with permission from Elsevier.)

Furthermore, Takai *et al.* also investigated the water droplets on the films by environmental scanning electron microscope (ESEM). Water droplets exhibited perfect spherical shape on the superhydrophobic surface with contact angle larger than 150°; while on the CH_3-terminated SAM surfaces, they exhibited semispherical shape, with contact angle about 100°. Based on these researches, the mechanical properties and optical transparence are improved by combining the process of microwave plasma-enhanced CVD (MPECVD) and oxygen-plasma treatment, and using CO_2 as an additive gas instead of Ar.

A. Amirfazli *et al.* reported the fabrication of superhydrophobic surfaces of n-hexatriacontane in a single-step physical vapor deposition process.[39] The low surface energy of n-hexatriacontane together with the randomly distributed micro/nano structures led to the very large contact angles and small roll-off angles for water droplets. The n-hexatriacontane–modified superhydrophobic surfaces had robust chemical stability and stable wetting characteristics.

4.1.6 Electrochemical method

By utilizing electrochemical deposition method, the authors' group have prepared conductive hydrophobic zinc oxide (ZnO) thin films with rough structures.[40] The as-prepared thin films showed a contact angle (CA) of 128.3°, whereas the superhydrophobic surface with a water contact angle of 152.0° was obtained by fluoro-alkyl silane modification. Moreover, the electrical properties of the films were measured by conducting atomic force microscopy (AFM) (Fig. 4.13). It turned out that the prepared ZnO films were semiconductors. Further study found that ZnO films could be superhydrophobic after heating treatment, without any extra modification. Professor Shi *et al.* from Tsinghua University, obtained superhydrophobic poly-thiophene films by direct electropolymerization of thiophene in boron trifluoride–diethyl etherate (BFEE), of which the contact angle was about 134°.[41]

Furthermore, by making use of simple electrohydrodynamics (EHD) technique and employing the low-cost polystyrene (PS) as

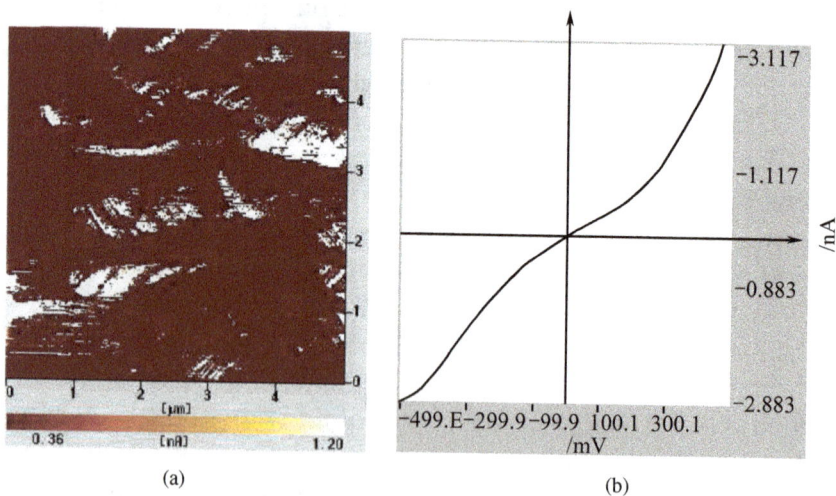

(a)   (b)

Fig. 4.13   (a) Conducting AFM image of the prepared ZnO films. (b) Typical I-V curve of the thin films. (Reprinted with permission from Ref. 40. Copyright 2003, American Chemical Society.)

Fig. 4.14 Composite PS films prepared by electrohydrodynamics (EHD) technique. (a)–(c): SEM images of the prepared PS films. (d) Water droplet on the film (CA = 160.4°).[42] (Copyright Wiley–VCH Verlag GmbH & Co. KGaA. Reproduced with permission.)

raw material, the authors' group prepared a superhydrophobic film with a novel composite structure consisting of porous microspheres and nanofibers.[42] The porous microspheres contributed to the superhydrophobicity by increasing the surface roughness; while nanofibers interweaved to form a 3D network, binding the porous micropheres, which reinforced the composite film. Besides, PS films with different morphology could be obtained by adjusting the concentration of the starting solution (Fig. 4.14).

4.1.7 Layer-by-layer (LBL) method

Zhang *et al.* from Tsinghua University fabricated polyelectrolyte multilayer films on indium tin oxide (ITO) electrode by utilizing layer-by-layer (LBL) technique, and subsequently covering it with dendritic gold clusters by electrochemical deposition (Fig. 4.15).

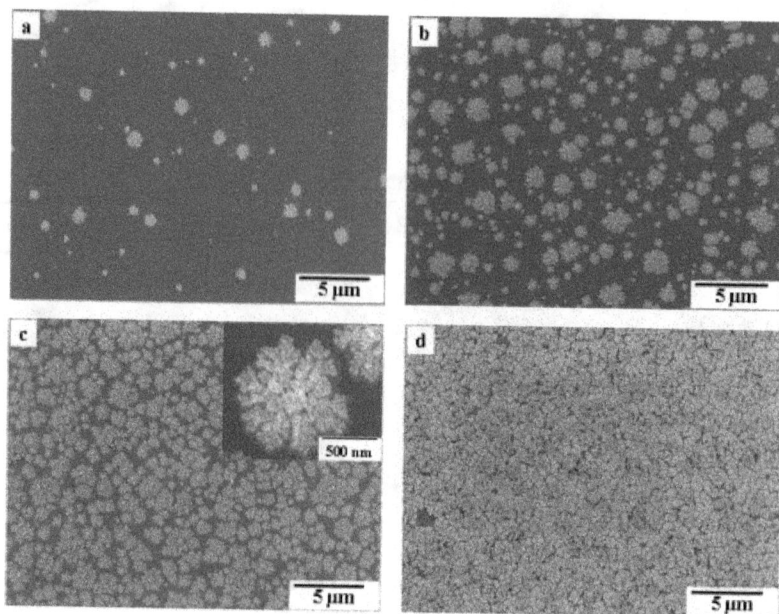

Fig. 4.15 SEM images of dendritic gold clusters formed on an ITO electrode modified with a polyelectrolyte multilayer by electrochemical deposition. Deposition time was (a) 2 s, (b) 50 s, (c) 200 s, and (d) 800 s. (Reprinted with permission from Ref. 43. Copyright 2004, American Chemical Society.)

After further chemisorption of a self-assembled monolayer of n-dode-canethiol, it showed superhydrophobic properties.[43] The contact angles of the modified-gold-cluster films remarkably depended on the duration of the electrochemical deposition (Fig. 4.16). The contact angles increased gradually with the time of electrochemical deposition. When the deposition time exceeded 1000 s, the contact angle reached a constant value as high as 156°. After exposing to ambient environment for 40 min, the contact angle of the film changed from 156° to 173°, as shown in Fig. 4.16. This apparent change should be related to the shrinkage of water droplet because of the evaporation and the reduction of gravitational effect. On the other hand, this phenomenon suggested that the superhydrophobic surface fabricated by this method was rather stable.

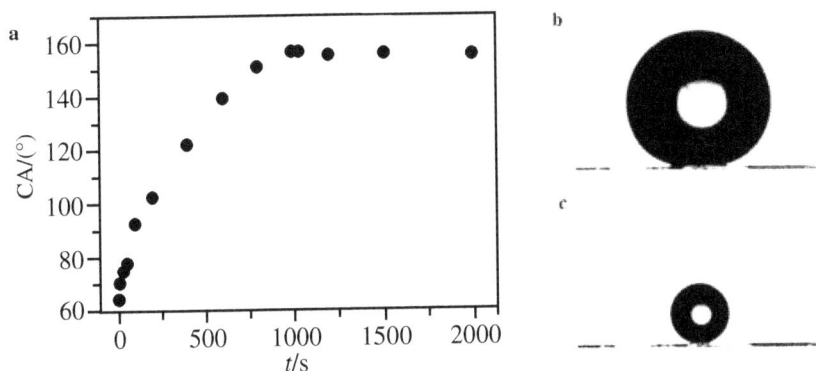

Fig. 4.16 (a) Dynamic water contact angle measurements on the surface of dendritic gold clusters as a function of the duration of electrochemical deposition at –200 mV (vs Ag/AgCl) by single potential time base mode. (b) Shape of a water droplet on the surface of dendritic gold clusters (drop weight: 5 mg). (c) Shape of the water droplet in (b) after 40 min of exposure to ambient environment. (Reprinted with permission from Ref. 43. Copyright 2004, American Chemical Society.)

Research has demonstrated that stable superhydrophobic behavior with honeycomb-like structure could be prepared by choosing suitable acid treatment of polyelectrolyte multilayer film. Rubner *et al.* firstly achieved a honeycomb-like polyelectrolyte multilayer surface by LBL technique, and subsequently coated it with silica nanoparticles (diameter: 50 nm). They finally obtained stable superhydrophobicity after the modification of semifluorinated silane by CVD.[44] Figure 4.17(a) shows the SEM images of the porous (PAH/PAA)$_{100.5}$ (pH = 8.5/3.5) films after a single-acid treatment (immersing in the solution of pH = 2.3 for 6 h, without water washing); the diameter of the pores was 0.5–2 µm, and the roughness was about 100 nm. However, it was evidently different after a combined-acid treatment (immersing in solution of pH = 2.7 for 2 h, and subsequent immersion in solution of pH = 2.3 for 4 h, without water washing); the pores were in honeycomb-like array with diameter of 10 µm and roughness reaching 440 nm. If the porous films were washed by water (pH ≈ 5.5) after acid treatment, the surface roughness would decrease largely. Experiments have proved that the prepared films could keep their

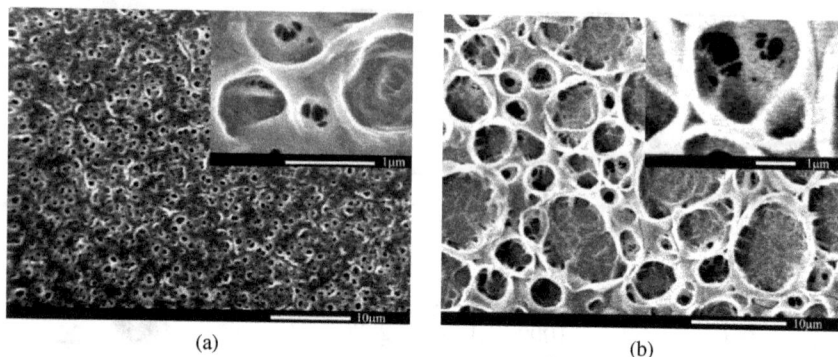

(a) (b)

Fig. 4.17 SEM images of porous (PAH/PAA)$_{100.5}$ films (a) after a single-acid treatment and (b) after a combined-acid treatment. (Reprinted with permission from Ref. 44. Copyright 2004, American Chemical Society.)

superhydrophobic property even after being immersed thoroughly in water for at least one week, or being exposed to humid environment for at least one month. In addition, they also found that superhydrophilic surface could be achieved by coating the polyelectrolyte multilayer film with 3.5 bilayers of silica nanoparticles. This means that it is possible to create coatings with identical surface morphologies but with dramatically different wetting characteristics, which was significant for preparation of surfaces with different wetting behaviors.

Shen *et al.* from Zhejiang University reported a method for developing superhydrophobic surfaces from a fluorinated exponential-growth multilayer.[45] They alternately immersed the 3-aminopropyltriethoxysilane-coated substrates in an polyelectrolyte solution of poly(acrylic acid) (PAA, 3 mg/mL) with negative ions and polyethyleneimine (PEI, 1 mg/mL) with positive cations, to form the multilayer after several cycles. The multilayer growth represented a typical exponential growth behavior. The thickness of the film was reported to reach several hundreds of nanometers or even micrometers after only 10 to 20 assembly cycles. Most importantly, the addition of Ag$^+$ into the PEI solution enhanced exponential growth of the multilayer significantly. This is because Ag$^+$ not only can form complex with PEI by the amino group, but also can chemical bond with PAA by ion exchange. The thickness of film grown with enhanced exponential was reported to reach 12 μm after about 15 layers. Further study showed that this enhanced grown-exponential could induce the multilayer film to form micro/nano-hierarchical structure, leading to lotus effect. After the chemical vapor deposition of (tridecafluoroctyl)-triethoxysilane on the multilayer film, it exhibited a contact angle as high as $172 \pm 1.5°$ and a slide angle as low as $6.5°$.

4.1.8 *Template method*

A template-extrusion method to construct rough surfaces was developed by the authors' group. Using porous anodic aluminum oxide membrane as template, aligned polymer nanofibers were prepared by extruding the polymer solution at certain concentration under certain pressure, with subsequent solidification. We firstly chose superhydrophobic polyacrylontitrile (PAN) as the precursor (having a contact angle of $100.8°$ on its smooth surface), and obtained aligned PAN nanofibers by the template-extrusion method. The prepared surfaces demonstrated superhydrophobic behavior without any modification, with the water contact angle as high as $173.8°$.[46] Besides, by employing amphililic polyvinyl alcohol (PVA) as the precursor, we fabricated superhydrophobic aligned PVA nanofiber surface[47] (Fig. 4.18) by the

(a)   (b)

Fig. 4.18 PVA nanofibers prepared by template extrusion method. (a) Cross-sectional SEM image. (b) Shape of a water droplet on the surface of nanofiber films.[47] (Copyright Wiley–VCH Verlag GmbH & Co. KGaA. Reproduced with permission.)

same method. It was proposed that when the precursor of the PVA solution was compressed into the nanostructures, molecules on the surface were recombined and intermolecular hydrogen bonds were created. Since air is a hydrophobic medium, the hydrophobic groups might be exposed to the surface, which would minimize the surface free energy of the system. This has been proven by the results of angle-resolved X-ray photoelectron spectroscopy (XPS).

The authors' group also reported that PC nanopillars copied from the structure of anodic alumina pore could be obtained by the following procedure (Fig. 4.19(a)).[48] Firstly, the film of polycarbonate (PC) was covered with anodic alumina template, then after heating and stressing, the porous anodic template could be mechanically stripped off, leaving the well-aligned polymer nanopillar arrays. Large-scale aligned PC nanopillars could be prepared by using tubular

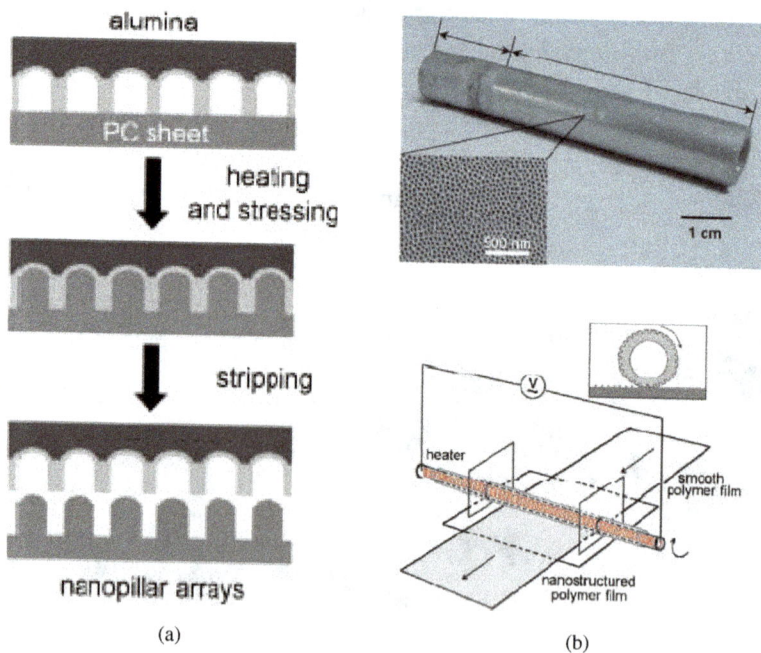

Fig. 4.19 (a) Schematic outline of the experimental procedure used to fabricate PC nanopillar arrays. (b) Equipmental scheme of preparation on a large scale.[48] (Copyright Wiley–VCH Verlag GmbH & Co. KGaA. Reproduced with permission.)

template rolling on polymer films with suitable temperature and pressure (Fig. 4.19(b)).

4.1.9 *Self-assembly method*

By making use of mechanically assembled monolayers (MAMs),[49] Genzer *et al.* firstly prepared HO-silicon substrate by treating the PDMS with UV light and O_3, then grafted the semi-fluorinated (SF) molecules on the surface, and finally created long-lived superhydrophobic polymer surface (Fig. 4.20).

Whitesides *et al.*[50] prepared monolayers or multilayers of silica colloid by firstly drop-casting aqueous suspensions of particles onto glass slides. After depositing an adhesion layer of titanium or nickel (0.5–0.8 nm) and a thin film of metal (gold, platinium, or oalladium: 8.0–15.0 nm) on the templates by electron-beam evaporation, the colloid particles were released from the substrate by sonication. After further dissolving the silicon core and adhesion layer by etching (aqueous HF), intact metallic half-shells were obtained, showing superhydrophobic property with water static contact angle of 151°. However, the contact angle hysteresis was remarkable for the surface, as the water droplet did not slide off even when the substrate was tilted up to 90°. When the gold half-shell was modified with hexadecanethiolate self-assembled monolayers (SAMs), superhydrophobic surface with static water contact angle which was equal to 163° was realized, and its sliding angle was about 1°.

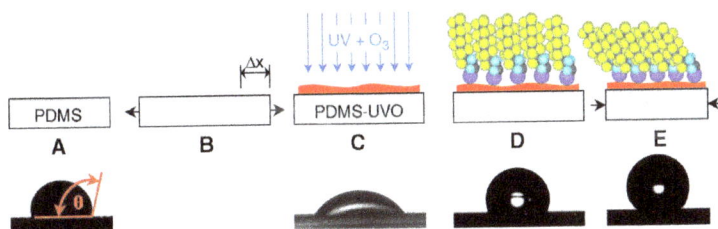

Fig. 4.20 Procedure of the creation of superhydrophobic MAMs, and their corresponding contact angle profiles. (From Ref. 49. Reprinted with permission from AAAS.)

Xiao *et al.*[51] firstly obtained polyethyleneimine (PEI) coated layer in the aqueous solution by immersing the hydroxylation-roughened aluminum substrates into the solution of PEI. Then, the PEI-coated aluminum substrates were put into the mixed solution of stearic acid (STA) and *N,N'*-dicyclohexylcarbodiimide (DCCD). Due to the chemical interaction between the carboxyl group in STA and the amine group in PEI, self-assembled monolayers of STA were adsorbed on the coating, and finally superhydrophobic film was obtained with contact angle reaching 166°. This work also demonstrated that if the same kind of PEI-coated Al substrate was immersed in STA solution without DCCD, the contact angle of prepared monolayer film was only about 120°, indicating that STA might adsorb on the coated surface only through hydrogen-bonding or electrostatic interaction in the absence of DCCD. However, the addition of DCCD as dehydration reagent into the STA solution could enhance the chemical reaction between PEI and STA, forming covalent amide bond (Fig. 4.21).

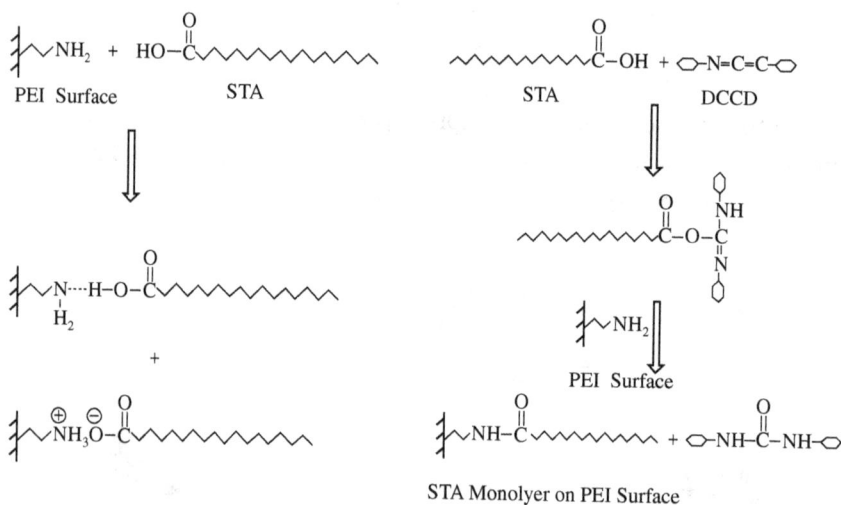

Fig. 4.21   Interaction between PEI and STA. (a) In the absence of DCCD. (b) In the presence of DCCD. (Reprinted from Ref. 51. Copyright 2003, with permission from Elsevier.)

4.1.10 *Solvent-nonsolvent methods*

Solvent–nonsolvent method is a facile way to construct superhydrophobic surface with rough structure. Erbil *et al.* achieved gel-like porous films on different substrates (such as glass slides, Al foils, stainless steel sheets, Teflon, high-density polyethylene, and polypropylene) via vacuum heating, by using polypropylene (PP) as the raw material, p-xylene as the solvent, and using methyl ethyl ketone (MEK), cyclohexanone, and isopropyl alcohol as the nonsolvent.[52] Detailed procedure is described as follows. Firstly, the polypropylene was dissolved into the mixed solution of solvent and nonsolvent (certain ratio), and then a few drops of that polypropylene solution was dropped onto the substrate and the solvent was evaporated at certain temperature, and then white gel-like porous films were formed. Investigation showed that the polymer concentration, film formation temperature and the nonsolvent had certain influence on the roughness of the film surface.

In the absence of nonsolvent, it is important to understand the mechanism of porous structure formation by investigating the film formation at different dry temperatures (Fig. 4.22). At low temperature, there is an increase in the inhomogeneity and size of the pores,

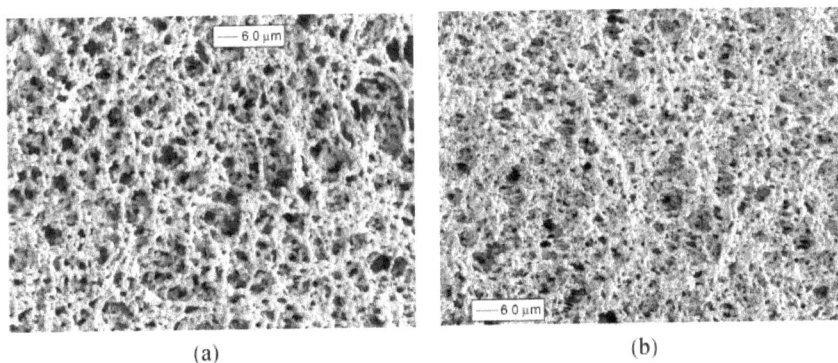

(a)   (b)

Fig. 4.22 SEM images of porous PP films on glass slides. (a) At the drying temperature of 30°C. (b) At the drying temperature of 60°C. (From Ref. 52. Reprinted with permission from AAAS.)

which directly influences the surface roughness and water contact angle (change between 138° and 155°). As the temperature decreases, the solvent evaporation rate slows down, and the crystallization time thus increases, giving a higher overall crystallinity. Moreover, the rates of nucleation and pore formation increase at the lower temperature, resulting in a loose network.

Experiments have proved that by choosing MEK as nonsolvent, the films with the best homogeneity and largest water contact angle could be realized. MEK's high polar solubility parameter caused stronger interaction with the hydroxyl groups on the glass surface and resulted in better spreading of the PP solution. Compared to that without nonsolvent, PP coatings from the solvent–nonsolvent system of p-xylene/MEK had smaller crystallites and narrower cylindrical bridges. On this superhydrophibic surface, the fraction of air in touch with the water drop was larger, leading to the increase of water contact angle from 149° to 160°.

There are three impacts of nonsolvents for the formation of porous film: (1) Nonsolvent acts as a polymer precipitator during the polymer phase-separation between the two phases of *nonsolvent plus p-xylene* and *polymer plus small amount of p-xylene*. This process decreases the crystallization time and gives rise to smaller aggregates. Two macroscopic phases are formed after the completion of this process, one polymer-rich and the other polymer-poor. Crystallization happens in the polymer-rich phase by forming crystal nuclei, which further develop into spherulites, fibrillates, and other crystal shapes. (2) The presence of nonsolvents speed up the nucleation rate, giving rise to smaller spherulites. At the same time, because the nonsolvents are more volatile than p-xylene, they increase the evaporation rate and decrease the time of crystal formation. (3) The addition of nonsolvents containing oxygen increases the wettability of the polymer solution on the glass slides, which is due to the presence of –OH groups on the substrate surface. This surface firstly gives rise to a homogeneous initial precipitate layer, then grows a network over it, and finally forms a homogeneous coating.

Based on the work by Erbil *et al.*, Han and co-workers[53] achieved low-density polyethylene (LDPE) superhydophobic surface with different structures by controlling crystallization behavior of

LDPE via adjusting the crystallization time and nucleation rate. It was indicated that lower solvent evaporation temperature could increase the time of crystallization and the nucleation rate, thus increasing the contact angle. By adding nonsolvent (cyclohexane) and drying at room temperature, LDPE superhydrophobic surface with floral structure could be realized; its water contact angle was 173.0° (Fig. 4.23(a) and (b)). Besides, Ma *et al.*[54] also reported the preparation of super-hydrophobic polyvinylchloride (PVC) film with micro/nano-binary structure on glass substrate (Fig. 4.23(c) and (d)).

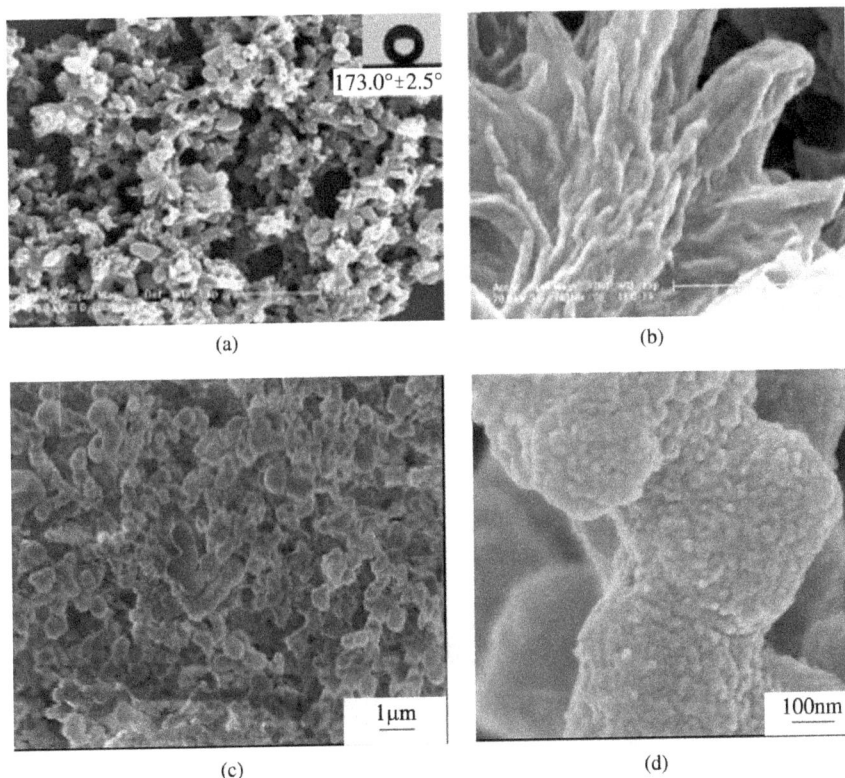

(a)

(b)

(c)

(d)

Fig. 4.23   (a) SEM pictures of porous LDPE films. (b) Enlarged view of a single floral structure.[53] (c) The SEM image of super-hydrophobic PVC films by coating a glass plate with a mixture of 2:1 (v/v) ethanol and H_2O as nonsolvent. (d) Highly-magnified nanostructures of (c).[54]

4.1.11 One step method

A two-block copolymer of polypropylene and poly(methyl methacrylate) (PP-PMMA) was used as film materials to construct polymer surfaces in one step. The surfaces possess the 3D micro/nano-multiscale structure (Fig. 4.24). A block copolymer could form a multi-molecular polymer micelle in a selective solvent, and the diameter of a single micelle was in the range of 50–200 nm. Micelles aggregated together to minimize the surface energy and to form microscale aggregate. The diameters of the microscale-aggregate sphere were mostly around 1–2 μm. On the surface of every sphere, the nanoscale multimolecular micelle spheres could be observed. The structure of this copolymer surface was similar to that of the lotus leaf, leading to the property of self-cleaning. The contact and sliding angles of water drops on the surface were 160.5 ± 2.1° and 9.0 ± 2.1°, respectively.

The wettability of solid surfaces is governed by both the chemical composition and the geometrical microstructure of the surface. However, researchers believe that the latter is more important. The self-cleaning biomimetic surface, as lotus leaf, could be obtained without modification from low-surface-energy materials as long as suitable micro-structures exist. In addition, polymer materials are vital in the paint industry, so these research results could broaden the selective range of biomimetic self-cleaning materials, deepen the understanding

Fig. 4.24   SEM images of the micro/nano-structures in the bionic polymer PP/PMMA.[55] (Copyright Wiley–VCH Verlag GmbH & Co. KGaA. Reproduced with permission.)

of wettability, and have potential application in the development of painting technology.

The authors' group presented a very simple and universally feasible method for constructing a stable bionic superhydrophobic surface with morphogenesis at room temperature. Taking copper as a model system, a film of copper carboxylate with micro/nano-structure was successfully fabricated on the surface of copper by immersing a copper plate into a solution of fatty acids of certain chain length at proper concentration. On the process of immersion, a few nano-sheets and small clusters were formed at first. Further increase of the immersion time led to an increasing surface density of the clusters and nano-sheets, finally the flowerlike clusters grew much bigger, became continuous, and almost completely covered the copper surface (Fig. 4.25(a)–(c)). Figure 4.25(d) shows SEM image of a typical single micro- and nano-cluster with a special flowerlike structure. The surface wettability of the as-prepared substrates was studied during the whole immersion process. As shown in Fig. 4.25, the

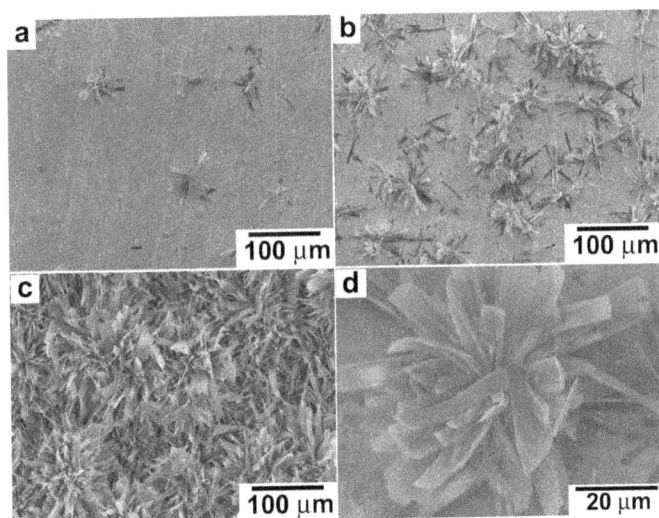

Fig. 4.25 (a)–(c): SEM images of Cu $(CH_3 (CH_2)_{12}COO)_2$ clusters formed on a copper plate by the solution-immersion process. (d) A high-magnification image of a single cluster showing its unique flowerlike structure.[56] (Copyright Wiley–VCH Verlag GmbH & Co. KGaA. Reproduced with permission.)

wettability of this surface varied from hydrophilic to hydrophobic finally to superhydrophobic with increasing immersion time. This kind of surface not only had a high water contact angle of about 162°, but also had a low sliding angle of about 2°. The superhydrophobic surface fabricated here was rather environmentally stable and was durable to solvent.

4.1.12 Other methods

Miller *et al.* reported that they controlled the roughness of surface with the ion-plated method by varying the voltage, and obtained superhydrophobic surfaces with contact angles in the range of 150° to 160°.[57] Polytetrafluorethylene (PTFE) thin films with nanometer-size roughness had also been prepared by vacuum-deposited method, which possessed the superhydrophobic property, and the contact angles to water droplets could reach as high as 150°.[58]

By means of an ion-assisted deposition method, a TiO_2 photo catalyst was prepared at the surfaces of superhydrophobic and porous Teflon sheets (PTS). UV light irradiation of TiO_2 photo catalyst on PTS led to the photo catalytic degradation of organic pollutants, which could keep the superhydrophobic properties of PTS for a long time. (Fig. 4.26).[59]

The Sn nanoflowers were formed on a titanium substrate by thermal-pyrolysis of a tin organometallic precursor.[60] Each flower was made of several dozens of porous nanopetals that connected with each other by self-assembly[60] (Fig. 4.27). After the rheotaxial growth

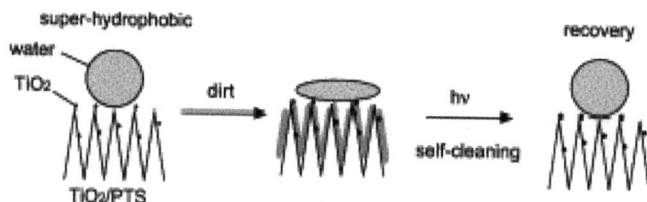

Fig. 4.26   The mechanism of the self-cleaning property on the TiO_2/PTS (TiO_2 content: 6.0×10^{-4} wt %). (Reprinted from Ref. 59. Copyright 2003, with permission from Elsevier.)

Fig. 4.27 SEM images of Sn nanoflowers (left) and SnO$_2$ nanoflowers (right) with different magnification. (From Ref. 60 — Reproduced by permission of The Royal Society of Chemistry.)

and thermal oxidation (RGTO), the assynthesized 3D Sn nanoflowers were completely converted into SnO$_2$ 3D nanoflowers without any obvious shape change. However, the SnO$_2$ nanopetal porosity greatly increased when compared with the assynthesized nanopetals of the Sn nanoflowers. After further investigation of the wetting properties of the Sn and SnO$_2$ nanoflowers, it was found that the water contact angle was 90° for the Sn nanoflowers and 155° for the SnO$_2$ flowers. The SnO$_2$ flowers had the property of being superhydrophobic. In contrast, it was found that the water contact angle was around 60° for a smooth Sn surface and about 90° for a flat SnO$_2$ surface.

Ming *et al.*[61] reported a procedure for preparing a rough surface with dual-size hierarchical structure through an epoxy-amine system that originated from well-defined raspberry-like silica oxygen particles which were chemically deposited onto epoxy films. During these procedures, loose particles could be flushed away easily with ethanol or water, leaving one layer of the particles covalently bonded to the epoxy-based films. Furthermore, these particles did not always have

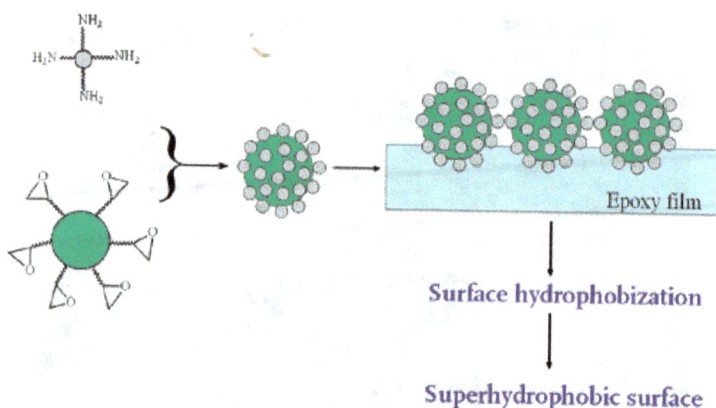

Fig. 4.28 Preparation of superhydrophobic films based on raspberry-like particles. (Reprinted with permission from Ref. 61. Copyright 2005, American Chemical Society.)

only one contact point with the epoxy-based film; some of them were embedded in the surface. The epoxy could be fully cured after the particle deposition. In the end, monoepoxy-endcapped PDMS was grafted onto the particles followed by hydrophobic treatment, then a solid hydrophobic surface was obtained (Fig. 4.28).

Superhydrophobic metal surfaces with micro/nano-structures were fabricated by using the simple electrochemical reaction of Cu or Cu–Sn alloy plated on steel sheets with sulfur gas, followed by subsequent perfluorosilane treatment. The water contact angles of the super-hydrophobic metal surfaces were found to be over 160°, and these surfaces exhibited a low contact angle hysteresis. The microstructure of these surfaces was obtained through nonelectric chemical plating of the copper onto steel sheets, and the nanotexturing of the surfaces was achieved via an electrochemical reaction of copper in a sulfur-containing environment at 150°C, resulting in the formation of a copper sulfide nanostructure on the microstructure.

A simple method for forming superhydrophobic surfaces is presented by ^{60}Co γ irradiation–induced hexafluoropropylene/ethyl methacrylate (HFP/EMA) vapor phase copolymerization under atmospheric pressure conditions. The resulting coral-reef-like micro-texture surface had a water contact angle of 153°.[63] The method could easily be extended to prepare superhydrophobic surface from a wide variety of materials. Xia *et al.*[64] reported that they had fabricated highly stable gold surface with micro/nano-composite structure through reaction of current exchange. It possessed the nature of superhydrophobicity after being modified by n-dodecanethiol (contact angle of 165°). A variety of topography of the gold surface could be prepared by changing the conditions, including the concentration of salt, light illumination, solution temperature, ultrasound or adding surfactant. In addition, a superhydrophobic thin film could be realized by a method of appending colloidal silica particles or polytetra-fluoroethylene (PTFE) particles.

4.2 Multi-functional Superhydrophobic Surfaces

4.2.1 *Stable superhydrophobic surface*

The authors' group obtained carbon nano-fiber arrays through a pyrolysis pathway, which was a typical preparation process of poly-acrylonitrile (PAN). Nanosturctured PAN films were used as precursors, which were prepared by template extrusion; we also found that the arrayed nano-structured carbon fiber had these properties without any modification of low energy substance, i.e. their water contact angle was larger than 150° for not only pure water but also corrosive liquids, such as acidic and basic solutions (Fig. 4.29). The as-prepared films showed superhydrophobic property in the whole pH range. It was found that the structures of these carbon nano fibers were similar to that of graphitelike. The nanostructure of the carbon fiber on the graphitelike structure and the endurability to the corrosion of acid and base were deemed to be the main reason for superhydrophobicity in the whole pH range.

The authors' group also prepared a novel polymer composite film which exhibited conductive and excellent superhydrophobic properties in a whole pH range by electrospinning method (Fig. 4.30).[70] We fabricated the PANI/PS composite film which had a self-cleaning effect by a mixture of conductive PANI and hydrophobic PS as their electrospinning materials. The exterior surface of this film was

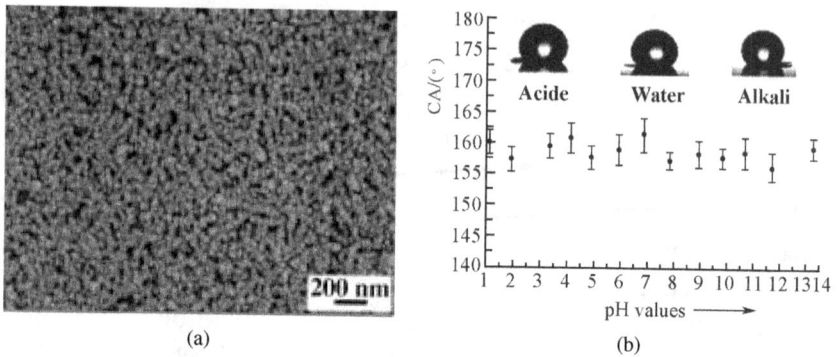

(a)   (b)

Fig. 4.29 (a) SEM image of the superhydrophobic nanostructured carbon fiber films in the whole pH range. (b) Graph showing the relationship between pH values and contact angles on the nanostructured carbon fiber films.[69] (Copyright Wiley–VCH Verlag GmbH & Co. KGaA. Reproduced with permission.)

composed of carbon element that possessed low surface energy, while the nitrogen atoms in the PANI which played an important role of doping and undoping were distributed in the internal membrane. Thus, this could maintain the superhydrophobicity and conductivity in the acidic and basic environment. Ordinary superhydrophobic surface could easily bring about static-charge accumulation, possibly causing a fire or an explosion under dry conditions, thus confining its practical application. So the present findings were of crucial significance for conductive coatings and for maintaining their stable conductivity and superhydrophobicity, which would be used effectively in many fields.

Yan *et al.* fabricated superhydrophobic films with Needle-like poly(alkylpyrrole) structures by electrochemical method. The films showed excellent environmental stability to both temperature and organic solvents or oils (Fig. 4.31).

4.2.2 *Superhydrophobic surfaces with special optics properties*

Transparency of the super-hydrophobic surface attracts much attention because of important applications on coatings of glasses, automobile glasses, windows, and so on. Hydrophobicity and transparency are

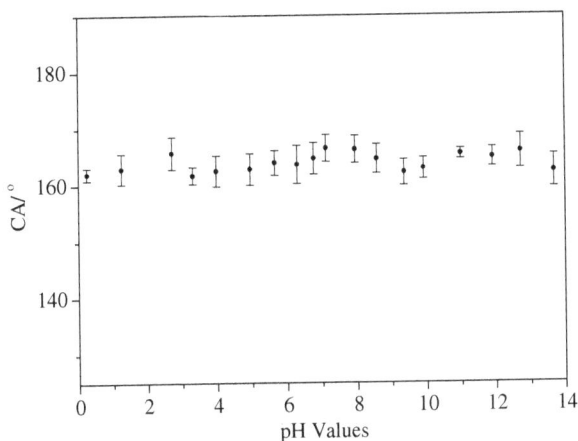

(a)

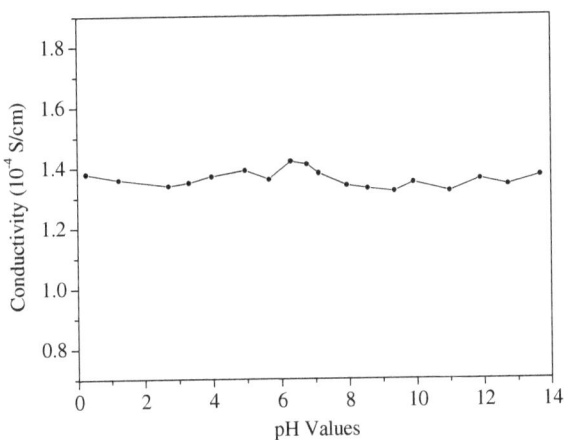

(b)

Fig. 4.30 (a) Relationship between pH and the CA on a PANI/PS composite film. (b) The relationship between pH and conductivity of a PANI/PS composite film.[70] (Copyright Wiley–VCH Verlag GmbH & Co. KGaA. Reproduced with permission.)

two competitive properties when taking into account the factors of roughness. The roughness of surface could enhance the property of hydrophobicity, while the light scattering would reduce its transparency. In general, transparency films have a minor roughness, while the hydrophobicity of surface decreases along with the reducing roughness. In order to prepare films that are both transparent and

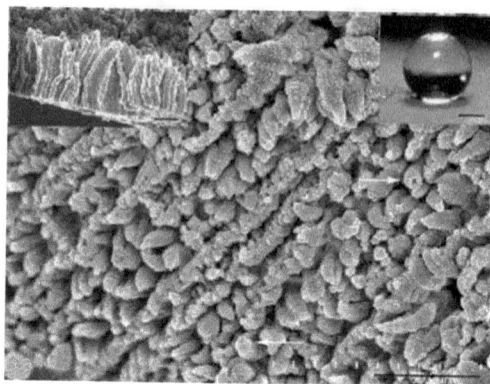

(a)

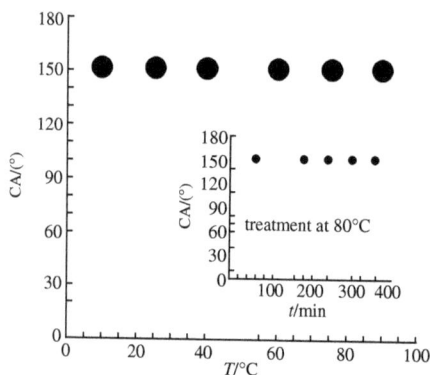

(b)

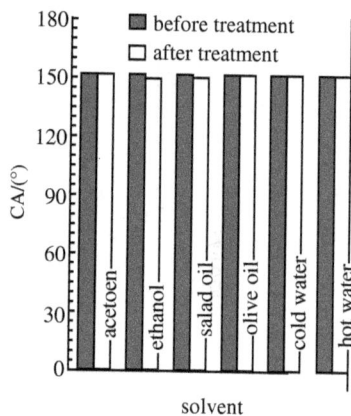

(c)

Fig. 4.31   (a) Scanning electron microscopic (SEM) image of the superhydrophobic poly(alkylpyrrole) film. (b) The relationship between temperature and contact angle. (c) The relationship between organic solvent and contact angle of the film.[71] (Copyright Wiley–VCH Verlag GmbH & Co. KGaA. Reproduced with permission.)

hydrophobic, effective control of the roughness of surface and building of a suitable rough surface are important. As mentioned above, a rough surface could be prepared by methods such as plasma processing, sublimation, separation, sol-gel, and chemical vapor deposition. A transparent superhydrophobic surface could be fabricated by

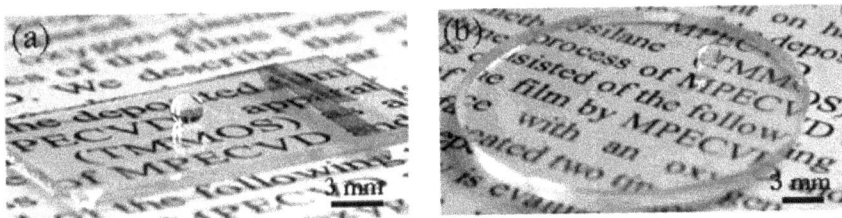

Fig. 4.32 Photographs of a water drop placed on the films prepared on (a) glass, and (b) organic substrates.[36] (Copyright Wiley–VCH Verlag GmbH & Co. KGaA. Reproduced with permission.)

adjusting the roughness and chemical modification (Fig. 4.32). For example, Watanabe and co-workers[72,73] succeeded in developing a new process to prepare the nanoscale transparent superhydrophobic AlO(OH), SiO_2 or TiO_2 films by using aluminum acetylacetonate $(Al(C_5H_7O_2)_3$, AACA, sublimation temperature 193°C) as a sublimation material. Each of the three kinds of films was roughened by the sublimation of aluminum acetylacetonate after calcination. Then, transparent superhydrophobic films were obtained by subsequent coating of fluoroalkyl silane. Their contact angles to water droplets were 161°, 150° and 157°, respectively. According to Watanabe *et al.*, the surface roughness is decided by the following factors: (1) the size of solid particles; (2) the concentration of sublimate and solid particles; and (3) the way of sublimate deposition.

Antireflectiveness is another very important optical property, which plays a crucial role in the glasses and the glass surfaces that cover the solar cells. A superhydrophobic and antireflective film was seldom reported, though the antireflective film has been developed widely. Xu *et al.*[74,75] fabricated a superhydrophobic film that possessed antireflective property by the method of methyl-modified SiO_2 sol. Its contact angle could reach 165°, and the minimum reflectivity could reach 0.03%.

4.2.3 *Superhydrophobic surfaces with other properties*

A superhydrophobic and electromagnetic material film could be prepared by filling carbon nano-fibers with the magnetic Fe_3O_4 particles.[76]

The specific process is that PVA/FeAc$_2$ nano fibers could be prepared by electro-spinning method from the mixture solution of PVA/FeAc$_2$. Then carbonization of PVA/FeAc$_2$ nano-fibers was completed in a high-temperature furnace, which has been heated up to 600°C under argon atmosphere for 8 h (Fig. 4.33(a)). Meanwhile, the existence of iron ions promotes the formation of carbon graphite in the carbonization process at low temperature. The carbon-fiber coating on the outer side of the iron oxide could prevent it from further oxidation so as to maintain the stability of the magnetic iron. In addition, the nature of the carbon nano-fibers could be controlled by changing the concentration of FeAc$_2$ electrical spinning solution (Fig. 4.33(b)).

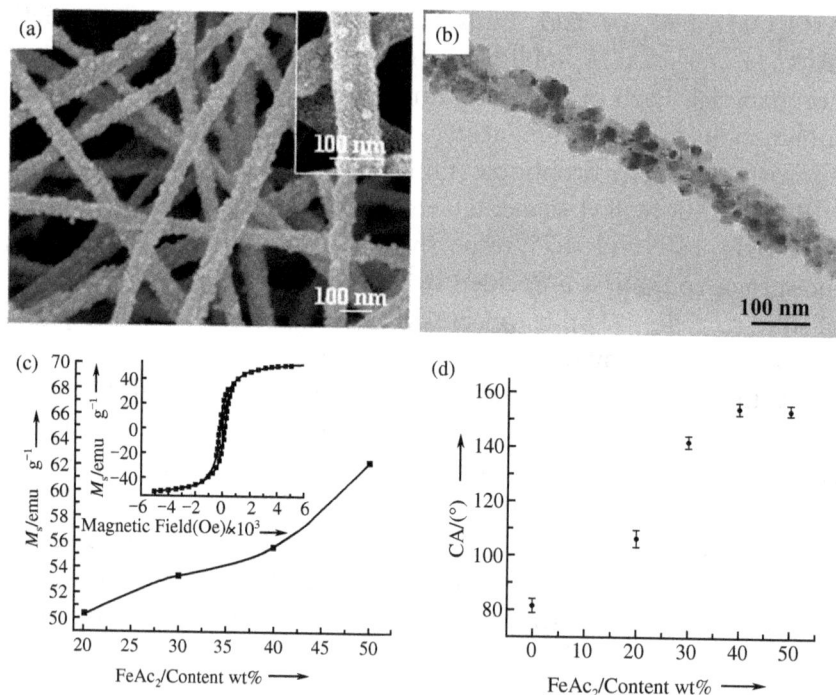

Fig. 4.33 (a) and (b): SEM and TEM image of Fe$_3$O$_4$-filled CNFs. (c) and (d): Relationship between the magnetization (Ms) or contact angle of carbon nano-fiber and the concentration of FeAc$_2$ in the electron spin solution.[76] (Copyright Wiley–VCH Verlag GmbH & Co. KGaA. Reproduced with permission.)

Nanostructured superhydrophobic surface with high blood compatibility was prepared by dip-coating fluorinated poly(carbonate urethane)s (FPCU) on aligned carbon nanotube (ACNT) films.[77] No platelet adhesion on the resulted superhydrophobic surfaces was observed in sharp contrast with that of the traditional smooth surfaces of FPCU (Fig. 4.34). Poly(carbonate urethane)s (PCUs) have many biomedical applications because of their long-term biostability and moderate blood compatibility. The fluorinated PCU may further lower the surface free energy and improve the properties of the polymers, which could be widely used as a new type of biological material. The face with nano-structure modified with FPCU could enhance their blood compatibility.

Superhydrophobic surfaces have a broad range of applications: when they are coated on the inner walls of micro-channels, microfluid will flow through under less flow resistance; on the satellite antennas,

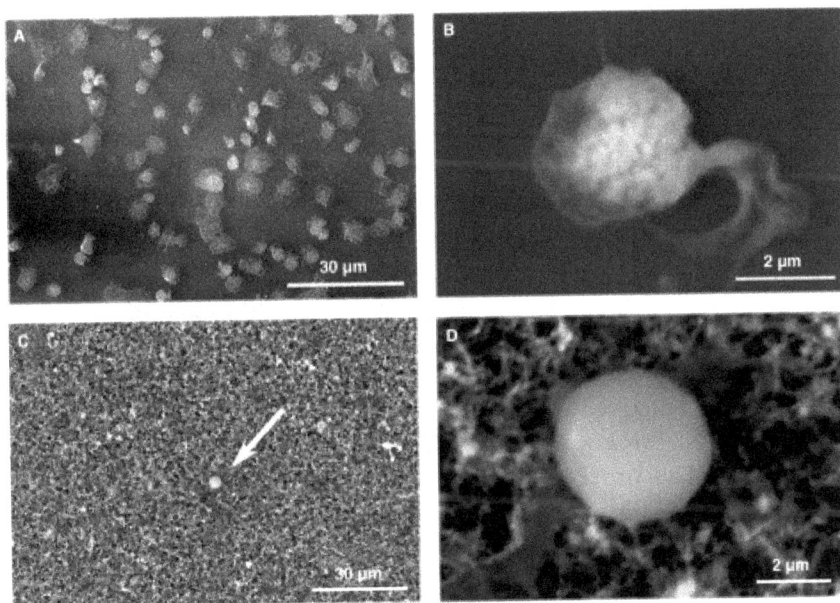

Fig. 4.34 SEM images of remarkable platelet-compatible superhydrophobic films of poly(carbonate urethane)s (PCUs) with fluorinated alkyl side chains.[77] (Copyright Wiley–VCH Verlag GmbH & Co. KGaA. Reproduced with permission.)

communication interference or interruption resulted by snow covering could be avoided; on the watercrafts like ships or submarines, increase in speed could be attained with reduced water resistance; on the needles of micro injector, wastage of expensive chemicals and contamination of the needles could be avoided. In summary, superhydrophobic surfaces have received wide attention due to their broad potential applications, which make it a hot topic in the current research on materials surface, and many effective preparation methods have been developed. Although some technical problems still exist in this field, e.g. economic practicality, wearability and durability of superhydrophobic surfaces, development of more high-performance superhydrophobic surfaces is expected to prosper with the collective effort of researchers in this promising field.

References

1. Onda T, Shibuichi S, Satoh N. *Langmuir*, 1996, **12**: 2125.
2. Shibuichi S, Onda T, Satoh N, Tsujii K. *J. Phys. Chem.*, 1996, **100**: 19512.
3. Chen W, Fadeev A Y, Hsieh M C, Öner D, Youngblood J, McCarthy T J. *Langmuir*, 1999, **15**: 3395.
4. Öner D, McCarthy T J. *Langmuir*, 2000, **16**: 7777.
5. Sacher E, Klemberg-Sapieha J, Schreiber H P, Wertheimer M R, McIntyre N S. In *Silanes, Surfaces, and Interfaces*, Leyden D. E. (Ed.), Gordon and Breach, New York, 1986, p. 189.
6. Washo B D. *Org. Coat. Appl. Polym. Sci. Proc.*, 1982, **47**: 69.
7. Youngblood J P, McCarthy T J. *Macromolecules*, 1999, **32**: 6800.
8. Matsumoto Y, Ishida M. *Sensors and Actuator*, 2000, **83**: 179.
9. Zhang Y, Kang E T, Neoh K G, Huang W, Huan A C H, Zhang H, Lamb R N. *Polymer*, 2002, **43**: 7279.
10. Fresnais J, Chapel J P, Poncin-Epaillard F. *Surface & Coatings Technology*, 2006, **200**: 5296.
11. Teare D O H, Spanos C G, Ridley P, Kinmond E J, Roucoules V, Badyal J P S, Brewer S A, Coulson S, Willis C. *Chem. Mater.*, 2002, **14**: 4566.
12. Woodward I, Schofield W C E, Roucoules V, Badyal J P S. *Langmuir*, 2003, **19**: 3432.
13. Favia P, Cicala G, Milella A, Palumbo F, Rossini P, d'Agostino R. *Surf. Coat. Technol.*, 2003, **169–170**: 609.

14. Teshima K, Sugimura H, Inoue Y, Takai O, Takano A. *Langmuir*, 2003, **19**: 10624.
15. Shiu J, Kuo C, Chen P, Mou C. *Chem. Mater.*, 2004, **16**: 561.
16. Yoshimitsu Z, Nakajima A, Hashimoto K, Watanabe T. *The 7th China-Japan Bilateral Symposium On Intelligent Electrophotonic Materials & Molecular Electronics*, Beijing, China, 2000, p. 110.
17. Bartell F E, Shepard J W. *J. Phys. Chem.*, 1953, **57**: 211.
18. Bico J, Marzolin C, Quéré D. *Europhys. Lett.*, 1999, **47**: 220.
19. Pozzato A, Zilio S D, Fois G, Vendramin D, Mistura G, Belotti M, Chen Y, Natali M. *Microelectronic Engineering*, 2006, **83**: 884.
20. Baldacchini T, Carey J E, Zhou M, Mazur E. *Langmuir*, 2006, **22**: 4917.
21. Jin M H, Feng X J, Xi J M, Zhai J, Cho K, Feng L, Jiang L. *Macromol. Rapid Commun.* 2005, **26**: 1805.
22. Tadanaga K, Katata N, Minami T. *J. Am. Ceram. Soc.*, 1997, **80**: 3213.
23. Tadanaga K, Katata N, Minami T. *J. Am. Ceram. Soc.*, 1997, **80**: 1040.
24. Tadanaga K, Kitamuro K, Morinaga J, Kotani Y, Matsuda A, Minami T. *Chem. Lett.*, 2000, **29**: 864.
25. Tadanaga T, Morinaga J, Matsuda A, Minami T. *Chem. Mater.*, 2000, **12**: 590.
26. Shirtcliffe N J, McHale G, Newton M I, Perry C C. *Langmuir*, 2003, **19**: 5626.
27. Yamanaka M, Sada K, Miyata M, Hanabusad K, Nakano K. *Chem Commun.*, 2006, 2248.
28. Han J T, Lee D H, Ryu C Y, Cho K. *J. Am. Chem. Soc.*, 2004, **126**: 4796.
29. Shang H M, Wang Y, Limmer S J, Chou T P, Takahashi K, Cao G Z. *Thin Solid Films*, 2005, **472**: 37.
30. Rao A V, Kulkarni M M, Amalnerkar D P, Seth T. *J. Non-Crystalline Solids*, 2003, **330**: 187.
31. Li S, Li H, Wang X, Song Y, Liu Y, Jiang L, Zhu D. *J. Phys. Chem. B*, 2002, **106**: 9274.
32. Li S H, Feng L, Li H J, Song Y L, Jiang L, Zhu D B. *Chemical Journal of Chinese Universities*, 2003, **24**: 340.
33. Song X Y, Zhai J, Wang Y L, Jiang L. *J. Colloid & Interface Sci.*, 2006, **298**: 267.
34. Lau K K S, Bico J, Teo K B K, Chhowalla M, Amaratunga G A J, Milne W I, McKinley G H, Gleason K K. *Nano Lett.*, 2003, **3**: 1701.
35. Hozumi A, Takai O. *Thin Solid Films*, 1997, **303**: 222.
36. Wu Y, Sugimura H, Inoue Y, Takai O. *Chem. Vap. Deposition*, 2002, **8**: 47.

37. Wu Y, Inoue Y, Sugimura H, Takai O, Kato H, Murai S, Oda H. *Thin Solid Films*, 2002, **407**: 45.
38. Wu Y, Bekke M, Inoue Y, Sugimura H, Kitaguchi H, Liu C, Takai O. *Thin Solid Films*, 2004, **457**: 122.
39. Tavana H, Amirfazli A, Neumann A W. *Langmuir*, 2006, **22**: 5556.
40. Li M, Zhai J, Liu H, Song Y, Jiang L, Zhu D. *J. Phys. Chem. B*, 2003, **107**: 9954.
41. Zhang Z, Gu L, Shi G. *J. Mater. Chem.*, 2003, **13**: 2858.
42. Jiang L, Zhao Y, Zhai J. *Angew. Chem. Int. Ed.*, 2004, **43**: 4338.
43. Zhang X, Shi F, Yu X, Liu H, Fu Y, Wang Z, Jiang L, Li X. *J. Am. Chem. Soc.*, 2004, **126**: 3064.
44. Zhai L, Cebeci F C, Cohen R E, Rubner M F. *Nano Lett.*, 2004, **4**: 1349.
45. Ji J, Fu J H, Shen J C. *Adv Mater.*, 2006, **18**: 1441.
46. Feng L, Li S H, Li H J, Zhai J, Song Y L, Jiang L, Zhu D B. *Angew. Chem. Int. Ed.*, 2002, **41**: 1221.
47. Feng L, Song Y L, Zhai J, Liu B Q, Xu J, Jiang L, Zhu D B. *Angew. Chem. Int. Ed.*, 2003, **42**: 800.
48. Guo C, Feng L, Zhai J, Wang G, Song Y, Jiang L, Zhu D. *ChemPhysChem*, 2004, **5**: 750.
49. Genzer J, Efimenko K. *Science*, 2000, **290**: 2130.
50. Love J C, Gates B D, Wolfe D B, Paul K F, Whitesides G M. *Nano Lett.*, 2002, **2**: 891.
51. Ren S, Yang S, Zhao Y, Yu T, Xiao X. *Surf. Sci.*, 2003, **546**: 64.
52. Erbil H Y, Demirel A L, Avci Y, Mert O. *Science*, 2003, **299**: 1377.
53. Lu X, Zhang C, Han Y. *Macromol. Rapid Commun.*, 2004, **25**: 1606.
54. Li X H, Chen G M, Ma Y M, Feng L, Zhao H Z, Jiang L, Wang F S. *Polymer*, 2006, **47**: 506.
55. Xie Q D, Fan G Q, Zhao N, Guo X L, Xu J, Dong J Y, Zhang L Y, Zhang Y J, Han C C. *Adv. Mater.*, 2004, **16**: 1830.
56. Wang S T, Feng L, Jiang L. *Adv. Mater.*, 2006, **18**: 767.
57. Veeramasuneni S, Drelich J, Miller J D, Yamauchi G. *Prog. Org. Coat.*, 1997, **31**: 265.
58. Miller J D, Veeramasuneni S, Drelich J, Yalamanchili M R, Yamauchi G. *Polym. Eng. Sci.*, 1996, **36**: 1849.
59. Yamashita H, Nakao H, Takeuchi M, Nakatani Y, Anpo M. *Nucl. Instr. Meth. Phys. Res. B*, 2003, **206**: 898.
60. Chen A X, Peng X, Koczkur K, Miller B. *Chem. Commun.*, 2004, **17**: 1964.

61. Ming W, Wu D, van Benthem R, de With G. *Nano Lett.*, 2005, **5**: 2298.

62. Han J T, Jang Y, Lee D Y, Park J H, Song S H, Ban D Y, Cho K. *J. Mater. Chem.*, 2005, **15**: 3089.

63. Wang Q, Quan Y, Zhang J, Chen Q. *Surface & Coatings Technology*, 2006, **200**: 5493.

64. Wang C H, Song Y Y, Zhao J W, Xia X H. *Surf. Sci.*, 2006, **600**: L38.

65. Nakajima A, Saiki C, Hashimoto K, Watanabe T. *J. Mater. Sci. Lett.*, 2001, **20**: 1975.

66. Takeda K, Sasaki M, Kieda N, Katayama K, Kako T, Hashimoto K, T Watanabe T, Nakajima A. *J. Mater. Sci. Lett.*, 2001, **20**: 2131.

67. Murase H, Nanishi K, Kogure H, Fujibayashi T, Tamura K, Haruta N. *J. Appl. Polym. Sci.*, 1994, **54**: 2051.

68. Yamauchi G, Miller J D, Saito H, Takai K, Ueda T, Takazawa H, Yamamoto H, Nislhi S. *Colloid Surfaces A*, 1996, **116**: 125.

69. Feng L, Yang Z L, Zhai J, Song Y L, Liu B Q, Ma Y M, Yang Z Z, Jiang L, Zhu D B. *Angew. Chem. Int. Ed.*, 2003, **42**: 4217.

70. Zhu Y, Zhang J C, Zheng Y M, Huang Z B, Feng L, Jiang L. *Adv. Funct. Mater.*, 2006, **16**: 568.

71. Yan H, Kurogi K, Mayama H, Tsujii K. *Angew. Chem. Int. Ed.*, 2005, **44**: 3453.

72. Nakajima A, Fujishima A, Hashimoto K, Watanabe T. *Adv. Mater.*, 1999, **11**: 1365.

73. Nakajima A, Hashimoto K, Watanabe T, Takai K, Yamauchi G, Fujishima A. *Langmuir*, 2000, **16**: 7044.

74. Xu Y, Fan W H, Li Z H, Wu D, Sun Y H. *Appl. Optics*, 2003, **42**: 108.

75. Xu Y, Wu D, Sun Y H, Huang Z X, Jiang X D, Wei X F, Wei Z H, Dong B Z, Wu Z H. *Appl. Optics*, 2005, **44**: 527.

76. Zhu Y, Zhang J C, Zhai J, Zheng Y M, Feng L, Jiang L. *ChemPhysChem*, 2006, 7: 336.

77. Sun T L, Tan H, Han D, Fu Q, Jiang L. *Small*, 2005, **1**: 959.

Chapter 5

Smart Nanoscale Interfacial Materials with Special Wettability

Some peculiar abilities of organisms are derived from the special microstructures on their surfaces. Inspired by these findings, lots of smart materials have been designed and fabricated. Bioinspired interfacial materials produce a new way to create intelligent materials. By combining the interface, structure, functionality, and multilevel weak interactions, multiple binary cooperative complementary systems will be realized.

Based on the fundamental wetting properties, that is, superhydrophobicity, superhydrophilicity, superoleophobicity, and superoleophilicity, many surface functions can be obtained by combining any two properties (illustrated in Fig. 5.1). For a solid substrate with micro/nanostructures and very high surface free energy, superhydrophilicity and superoleophilicity can coexist, and the film will show superamphiphilicity (Fig. 5.1, right). For a rough surface with very low surface free energy, superhydrophobicity and superoleophobicity can coexist, and the film will show superamphiphobicity (Fig. 5.1, left). Analogously, when superhydrophobicity/superoleophilicity or superhydrophilicity/superoleophobicity coexist (Fig. 5.1, center), separation of water from oil (or oil from water) can be realized. In addition to surfaces with these static wetting properties, in some cases the surface chemical composition or geometrical structure of a rough surface can be tuned dynamically. In such a case, a smart surface whose wettability can be modulated reversibly between superhydrophobicity and superhydrophilicity (Fig. 5.1, top) or superoleophobicity and

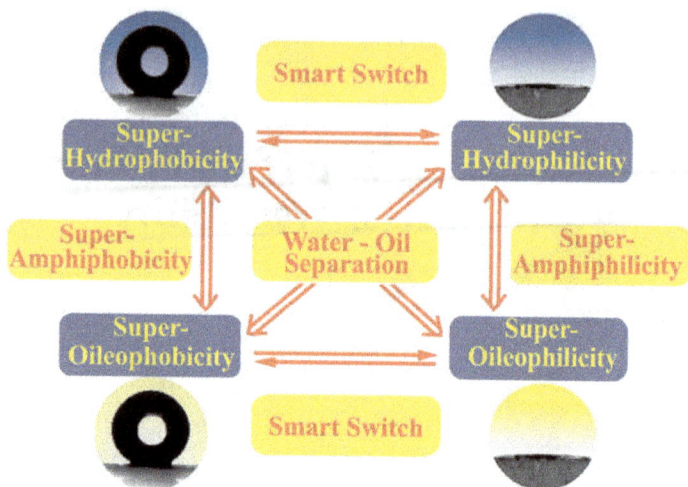

Fig. 5.1   Illustration of the relationships between the four kinds of fundamental superwetting/antiwetting properties (in blue boxes) and the further special surface superwetting/antiwetting functions (in yellow boxes) that are obtained by combining any two of the fundamental properties. Herein, the double arrow indicates coexistence of the two properties, and the reversible arrow indicates switching of the two properties.[1]

superoleophilicity (Fig. 5.1, bottom) is obtained. These novel materials will potentially be used in some important fields, such as catalysis, rheology, and so on.

5.1 Superamphiphobic Surface

Surfaces with both water contact angle and oil contact angle larger than 150° are defined as superamphiphobic surfaces, which have wide prospect in industries and daily lives. For example, boats with such surface will have anticorrosive and antipollution properties. The waste of petrol can be decreased by using conduits with such inner surface during transportation. Meanwhile, these superamphiphobic materials can also be used to modify textile to produce novel cloth with rainproof and antipollution abilities.

Polytetrafluoroethylene (PTFE) is a representative material with low surface energy (18.5 mN/m), which has attracted much attention for its thermal stability, chemical stability, and low friction coefficient. Although it is a well-known hydrophobic material, many applications were restricted for its lack of oleophobic property. By deposition of a fluorinated polymer layer with low surface energy onto microroughened PTFE substrate, Coulson *et al.*[2] fabricated a surface showing high repellency toward both polar and nonpolar liquid (Table 5.1). Furthermore, they prepared a novel surface with ultralow surface energy (4.3 mN/m) through room-temperature pulsed plasma polymerization of long-chain perfluoroacrylates, which is hydrophobic and oleophobic.[3]

Kiuru *et al.*[4] reported a diamond-like carbon–polymer hybrid (DLC–p-h) coating prepared with filtered pulsed arc discharge (FPAD) system using a graphite–polymer cathode. The polymers used in the hybrid coatings were polydimethylsiloxane (PDMS) and polytetrafluoroethylene (PTFE). It is worth noting that the surface is hydro- and oleo-phobic, on which water and oil can roll easily without any trace. Another example is reported by Nicolas *et al.*[5] In their work, a series of fluorinated conjugated polythiophene with low surface energy was prepared by incorporation of perfluoroalkyl moieties into a polythiophene backbone, which can reduce the surface energy significantly compared to the alkyl analogue due to a high percentage of fluorine at the outermost layer of the polymer. In addition, such polymers were electrodeposited on ITO plate to produce new surfaces with both excellent and stable water and oil repellency. Ma *et al.*[6]

Table 5.1   Probe liquid contact angles for flat and porous PTFE substrates.

| | Contact Angle/° | | | |
| | Water | | Decane | |
Treatment	Flat	Porous	Flat	Porous
Untreated	116 ± 2°	146 ± 1°	46 ± 2°	wicks
O$_2$ plasma	131 ± 4°	153 ± 2°	55 ± 2°	wicks
Plasma polymer	131 ± 3°	144 ± 1°	83 ± 1°	115 ± 1°
O$_2$ plasma + plasma polymer	148 ± 1°	152 ± 1°	92 ± 3°	133 ± 2°

reported a versatile method to produce superhydrophobic fabrics by combining electrospinning and initiated chemical vapor deposition (iCVD). Briefly, poly(caprolactone) (PCL) was first electrospun and then coated with a thin layer of hydrophobic polymerized perfluoroalkyl ethyl methacrylate (PPFEMA) by iCVD. By combining the hierarchical surface roughness of the PCL film and the low free energy of the coating layer, the surface is superhydrophobic with a contact angle of 175° and a sliding angle of less than 2.5° for a 20 mg droplet. This PPFEMA-coated PCL mat was also shown to exhibit excellent oleophobicity.

As mentioned above, surfaces which are both hydrophobic and oleophobic have been realized in some extent, however, reports about superamphiphobic surfaces are limited. Tsujii and co-workers[7,8] have reported about preparation of superlipophobic surfaces. Aluminum plates were first anodically oxidized to obtain hierarchical structures on the aluminium surface, and then modified by the fluorinated monoalkylphosphates $[CF_3(CF_2)_9CH_2OP(=O)(OH)_2$, F10-MAP]. The as-prepared surface is superlipophobic with a contact angle of 150° for rapeseed oil. According to the relation between contact angle and the surface tension, it can be deduced that the surface is superhydrophobic, and the water contact angle is about 160°. Thus, such surface should be superamphiphobic. Through compressing pellets of submicrometer and variable-diameter spherical particles of PTFE oligomers, McCarthy and co-workers[9] fabricated a surface exhibiting water contact angles of $\theta_A/\theta_R = 177°/177°$, methylene iodide contact angles of $\theta_A/\theta_R = 140°/138°$, and hexadecane contact angles of $\theta_A/\theta_R = 140°/125°$. Wang *et al.*[10] brought forward a method for producing superamphiphobic surfaces through plasma modification of benzoxazine films. Microroughening and fluorination of the benzoxazine films occur during the plasma-treatment process, and a rugged surface with a micro/nano-binary structure is formed, which results in high advancing contact angles (157° for water, 152° for di-iodomethane) and low contact angle hysteresis.

In fact, the concept of "superamphiphobic" was first proposed by our group in 2000.[11] Subsequently, we successfully prepared the superamphiphobic aligned carbon nanotube films, and studied the wettability of the films.[12,13] The as-prepared surface without any treatment is superhydrophobic with a water contact angle of 158.5°, after being modified with the low-surface-energy material fluoroalkylsilane, the surface is superamphiphobic and the contact angles for water and rapeseed oil on the film are 171° and 161°, respectively.

Lotus leaf has aroused much attention for its self-cleaning ability, and experimental results show that the hierarchical micro/nanostructures on the surface are the key factors to induce such property. Inspired by this finding, Xu and co-workers[14] reported a simple casting process for creating a superamphiphobic polymer coating from two polymer materials — poly(methyl methacrylate) (PMMA) and fluorine-end-capped polyurethane (FPU) — in air. SEM study of the surfaces shows that the polymer surface possesses natural lotus-like micro/nanostructures as a result of the different solubility of the two polymers in a solvent (Fig. 5.2(a) and (b)). The surface is superamphiphobic, the contact angles for water and oil are 166° and 140° (Fig. 5.2(c) and (d)), respectively. Similar to the lotus leaf, the water droplet can roll easily on the surface. Meanwhile, the superhydrophobic and self-cleaning properties can be retained even though the outmost surface is destroyed, indicating the self-repairing ability of the surface.

5.2 Surface with Superhydrophobicity and Superoleophilicity

By combining superhydrophobicity with superoleophilicity, we previously fabricated a surface that can be used to separate oil from water, or water from oil.[15] Such films were prepared by coating polytetrafluoroethylene (PTFE), a hydrophobic and oleophilic material, onto a stainless steel mesh via a spray-and-dry method. Detailed SEM image indicates that the as-prepared mesh films exhibit micro/nanoscale composite rough surface structures (Fig. 5.3(a)).

Figure 5.3(b) shows the shape of a water droplet on the surface, of which the contact angle and the sliding angle are 156.2° and 4°, respectively, indicating that the surface is superhydrophobic. More importantly, the hard superhydrophobic coating mesh film has a

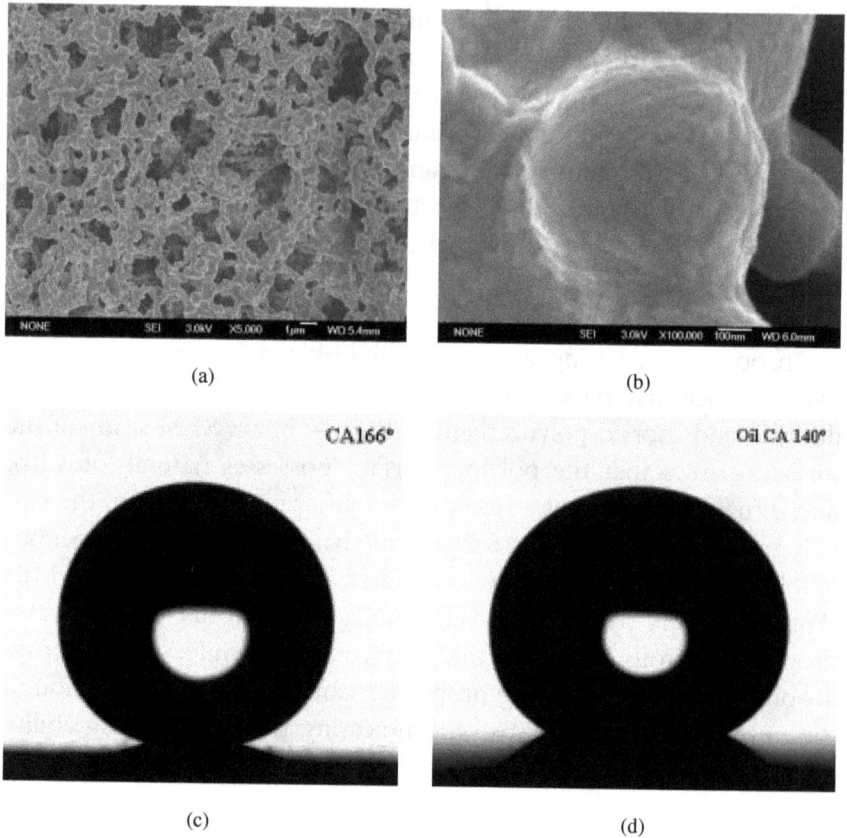

(a)

(b)

(c)

(d)

Fig. 5.2   (a) SEM image of coated surface of PMMA/FPU mixture. (b) micro/ nano-binary structures on a single micro-papilla surface from (a). (c) and (d): Contact angle of (c) water and (d) oil on the PMMA/FPU-mixture coated surface, respectively.[14]

practical application: the prepared film shows superoleophilic properties, with a contact angle of about $0° \pm 1.3°$. Figure 5.3(c) shows the spreading and permeating behavior of a diesel-oil droplet on the coating mesh film. Oil spreads quickly on the film and permeates thoroughly within only 240 ms. The coating films not only retain the intrinsic properties of stainless steel mesh, such as high strength, ventilation property, and anticorrosion, but also possess the new superhydrophobicity and superoleophilicity as a result of the hierarchical

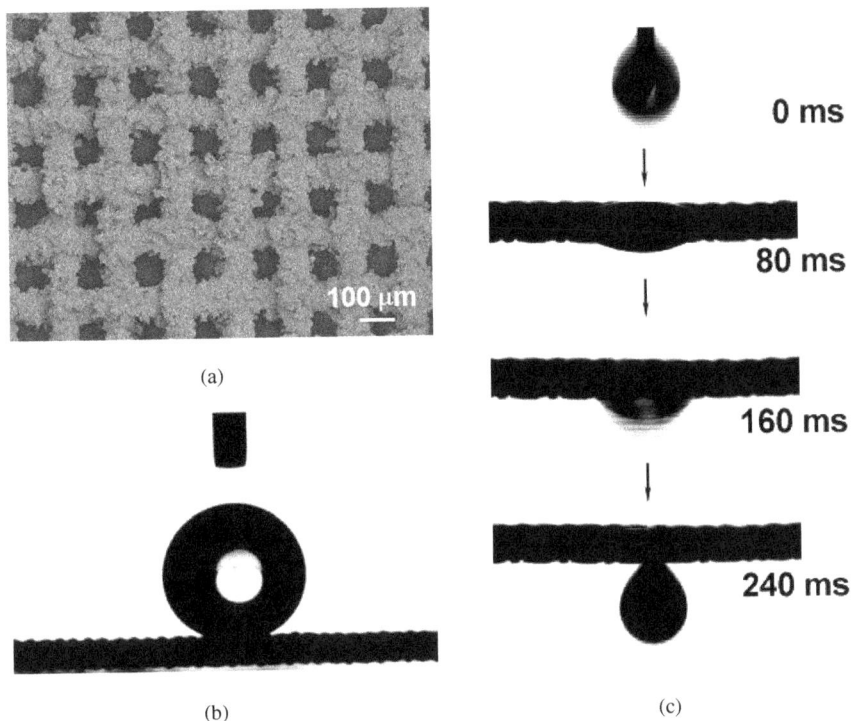

Fig. 5.3 (a) SEM image of the coating mesh film prepared from a stainless steel mesh. (b) Shape of a water droplet on the coating mesh film with a contact angle of $156.2° ± 2.8°$ and sliding angle of $4°$. (c) Spreading and permeating behavior of a diesel-oil droplet on the coating mesh film. The diesel oil spreads quickly on the coating mesh film and flows through (within only 240 ms).[15]

micro/nanostructures on the surface, which will enable the surfaces to be applied in different fields.

5.3 Smart Surface with Reversible Superhydrophilicity and Superhydrophobicity

Responsive materials have attracted much attention because of their variable performances triggered by external stimuli.[16] To date, many responsive materials with their surface free energy or morphology sensitive to the external environment are being used to alter the surface wettability.[17] However, the range of wettability transition is usually very limited, which restricts many important applications. Research results show that increasing the surface roughness is an effective way to amplify the surface wettability. Therefore, reversible transition between superhydrophilicity and superhydrophobicity can be realized by combining the responsive materials and the surface roughness.

5.3.1 *Surfaces with controllable wettability*

According to theories mentioned in Chapter 3, the wettability of solid surfaces is governed by both the chemical composition and the geo-metrical microstructures of the surface. Thus, the wettability of surfaces can be controlled by tuning the chemical composition or the microstructures of the surfaces. For examples, Chen and co-workers[18] reported that the wettability can be controlled by simultaneously manipulating surface topography and surface chemical structure. Two steps are included: firstly, charged submicrometer polystyrene latex particles are adsorbed to oppositely charged poly(ethylene terephthalate) (PET) film. By controlling the adsorption time, solution pH, solution ionic strength, latex particle size, and substrate charge density, surfaces with different topographies/roughness can be obtained. Then, the discrete functional groups are introduced to the surfaces through organic transformations. Amine groups ($-NH_2$) and alcohol groups ($-OH$) were introduced onto smooth PET surfaces by amidation with poly(allylamine) and adsorption with poly(vinyl alcohol) (PVOH), respectively. On latex particle-adsorbed surfaces, before the chemical transformation, a thin layer of gold was evaporated first to prevent particle redistribution. Reactions with functionalized thiols and adsorption with PVOH on patterned gold surfaces successfully enhanced surface hydrophobicity and hydrophilicity. Both hydrophobicity and hydrophilicity are affected by particle size and biomodal particle size distribution. By absorption of materials with different free energy, surfaces with extremely hydrophobic and hydrophilic properties can be prepared. For example, a very hydrophobic surface exhibiting water contact angles of 150°/126° (θ_A/θ_R) can be prepared by adsorption of 1-octadecanethiol, while a hydrophilic surface with water contact angles of $18°/8°$ (θ_A/θ_R) can be obtained by adsorption of PVOH on gold-coated surfaces. Another example is reported by Wu and Shi.[19] They reported a simple method for fabricating a lotus-like micro/nanoscale binary structured surface of copper phosphate dihydrate. The wettability of this surface can be changed from super-hydrophilic to highly hydrophobic or superhydrophobic by heating or modifying it with an *n*-dodecanethiol monolayer.

We previously proposed a facile method to fabricate colloidal crystal films with tunable wettability from an amphiphilic-material polystyrene-block-poly(methyl methacrylate)-block-poly-(acrylic acid) (Fig. 5.4).[20] The wettability of the film can be tuned from superhydrophilic (CA, 0°) to superhydrophobic (CA, 150.2°) by varying the assembly temperature, while the position of the photonic bandgap of the colloidal crystal films remains virtually unchanged. The tunable wettability is ascribed to phase inversion of the flexible groups of the latex surface during the assembly procedure. The latex sphere has a hydrophobic PS core and a hydrophilic P(MMA/AA) shell with carboxy and ester groups anchored on the surface. Thus, the resulting colloidal crystal film is hydrophilic when fabricated at a lower temperature. When the assembly temperature is increased, thermodynamics drives the polymer chains towards a common minimum energy: the hydrophilic groups tend to shield from the apolar air, and shrink towards the interior of the latex sphere, while hydrophobic CH_2 groups prefer to extend toward the apolar air as the water evaporates during the assembly procedure, so the resulting film is water repellent. The results indicate that the wettability of the films is controllable by the assembly temperature.

5.3.1.1 *Control the wettability through the morphology*

By a simple extrusion process with an anodic aluminum oxide membrane as the template, we previously prepared a superhydrophobic aligned polyacrylonitrile nanofibers film.[21] Furthermore, through an idealized model of the surface structure of the PAN nanofibers,

(a)　　　　　　　　　　　　　　　　　　(b)

(c)

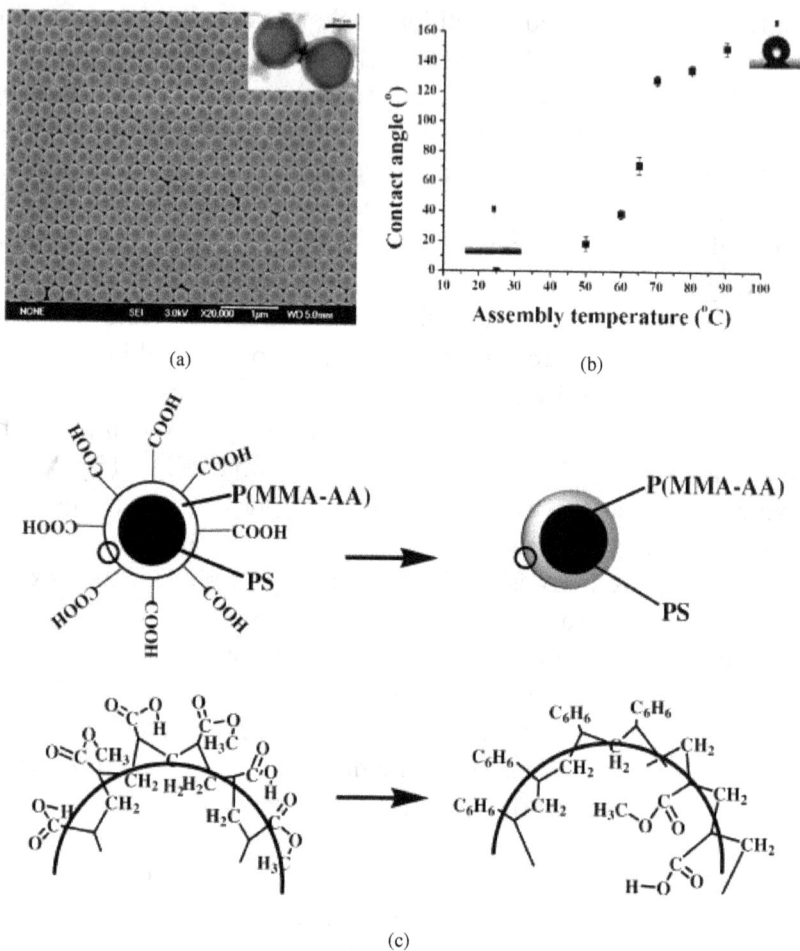

Fig. 5.4 (a) Typical SEM image of a colloidal crystal film assembled at 50°C (inset: typical TEM image of core-shell particles of P(St-MMA-AA), the bar is 200 nm). (b) The relationship of assembly temperature with water CA (inset is the water droplet profile of the relative water CA). (c) Schematic illustration of the phase inversion of the latex sphere during the assembly procedure.[20]

a simple equation was put forward to calculate the fraction of air in the troughs between individual nanofibers:

$$f_2 = 1 - k \cdot \left(\frac{d}{D}\right)^2. \tag{5.1}$$

Here, d and D are the average tip diameter of a nanofiber and an average distance interfiber distance, respectively, $k = \dfrac{\pi}{2\sqrt{3}}$.

According to the formula proposed by Cassie (Eq. (5.2)), θ_r can be increased with increasing the f_2; that is, increasing the fraction of air in the surface will increase the hydrophobicity of the surface.

$$\cos \theta_r = f_1 \cos \theta - f_2. \qquad (5.2)$$

Here, θ_r and θ are the contact angles on the PAN nanofibers with a rough surface and on the native PAN film with a smooth surface, respectively; f_1 and f_2 are the fractional interfacial areas of the PAN nanofibers and of the air in the troughs between individual nanofibers, respectively (i.e. $f_1 + f_2 = 1$). It is easy to deduce from this equation that increasing f_2 increases θ_r, i.e. the fraction of air in the surface is important to the superhydrophobicity.

According to Eq. (5.1), increasing D, or decreasing d will increase f_2, i.e. the water contact angle will be increased. The results offered us important theoretical elements for fabrication surfaces with controllable wettability.

Han and co-workers[22] proposed a simple way to achieve a superhydrophobic film by extending a Teflon film; the water contact angle can be increased from 118° to 165° by increasing the axial extension ratio from 5% to 190% of the Teflon tape. Figure 5.5 shows water contact angle and the corresponding shapes of water droplets as a function of extension ratio (ε) of Teflon tape: (1) $\varepsilon = 5\% \pm 5\%$, CA $= 118.2° \pm 6.0°$; (2) $\varepsilon = 35\% \pm 5\%$, CA $= 141.8° \pm 5.1°$; (3) $\varepsilon = 60\% \pm 5\%$, CA $= 147.1° \pm 5°$; (4) $\varepsilon = 90\% \pm 5\%$, CA $= 160.4° \pm 4.5°$; (5) $\varepsilon = 140\% \pm 5\%$, CA $= 165.1° \pm 4.1°$; and (6) $\varepsilon = 190\% \pm 5\%$, CA $= 165.0° \pm 4.9°$. The results indicate that decreasing the density improves the fraction of air, and further improves water repellency.

Well-ordered nanostructured PS surfaces were fabricated by utilizing nanosphere lithography and oxygen plasma firstly, and then coated with a 20 nm–thick gold film and further modified with octadecanethiol.[23] The apparent water contact angle (132°–168°) of

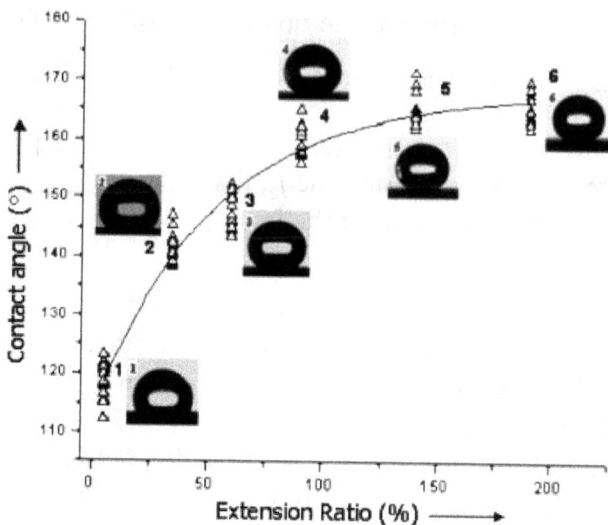

Fig. 5.5 Water-contact angle and the corresponding shapes of water droplets as a function of extension ratio of Teflon tape.[22] (Copyright Wiley–VCH Verlag GmbH & Co. KGaA. Reproduced with permission.)

the modified polystyrene surfaces can be tuned by changing the diameters (440–190 nm) of the PS arrays (Fig. 5.6).

By combining a phase separation of tetraethyl orthosilicate (TEOS) induced by the addition of an acrylic polymer, and subsequent fluoroalkylsilane (FAS) coating. Nakajima *et al.*[24] prepared a novel hard superhydrophobic thin film with high transmittance in the visible wavelength range. The crater-like structure was obtained by the addition of acrylic polymer to the system, and its size increases with increasing polymer concentration (Fig. 5.7(a)). Further investigations revealed that the silica sol had little effect on the phase relation. The water contact angle increases with increasing polymer concentration; while the sliding angle of water droplet initially increases, when the polymer concentration is larger than 0.4%, it decreases (Fig. 5.7(b)). Results revealed that the contact angle hysteresis on hydrophobic rough surfaces decreased with increasing roughness factor from a critical value, due to the switching of the dominant sliding mode from Wenzel to Cassie. In their work, the critical concentration for

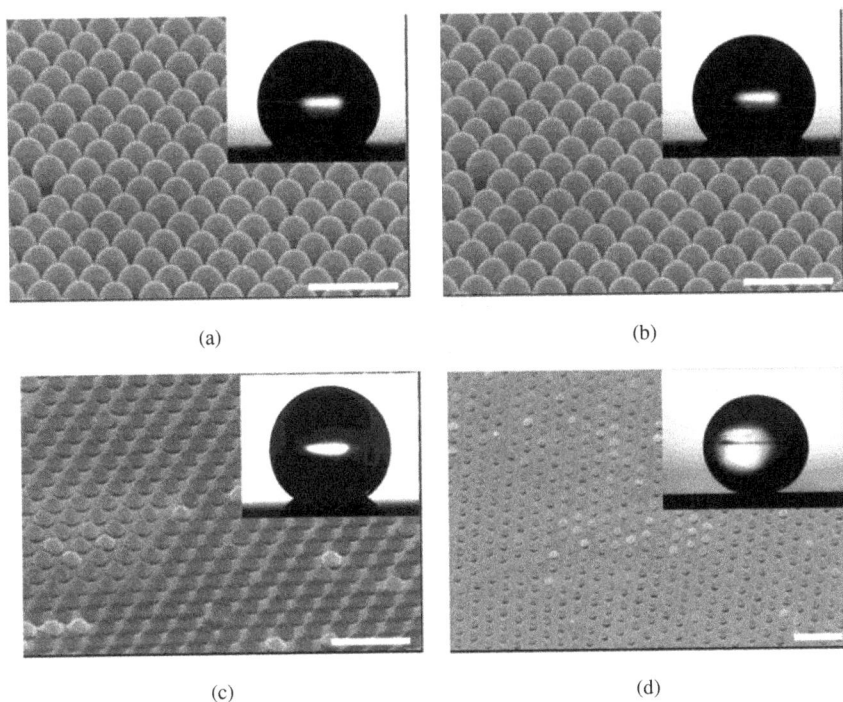

Fig. 5.6   SEM images (60°) of the size-reduced polystyrene beads and the water contact angle measurement on the corresponding modified surfaces (insets). The diameters of polystyrene beads and water contact angles on these surfaces were measured to be (a) 400 nm, 135°; (b) 360 nm, 144°; (c) 330 nm, 152°; (d) 190 nm, 168° (The bar is 1 μm). (Reprinted with permission from Ref. 23. Copyright 2004, American Chemical Society.)

the switching is in the range 0.2–0.4 wt.%. When the polymer concentration is lower than the critical value, surface roughness will be a resistance for the sliding of water droplet. When the polymer concentration is higher than the critical concentration, the sliding resistance is reduced because of the increase in air ratio at the solid–water interface, and the water will slide easily. This result suggests that a proper rough microstructure is necessary to satisfy both high contact angles and low sliding angles for rough hydrophobic surfaces.

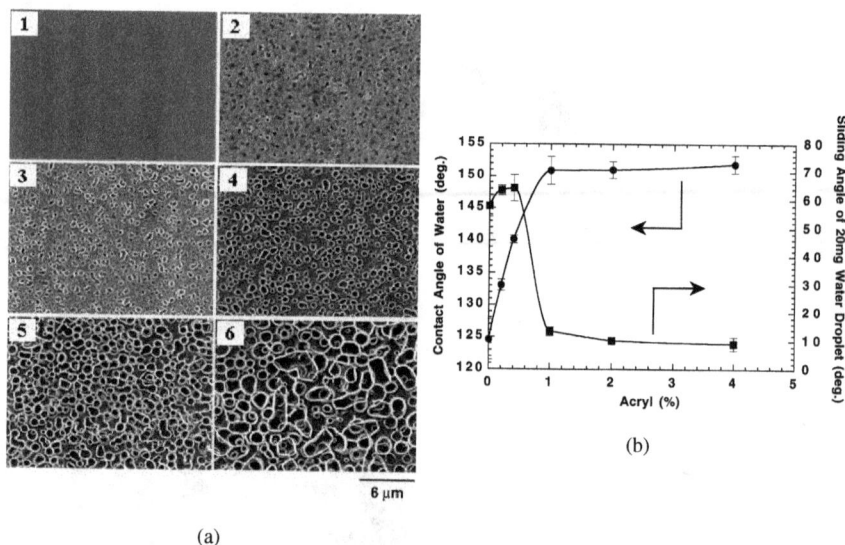

Fig. 5.7 (a) SEM micrographs of prepared films. Acrylic polymer concentration: (1) 0, (2) 0.2, (3) 0.4, (4) 1, (5) 2, and (6) 4%. (b) Contact angles and sliding angles of water droplet on the prepared films. (Reprinted from Ref. 24. Copyright 2000, with permission from Elsevier.)

Based on a monolayer-polystyrene (PS) colloidal crystal, Cai and co-workers[25] fabricated large-scale two-dimensional (2D) hierarchical porous silica film (orderly arranged macropores and disordered mesopores in its skeleton) with a highly specific surface area using the sol–gel technique. The surface is superhydrophilic with a water CA of 5° and superhydrophobic with a water CA of 154° after modification with fluoroalkylsilane. Noticeably, the water CA can be increased to 165° and the sliding angle can be decreased by using a heat-deformed PS template, indicating that the superhydrophobicity can be controlled by the template with different heat-deformed extents.

5.3.1.2 *Control the wettability on rough surface through the chemical composition*

The copper films with double scale micro- and nanostructures were prepared via a two-step electrochemical deposition (ECD) approach

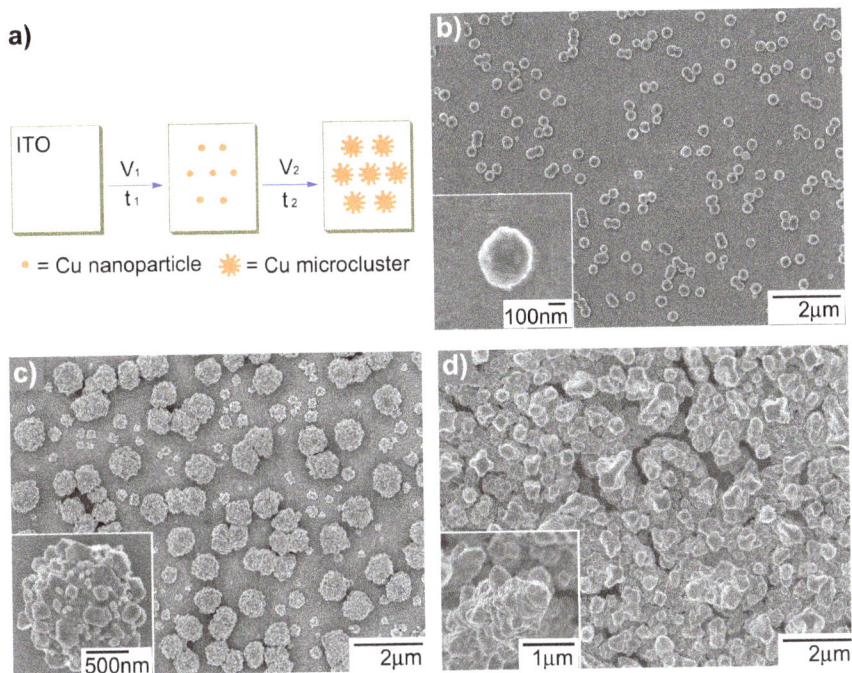

Fig. 5.8 (a) Simplified two-step electrochemical deposition process for preparing copper films with hierarchical structures. (b)–(d): SEM images of the copper films on ITO glass at different deposition times. (b) the first ECD step (t_1 = 20 s); the second ECD step with (c) t_2 = 80 s and (d) t_2 = 280 s. The inset in (b) shows a globe-shaped copper particle. The enlarged views of copper clusters in (c) and (d) show that they have micro/nanohierarchical structures.[26]

by our group previously.[26] The typical ECD process is shown in Fig. 5.8(a). Firstly, a small potential (−0.8 to −1.0 V) was used to create nucleation sites on indium tin oxide (ITO)–coated glass. As shown in Fig. 5.8(b), copper nucleation sites form in random distribution after the first deposition step (t_1 = 20 s), and every copper particle shows a bare globe-shaped surface (see the insert in Fig. 5.8(b)). Secondly, an overpotential (−2.0 to −2.5 V) helped the particles to grow larger and produce extended metal nanostructures. Figure 5.8(c) shows the SEM image of copper films obtained after the second deposition time (t_2) of 80 s. The dominant copper clusters have sizes of 0.7–1.5 μm, among which there are random ones which are less than 0.5 μm as well as a few new particles probably produced in the second step. Clearly, many protuberant nanostructures (40–300 nm) are created on copper microclusters (see the inset in Fig. 5.8(c)). Upon extension of the second deposition time, these protuberant nanostructures become larger and taller (Fig. 5.8(d)), with a height of 2–7 μm. The enlarged view of a copper cluster in Fig. 5.8(d) shows that it has a special micro/nanohierarchical structure, which is similar to that found on lotus leaves.

The as-prepared films were immersed in an ethanol solution of *n*-octanoic acid or *n*-propanoic acid for 12 h, and the surface-active molecules adsorb onto the native oxide surfaces of copper clusters and form a self-assembled monolayer through the acid–base reaction between *n*-alkanoic acids and copper oxide on the metal surface. Interestingly, the modified surface with *n*-octanoic acid shows super-hydrophobicity with a CA of $153.4° \pm 3.0°$ (Fig. 5.9(a), left), while that with *n*-propanoic acid is superhydrophilic with a CA of about $0°$ (Fig. 5.9(a), right). It can be concluded that the chain length of acids has an important influence on the wettability.

To investigate in detail the effect of chain length of acids on the surface wettability of copper films, we systematically modified as-prepared films ($t_1 = 20$ s, $t_2 = 360$ s) with *n*-alkanoic acids of different chain lengths, $CH_3(CH_2)_nCOOH$ ($n = 1, 2, 3,..., 16$). Figure 5.9(b) shows that the CAs are remarkably different depending on the chain length. On the as-prepared films, the CA increases from about $0°$ for the C_3 acid ($n = 1$) and levels off to a value of about $160°$ for acids longer than C_{11}, which indicates that surface wettability on the rough surface can be controlled to a large extent by adjusting the chain length of *n*-alkanoic acids. For comparison, smooth copper films prepared by vacuum evaporation were modified by the same process. However, the CA on the smooth films changed just in a limited range from $68.3°$ to a plateau of about $113°$. Results indicate that controllable transition from superhydrophilicity to superhydrophobicity can be realized on hierarchical micro/nanostructured surfaces.

(a)

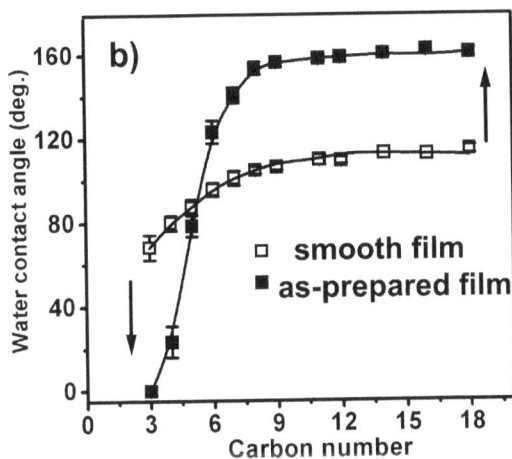

(b)

Fig. 5.9 (a) Shapes of water droplets on the modified as-prepared films with *n*-octanoic acid and *n*-propanoic acid. (b) Water contact angle on the modified copper films as a function of chain length of *n*-alkanoic acids.[26]

5.3.1.3 *Gradient wettability surfaces*

Gradient surfaces with a gradually changing surface wettability along their length are of great interest, and can be divided into two groups according to the factors governing the changing of wettability: chemical gradient surfaces and microstructured gradient surfaces.

Chemical gradient surfaces are the foremost among numerous gradient surface materials. In 1992, Chaudhury and Whiteside[27] firstly reported a surface having a spatial gradient in its surface free energy, which was capable of causing drops of water placed on it to move uphill. The surface with gradient free energy was achieved by exposing the surface of a polished silicon wafer to the diffusing front of the vapor of decyltrichlorosilane, $Cl_3Si(CH_2)_9CH_3$. The gradient reported is easy to prepare, and should be useful in the study of the motion of liquid drops induced by chemical gradients and of the interplay of chemical and thermal gradients.

Microstructured gradient surfaces can be obtained by controlling some factors during the fabrication process, such as potential, electric current, and temperature. Han and co-workers[28] fabricated the gradient surfaces by changing polystyrene (PS) microsphere topography in a temperature gradient field (Fig. 5.10). The wettability of the film has an obvious gradient change from 88.7° to 148.1°. Similarly, they

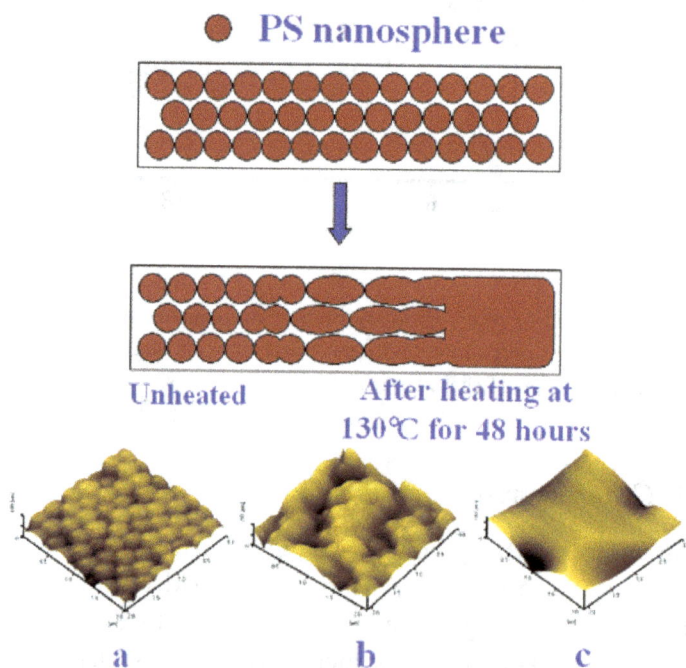

Fig. 5.10 AFM images at different locations of the film. (a) Unheated side. (b) Temperature ~ T_g. (c) Temperature ~ 130°C. (Reprinted with permission from Ref. 28. Copyright 2005, American Chemical Society.)

also reported an interesting approach to the fabrication of wettability gradients from hydrophobicity to superhydrophobicity in low-density polyethylene (LDPE) films.[29] Although the polymer used for the study is non-polar, its microporous layer provides sufficient superhydrophobicity. Lateral gradient heating of the layer results in partial melting of the polymer and correspondingly decreases porosity, thus decreasing hydrophobicity.

Zhang and co-workers[30] developed a simple method to fabricate a gradient surface that is from superhydrophobic to superhydrophilic. By controlling the self-assembled process of the monolayer consisting of a thiol $(HS(CH_2)_{11}CH_3)$ molecule on a rough gold surface prepared by electrochemical deposition, the density of molecules on the surface can be controlled effectively. By adding the dilute solution to the container holding the rough gold substrate, it will lead to a density gradient along the surface. After the complementary adsorption of $HS(CH_2)_{10}CH_2OH$, the surface exhibits a gradient from superhydrophobicity (CA > 150°) to superhydrophilicity (CA < 10°) (Fig. 5.11).

The superhydrophobic-to-superhydrophilic gradient will provide a larger oriented driving force for many important biological and physical processes and will have potential applications in water droplet movement. Daniel *et al.*[31] investigated the movement of liquid drops on a surface with a radial-surface-tension gradient. When saturated steam passes over a colder hydrophobic substrate, numerous water droplets nucleate and grow by coalescence with the surrounding drops. When a surface-tension gradient is designed into the substrate surface, the

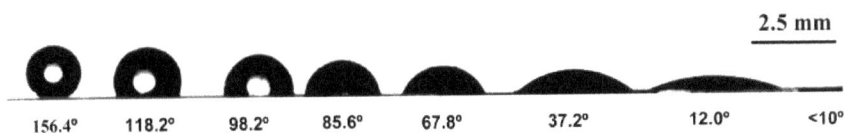

| 156.4° | 118.2° | 98.2° | 85.6° | 67.8° | 37.2° | 12.0° | <10° |

Fig. 5.11 Photograph of the water contact angles along the gradient surface. The photograph was combined with four continuous photographs along the substrate because the view angle of the contact angle measurement system is not wide enough. The volume of the water droplets was kept at ~3 µL. (Reprinted with permission from Ref. 30. Copyright 2006, American Chemical Society.)

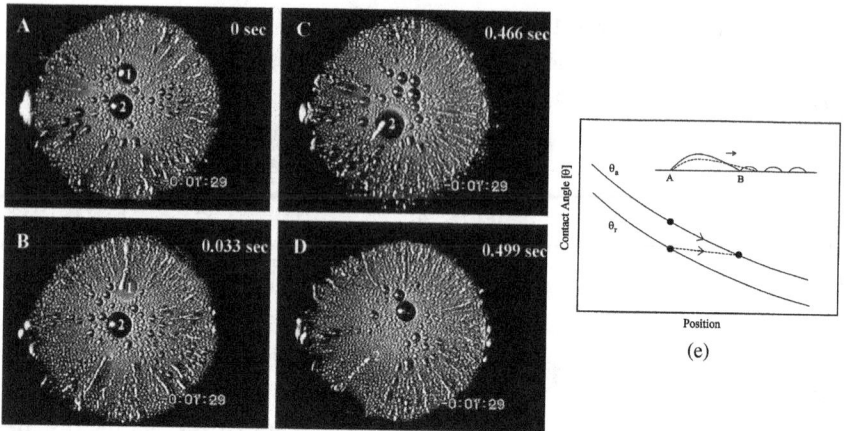

Fig. 5.12 Video prints showing fast movements of water drops (indicated by the plume- and streak-like appearances) resulting from the condensation of steam on a silicon wafer possessing a radial gradient (1 cm diameter) of surface energy. (a) and (b) show that drops 1 and 3 leave their original positions to cross the edge of the gradient in about 0.033 s. These movements leave behind streak-like appearances, which are due to the flashes of light reflecting from the drops during fast movements. The central drop 2 does not initially move, as it is on the weakest part of the gradient. However, after 0.47 s, as the drop becomes off-centered, it rapidly crosses the gradient zone ((c) and (d)). (e) Schematic drawing of the ID wettability gradient of a surface. (From Ref. 31. Reprinted with permission from AAAS.)

random movements of droplets are biased toward the more wettable side of the surface (Fig. 5.12). This effect has implications for passively enhancing heat transfer in heat exchangers and heat pipes.

It can be found that the surface roughness can increase both the surface hydrophobicity and hydrophilicity; that is, the surface roughness is a key factor in amplifying the surface wettability, which makes it possible to realize the reversible transition between superhydrophobic and superhydrophilic.

Combining the responsive materials and rough substrate, surfaces with reversible transition between superhydrophobicity and superhydrophilicity can be realized under different external stimuli. Such intelligent surfaces are often fabricated by modified roughness substrate with responsive materials, or constructed directly by responsive materials into rough structures. The following is some recent progress about the surfaces with responsive wettability, which will be discussed systematically according to different external stimuli, such as physical stimuli and chemical stimuli.

5.3.2 *Thermo-responsive surfaces*

Polymer as a thermo-responsive material has its unique advantages. For example, the spreading velocity of droplets on the surface of a liquid crystalline polymer can be extremely affected by a slight temperature change, resulting in a sharp change from rigid to soft behavior induced by a bulk transition between highly ordered smectic phase and isotropic phase, with consequent effects on the tack properties of the liquid crystalline polymer and on the dewetting dynamics of a liquid on its surface.[32] Owing to the flexible property, polymer chains exhibit conformational alteration in response to environmental stimuli.[33] Typically, thermo-responsive polymers show different characters above and below their lower critical solution temperature (LCST).

Poly-isopropylacrylamide is an excellent thermo-responsive polymer with a low LCST value. Below the LCST, it is soluble in water; while above the LCST, it becomes insoluble. The principle of such phenomenon lies in the reversible competitive process between intramolecular hydrogen bonds and intermolecular hydrogen bonds during temperature changes, which leads to the hydrophilic/hydrophobic changes of the molecule. In the following, poly-N-isopropylacryl-amide (PNIPAAm) is taken as an example to discuss the intelligent reversible wettability changes.

5.3.2.1 *Thermal stimuli–responsive poly-isopropylacrylamide*

Liang *et al.*[34] first grafted a photosensitizer, (N,N'-diethylamino) dithiocarbamoylpropyl(trimethoxy)-silane (DATMS), onto the substrate and then modified the surface of glass plate (Fig. 5.13) and quartz capillary tube (Fig. 5.14) with poly-N-isopropylacrylamide (PNIPAAm) by ultraviolet (UV) photopolymerization of NIPAAm monomers using N,N'-methylenebisacrylamide (BisAAm) as a cross-linker.

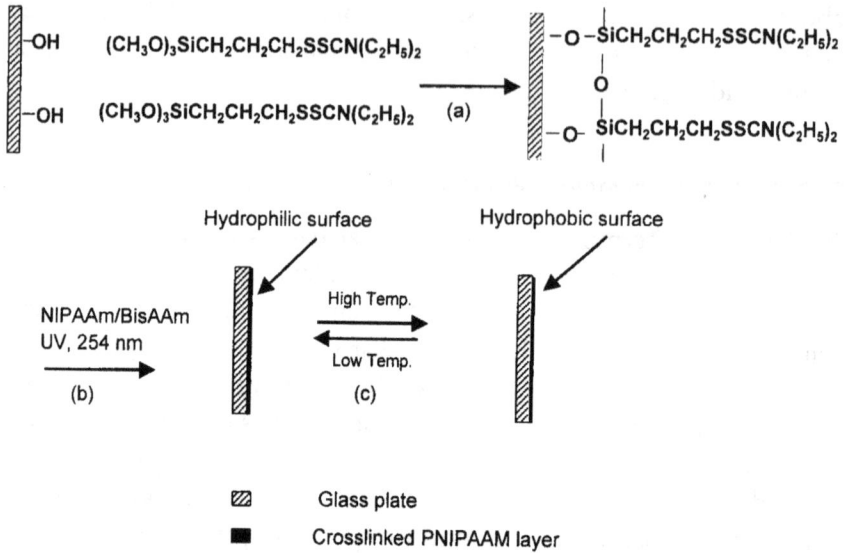

Fig. 5.13 PNIPAAm photopolymerization on the glass plate. (Reprinted with permission from Ref. 34. Copyright 1998, American Chemical Society.)

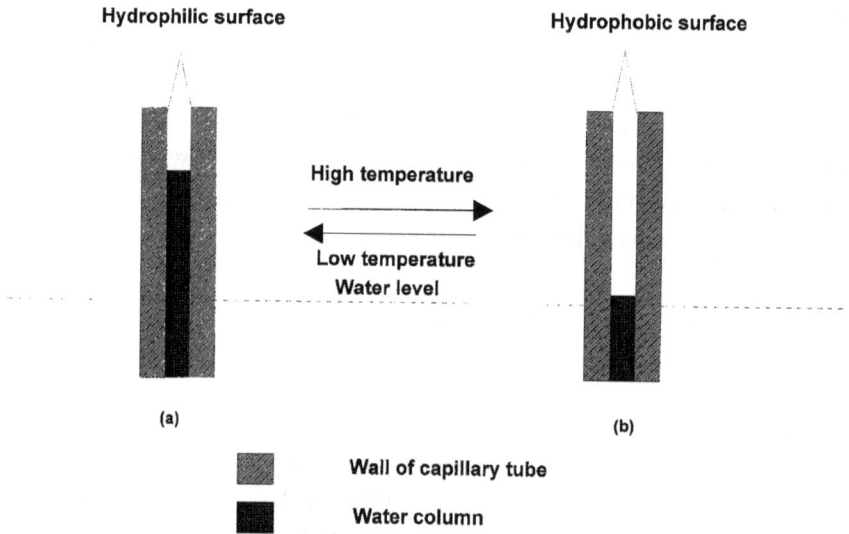

Fig. 5.14 Thermal-responsive PNIPAAm grafted in capillary tube. (Reprinted with permission from Ref. 34. Copyright 1998, American Chemical Society.)

Utilizing the property that PNIPAAm experiences a competition between intramolecular and intermolecular hydrogen bonds when temperature changes, thermal-responsive wettability was realized onto solid surface. On the surface of the glass plate, the thickness of the PNIPAAm film is less than 320 nm. The surface exhibits a hydrophilic property at a lower temperature (<30°C) and a hydrophobic property when temperature is higher than 40°C. The transition temperature is around 35°C, and the range of transition temperature change is approximately 8°C. The height of water meniscus in the capillary tube can be adjusted through temperature change: when temperature changed from 20°C to 40°C, a difference of about 7 mm was observed. With the same method, they also grafted PNIPAAm onto the surface of a silicone wafer,[35] and observed a remarkable change of advancing contact angle around 32°C. Grafting PNIPAAm or its copolymer hydrogels (Fig. 5.15) on solid substrate can also realize thermal stimuli–responsive wettability.[36–38]

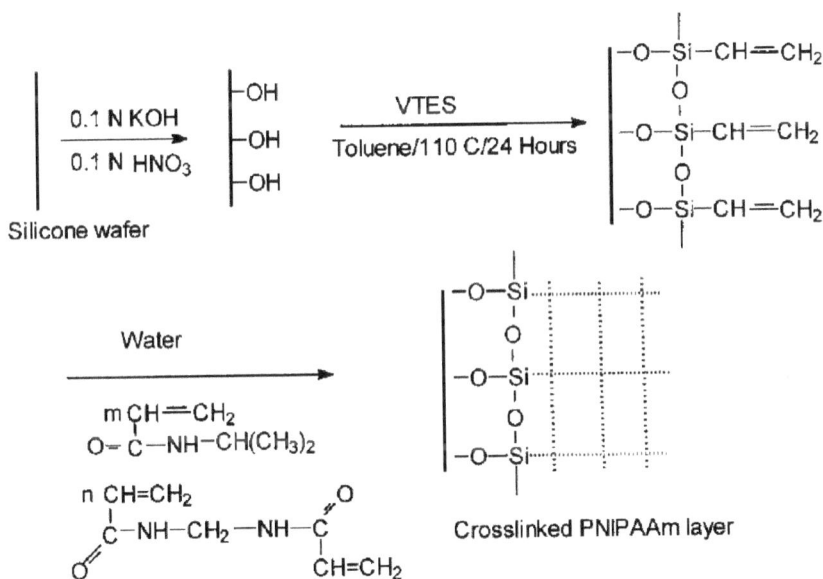

Fig. 5.15 Thermal-responsive PNIPAAm gels grafted on silicon substrate. (Reprinted with permission from Ref. 36. Copyright 2000, American Chemical Society.)

The fact that PNIPAAm thin films exhibit thermally responsive switching between hydrophilicity and hydrophobicity during temperature change is explained by the competition between intermolecular and intramolecular hydrogen bonding of PNIPAAm chain. Such effect is also the intrinsic mechanism of other properties of PNIPAAm such as reversible contraction and expansion, as well as solubility. At temperatures below the LCST (32–33°C), PNIPAAm chains present a loosely coiled conformation that leads to the predominantly intermolecular hydrogen bonding between –C=O, –NH groups and water molecules, contributing to the hydrophilicity of PNIPAAm film. At temperatures above the LCST, PNIPAAm chains present a compact and collapsed conformation that leads to intramolecular hydrogen bonding between –C=O and –NH groups, making it difficult to interact with water molecules and contributing to the hydrophobicity.[34–40] Our group[40] reported a PNIPAAm-modified surface with smooth silicon substrate, and measured the contact angle of the surface at different temperatures. The results show that when temperature was controlled at 25°C, the surface had a hydrophilic state with a contact angle of 63.5°; and when the temperature was increased to 40°C, the surface became hydrophobic with a contact angle of 93.2°. Repeat cycles of the experimental temperature showed that this effect was reversible. Such evidence reveals that by controlling the environmental temperature, the surface wettability can be easily controlled (Fig. 5.16).

5.3.2.2 *Thermal-responsive hydrophobic/hydrophilic reversible "switch"*

Our group[40] used surface-initiated atom-transfer radical polymerization (SI-ATRP) to fabricate thermally responsive PNIPAAm thin films onto silicon substrate, achieving a reversible switch between superhydrophilicity and superhydrophobicity in a narrow temperature range of about 10°C through the control of surface roughness. Such reversible switching property of surface results from the combined effect of the chemical variation of the surface, surface roughness, and their response to changing temperature. Hydrophilicity and hydrophobicity are two opposite properties in wettability, just like "Yin" and "Yang", which are the two opposite concepts of ancient Chinese philosophy. Due to the

a)

b)

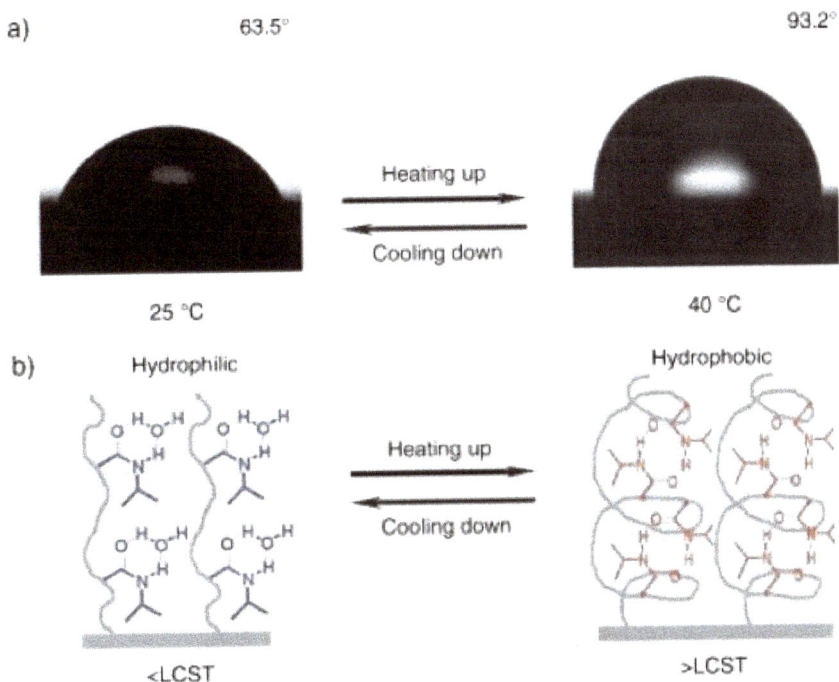

Fig. 5.16   The mechanism of thermal-responsive wettability-changing of PNIPAAm-grafted rough substrate.[40]

excellent responsibility, reversibility and stability, such switchable surfaces can be considered to have wide applications in industry fields such as intelligent microfluidic switching, functional textiles, thermally responsive filters, and controllable drug release.

Inspired by the above research results, it can be imagined that a kind of magic cloth with the effect of being "warm in winter and cool in summer" may soon be realized. We all known that hydrophilicity and hydrophobicity are two important features of a solid surface. Usually, the clothes that people wear are hydrophilic and easily get wet by water; and plastic is hydrohphobic that cannot be wetted by water. If the temperature-responsive superhydrophobic/superhydrophilic "switch" material can be used in cloth production, clothes can be made with such property that they are hydrophilic and can absorb sweat to make people feel cool in summer; while in winter with low temperature, the clothes become hydrophobic and keep the body warm.

We first generated some rough silicon substrates with different roughness by a laser cutter, and then modified PNIPAAm thin films onto the surfaces. Figure 5.17(a) is the SEM image for the regular array of silicon microconvexes with PNIPAAm modification (left), compared with that of flat substrate (right). It can be observed from the picture that the rough substrate is composed of a regular array of square silicon microconvexes, and the width of the microconvex and the spacing between the grooves can be adjusted discretionarily. Further magnified SEM images show that the surface of the silicon microconvexes is rather rough (Fig. 5.17(b) and (c)) for they are composed of many irregular nanoparticles and pores. The nanoparticles were sputtered from adjacent microgrooves and the pores were formed by the aggregation of nanoparticles. Although the nanoporous structure of the surface was partially filled by PNIPAAm polymerization, the micro/nanostructures of the substrates remained after polymerization.

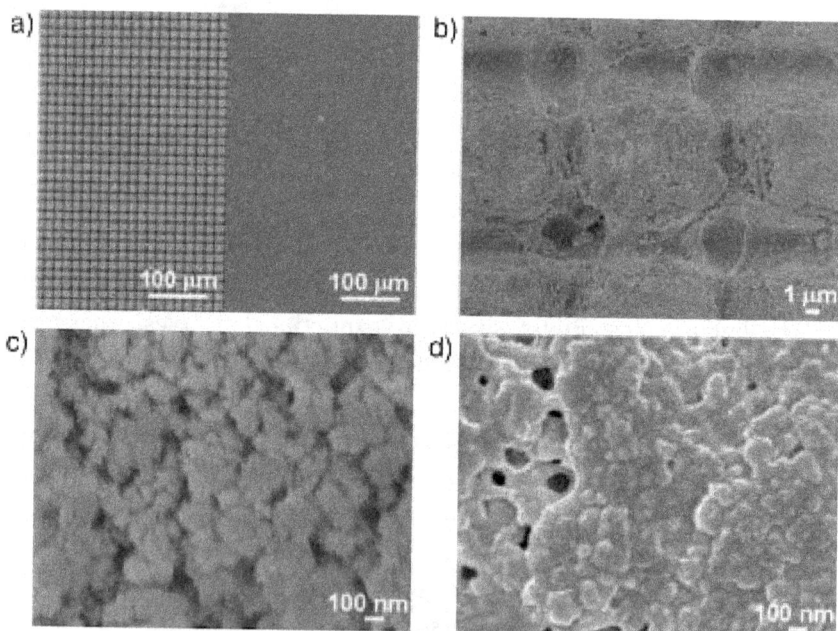

Fig. 5.17 SEM image of the regular array of silicon microconvexes with (a) PNIPAAm modification (left), compared with that of flat substrate (right). (b) and (c): Magnified roughness substrate. (d) PNIPAAm-grafted surface.[40]

As for the rough substrates with PNIPAAm modification, temperature-responsive wettability was significantly exhibited (Fig. 5.18). Compared with smooth substrate, the thermally responsive wettability was greatly enhanced by surface roughness. Figure 5.18(a) shows the relationships between the water CA and the surface roughness (presented as the groove spacing) at high temperature (40°C) and low temperature (25°C). It is noted from the picture that the CA changing range was amplified with the decrease of the groove spacing. The water

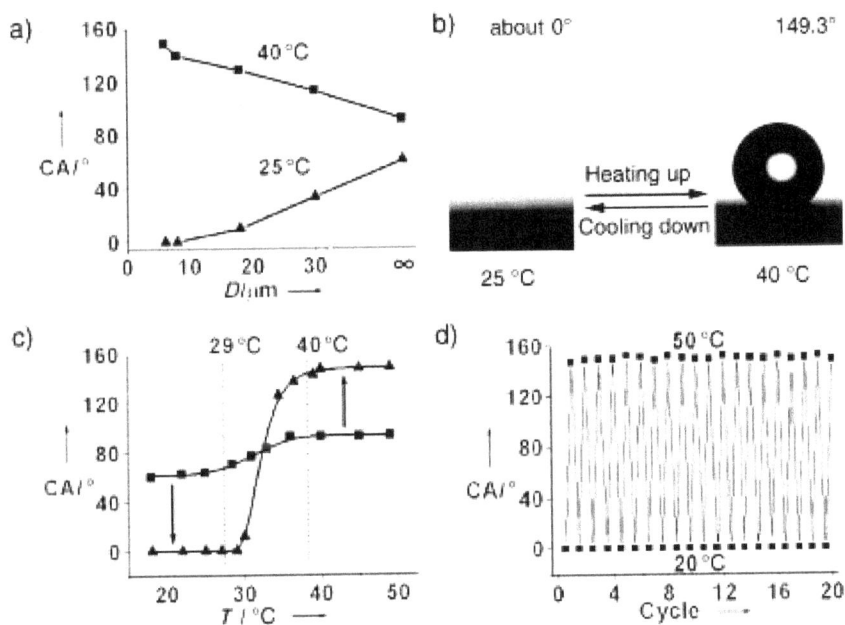

Fig. 5.18   The CA changes of PNIPAAm-grafted film on rough substrate. (a) The relationships between groove spacing D of rough surfaces and the water CAs at low temperature (triangles, 25°C) and at high temperature (squares, 40°C). The groove spacing of ∞ represents flat substrate. (b) Water drop profile for thermally responsive switching between superhydrophilicity and superhydrophobicity of a PNIPAAm-modified rough surface with groove spacing of about 6 μm, at 25°C and 40°C. The water CAs are about 0° and 149.3 ± 2.5°, respectively. (c) The dependence of temperature T on water CAs for PNIPAAm thin films on a rough substrate with groove spacing of about 6 μm (triangles) and on flat substrate (squares). (d) Water CA at two different temperatures for a PNIPAAm-modified rough substrate with groove spacing of 6 μm. Half cycles: 20°C; and integral cycles: 50°C.[40]

CA at 25°C decreased from 63° for a flat substrate to 34°, 10°, 0° and 0°; while at 40°C, the water CAs increased from 93° to 113°, 129°, 138°, and 149°. These results indicate that thermally responsive switching between superhydrophilicity and superhydropobicity can be realized when the groove spacing of the substrate is small, or in other words, when the substrate is sufficiently rough (Fig. 5.18(b)). The temperature dependency of water CA for the groove spacing of 6 µm rough substrate and the flat substrate after PNIPAAm modification was studied in detail and shown in Fig. 5.18(c). The range and changing point of wettability switching temperature for flat and rough substrates were almost the same, thus indicating that the chemical property of the PNIPAAm thin film did not change with the variation of substrate structure. For the rough substrate, the surface exhibited a water CA of about 0° below 29°C, whereas it was about 150° above 40°C, thus indicating that a thermally responsive switching between superhydrophilicity and superhydropobicity can be realized in the temperature range of 29–40°C. The experiment was repeated between cycles of low and high temperatures (the measurement of water CA was conducted on two sample stages of 20°C and 50°C with a single cycle that lasts about three minutes). The results show excellent reversibility for more than 20 cycles and a quick transformation between superhydrophilicity and superhydrophobicity (Fig. 5.18(d)). On the other hand, this reversibility remained after the samples had been laid aside without special protection for at least three months, which shows that the polymer film is stable.

5.3.2.3 PNIPAAm-grafted porous anodic aluminum oxide (AAO) membranes

Based on the above works, López' group[41] grafted PNIPAAm on porous anodic aluminum oxide (AAO) membranes, and discovered that the size of the surface pores played crucial effect on wettability (Fig. 5.19). At low temperature, increasing the pore size of the substrate led to a gradual decrease in the contact angles; while at high temperature, a dramatic increase in contact angles was observed. Thus, the difference in contact angles measured at low and high temperatures was increased (from 13° to 112°).

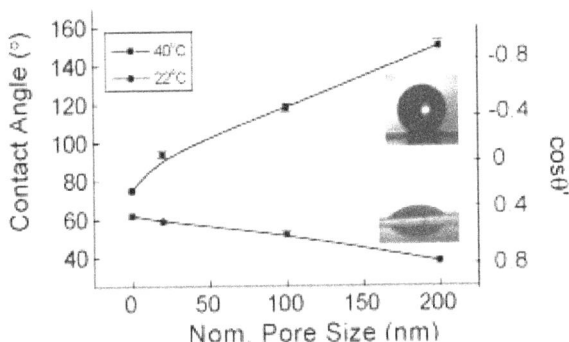

Fig. 5.19 The CA changes of PNIPAAm-grafted film on AAO substrates with different pore size. (Reprinted with permission from Ref. 41. Copyright 2004, American Chemical Society.)

They also used atomic force microscope (AFM) to examine the changes in the nanostructure of AAO surfaces before and after PNIPAAm modification, which include variations in pore size and temperature (below and above the LCST of PNIPAAm). From the images, the roughness of the PNIPAAm surfaces increased steadily as the pore size increased and increased significantly upon increase in temperature. For the pore size of 20 nm, the roughness factors of PNIPAAm-grafted AAO surface is 1.15 at 25°C, and increased to 1.24 at 40°C. When grafting PNIPAAm film on AAO surface with pore size of 100 nm, the roughness factors were 1.23 and 1.33 at temperatures of 25°C and 40°C, respectively. The 200-nm samples did not show a dramatic difference in roughness factor at low and high temperatures; such results reveal that changes in topography due to swelling and shrinkage of the thin polymer layer are less significant for larger pore sizes.

5.3.2.4 *Aligned carbon nanotube–enhanced thermal-responsive PNIPAAm*

Aligned carbon nanotubes (ACNTs) have significant appications in microdevices because of their superiority in easy integration into microdevices to realize functionalization. Surface modification of

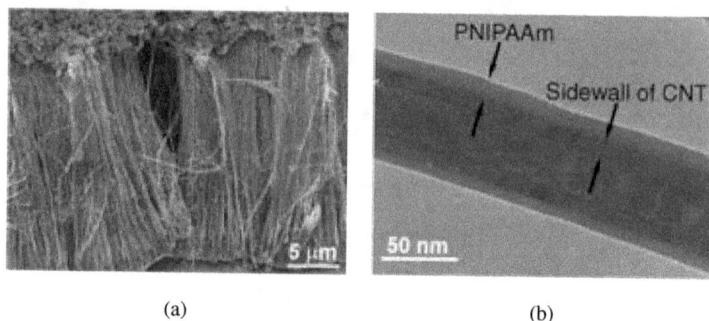

(a)   (b)

Fig. 5.20   Pictures of PNIPAAm-grafted ACNT. (a) SEM picture. (b) TEM picture.[42]

ACNTs by responsive molecules could help to realize better control over their properties, and thus may bring about more opportunities for their application. Our group have successfully grafted PNIPAAm onto ACNTs by using an SI-ATRP method (Fig. 5.20) and observed the temperature responsiveness of both the macroscopical (wettability) and the microcosmic (the diameter and rigidity of ACNTs) properties.

Figure 5.21 demonstrates the spreading behavior of a water droplet on a PNIPAAm-modified ACNT film under different temperatures. At 25°C, the superhydrophilicity with a contact angle of about 0° could be reached in about 10 seconds; at 40°C, however, the film remained superhydrophobic with a contact angle of about 150° (for about 1.5 seconds) before the water droplet started to spread to give a final CA value of 63.7 ± 1.8°. This result indicates that PNIPAAm-modified ACNT film has temperature-responsive wettability property.

Atomic force microscope (AFM) was used to investigate the temperature-induced contraction of the PNIPAAm layer on a single CNT (Fig. 5.22). It can be clearly observed that with an increase of temperature, the diameter of the nanotube reduced by about 4.3 ± 0.3 nm, which is a contraction of 21–25% in the film thickness of the PNIPAAm-modification layer compared to that of the original thickness. Meanwhile, accompanying the shrinking of PNIPAAm chains, the CNT exhibited a phase difference change from about 15.4° to 7.4° relative to the surrounding silicon substrate, when temperature

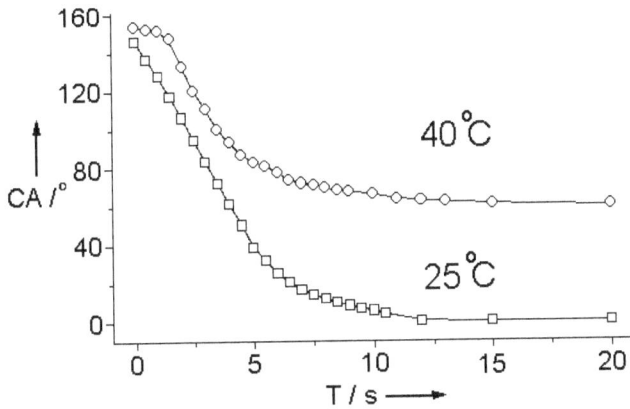

Fig. 5.21 The spreading behavior of a water droplet on a PNIPAAm-modified ACNT film under different temperatures.[42]

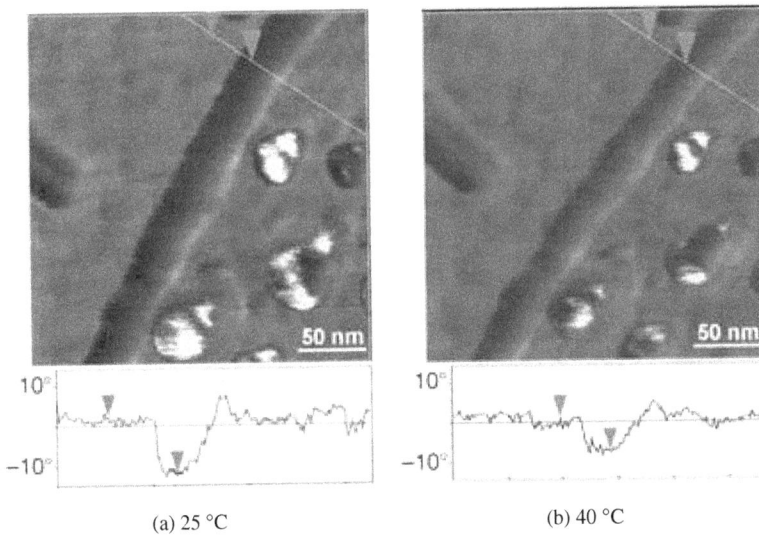

(a) 25 °C   (b) 40 °C

Fig. 5.22 AFM phase images of a single PNIPAAm-modified CNT.[42]

increased from 25°C to 40°C. As the phase difference directly reflects the rigidity of sample surface, thus, the above results indicate that not only the diameter of CNT is temperature-responsive, its rigidity is temperature-responsive as well.

5.3.2.5   *Irresponsive molecule enhanced responsive wettability of poly(N-isopropylacrylamide) film*

It has been reported that the PNIPAAm-modified surface can realize a reversibly responsive switching between superhydrophilicity and superhydrophobicity, resulting from the fact that the surface wettability can be enhanced by introducing roughness. However, in many application fields, especially in the microfluidic channel, it is hard to create roughness, so that this method is largely restricted. As we know, wettability is determined by both the chemical composition and surface roughness. In this case, we expect to realize the tunable responsive wettability on flat substrate by controlling the surface chemical composition. Based on this method, our group[43] obtained tunable thermal-stimuli responsive surface with scope from hydrophobic to hydrophilic by cooperating an irresponsive but very hydrophobic molecule, fluoroalkylsilane, with responsive molecule, PNIPAAm, as a composite film. In the experiments, different dosage ratios of polymers and silanes were examined to investigate the relationship between the dosage ratios and wettability response. The silanes used here were heptadecafluorodecyltrimethoxysilane (HTMS) and (3-aminopropyl)-trimethoxysilane (ATMS). Figure 5.23 shows the temperature-dependent curve of water CA on different

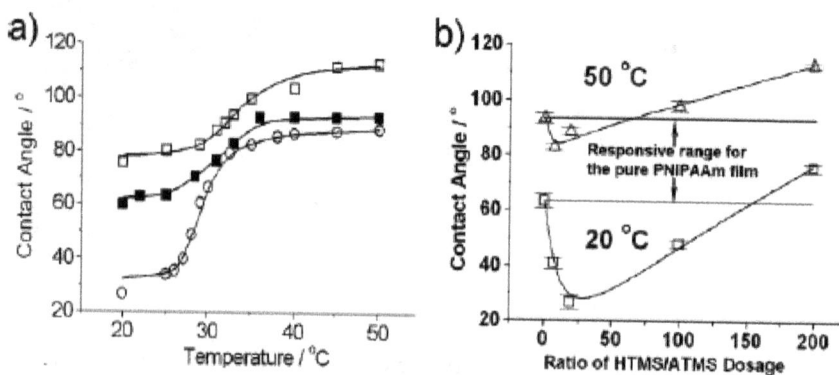

Fig. 5.23   (a) The temperature-dependent curve of water CA. (b) The CA values of different HTMS-dosage composite films. (From Ref. 43 — Reproduced by permission of The Royal Society of Chemistry.)

HTMS-dosage composite films, indicating that the water CA responsive range increases with the modification of HTMS, and more
interestingly, the CA range increases larger when the composite film
is composited by smaller HTMS dosage ratio.

This can be explained from the view of Cassie's equation, as it
anticipated that forming composite film with a lower surface energy
molecule can give rise to the increase of the hydrophobicity; however,
it is abnormal for the composite film to also bring about an increase
of hydrophilicity when measured at low temperature. On the other
hand, the contamination of the dissociative HTMS molecules on
air/water interface would lead to the reduction of the surface tension
of water drop and thus the decrease of CA, which confilcts with the
view of experiments. Thus, we consider that this effect is due to the
much looser configuration of PNIPAAm chains in the composite
films, which makes it easier for the hydrophilic groups, such as N–H
and C=O, to contact with water (Fig. 5.24). Comparing with the
extended one under low temperatures, the PNIPAAm chains may

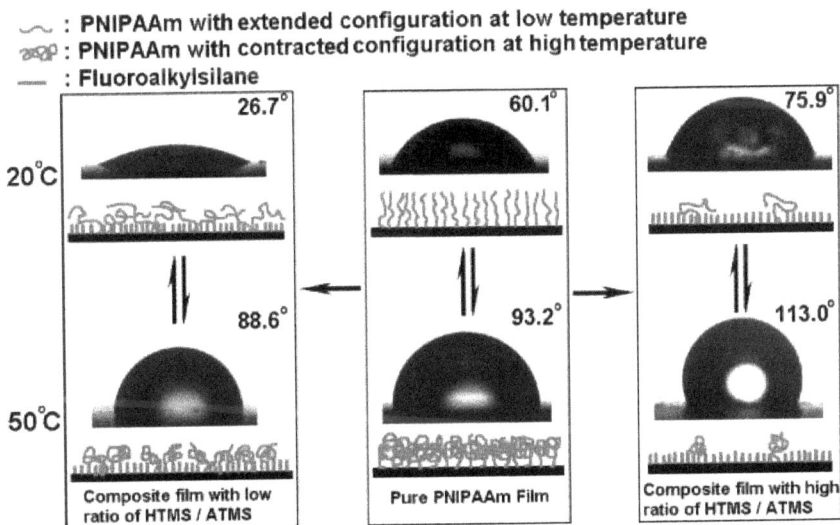

Fig. 5.24 Mechanism scheme of hydrophobicity-enhanced smooth surface modified with low surface energy material. (From Ref. 43 — Reproduced by permission
of The Royal Society of Chemistry.)

exhibit a contracted configuration under high temperatures because of the competition between the intermolecular and intramolecular hydrogen bonding. However, when the dosage ratio increases further, the surface density of PNIPAAm chains may be insufficient to cover the whole surface and part of the HTMS-modified surface may be exposed, and the upper surface of the composite film is heterogeneous. As anticipated by Cassie's equation, the CA will increase with the increase of the surface fraction of HTMS. Thus, an increase of hydrophobicity under both low and high temperatures is observed.

5.3.2.6 *Engineered thermally responsive surface patterns*

The polymer gels (hydrogels) may be responsive to a variety of environmental stimuli, such as changes in pH or temperature, or the presence of a specific chemical substrate, taking up or expelling water between the crosslinked polymer chains. These gels are being explored for numerous technological applications, particularly in the biomedical field. With the swelling or shrinking of the hydrogels, complex patterns may be generated on their surfaces. Hu and co-workers[44] reported the synthesis and controlled modulation of engineered patterns on the surface of environmentally responsive hydrogels. They modified the characteristics of a gel surface by selectively depositing another material with a mask. For example, they used sputter deposition to imprint a square array of gold thin films onto the surface of an N-isopropylacrylamide (NIPA) gel. Such periodical array can be continuously varied (altering the gel's volume) as a function of temperature or electric field. Figure 5.25(a) shows the surface array of a swollen NIPA gel film in water at 30°C. The dark area represents the gold-covered surface and the light area represents the bare gel surface. Because the spaces between the small gold squares in the gel surface are aligned in both the *x*- and *y*-directions, they can serve as slits for light diffraction. When monochromatic light passes through these slits, the diffraction pattern should consist of bright and dark spots instead of fringes (Fig. 5.25(b)). As the temperature was increased, the NIPA gel shrank owing to the volume phase transition (Fig. 5.25(c) and (d)), leading to the decrease of the array

(a)   30 °C   (b)

(c)   33.6 °C   (d)

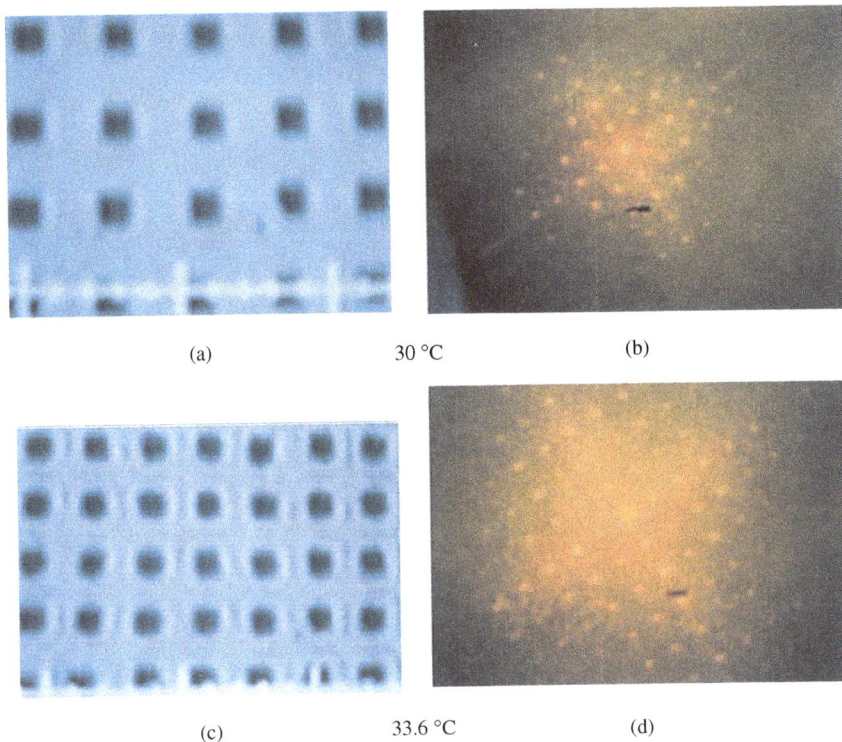

Fig. 5.25 (a) and (c): A square array of gold thin films on the surface of an (NIPAAm) gel. (b) and (d): The light diffraction pattern of the samples in (a) and (c).[44] (Reprinted by permission from Macmillan Publishers Ltd: *Nature*, copyright 1998.)

periodic constant (d). Similarly, many other structural arrays may be synthesized by using different masks, including strip, rectangular, hexagonal, and even non-periodic surface arrays. The periodic arrays generated by the "smart" gels can be used as raster sensors.

The NIPAAm pre-gel solution was deposited on the surface of the polyacrylamide (PAAM) gel. One sample was then covered by a mask of an outline map of Texas and the other a non-covered dinosaur picture. These two samples could both respond to temperature changes. At room temperature (lower than the LCST of NIPAAm), as both the PAAM substrate and the NIPAAm gel were transparent, no surface structure could be observed; but when heated to a higher

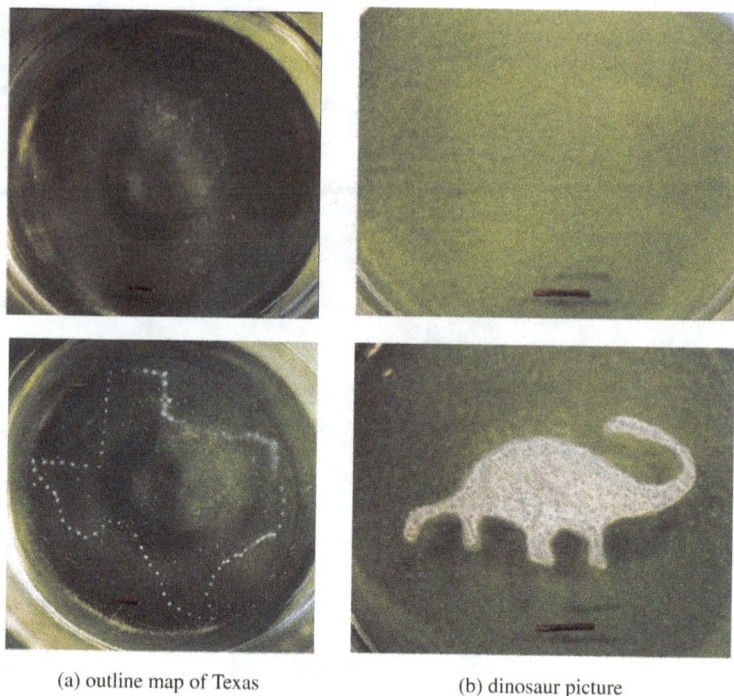

(a) outline map of Texas   (b) dinosaur picture

Fig. 5.26   The thermal-responsive arrays of NIPAAm-deposited PAAM gel (a) at room temperature; and (b) when heated to higher than 37°C.[44] (Reprinted by permission from Macmillan Publishers Ltd: *Nature*, copyright 1998.)

temperature (higher than the LCST of NIPAAm gel), the areas of polymerized NIPAAm became cloudy, while the PAAM gel remained in the transparent state. As a result, the image of the gel display could be clearly seen, and the two states could be reversibly switched (Fig. 5.26).

Except for the thermal-responsive polymers, porous materials also have the property of thermally induced contact angle change. For example, Shirtcliffe and co-workers reported that methyltriethoxysilane (MTEOS) sol-gel foams exhibited a superhydrophobic state after being heated in a furnace of 390°C and subsequently cooled to room temperature; while heated at 400°C and then cooled to room temperature, the foams became superhydrophilic (Fig. 5.27). The switch of wettability occurring in such narrow temperature range was caused

Fig. 5.27   The wettability performance of MTEOS sol-gel foams heated to different temperature. (a) Phenolphthalein in water. (b) Drop of water with brilliant blue G. (From Ref. 45 — Reproduced by permission of The Royal Society of Chemistry.)

by the cooperation of surface structure and chemical composition. On the one hand, the porous structure of MTEOS was produced by using a phase separation when dissolved in the mixture of organic solution and water. On the other hand, the increase of temperature induced the redistribution of surface organic groups, which resulted in the exposure of hydrophilic group to the surface.

5.3.3 *Photoresponsive surfaces*

5.3.3.1 *Light-induced amphiphilic surfaces of TiO$_2$*

In 1997, Fujishima's group (from Tokyo University, Japan) reported their new discovery of the photogeneration of a highly amphiphilic (both hydrophilic and oleophilic) titanium dioxide surface (Fig. 5.28).[46] This surface shows that after irradiation, water droplets spread out on the surface of film, resulting in a contact angle of 0°, and all the liquids spread across the surface upon ultraviolet irradiation (Fig. 5.29(a)). A longer storage period in dark induced a gradual increase in the water contact angle, revealing a surface wettability trend toward hydrophobicity. It is manifested as the reversible transformation of

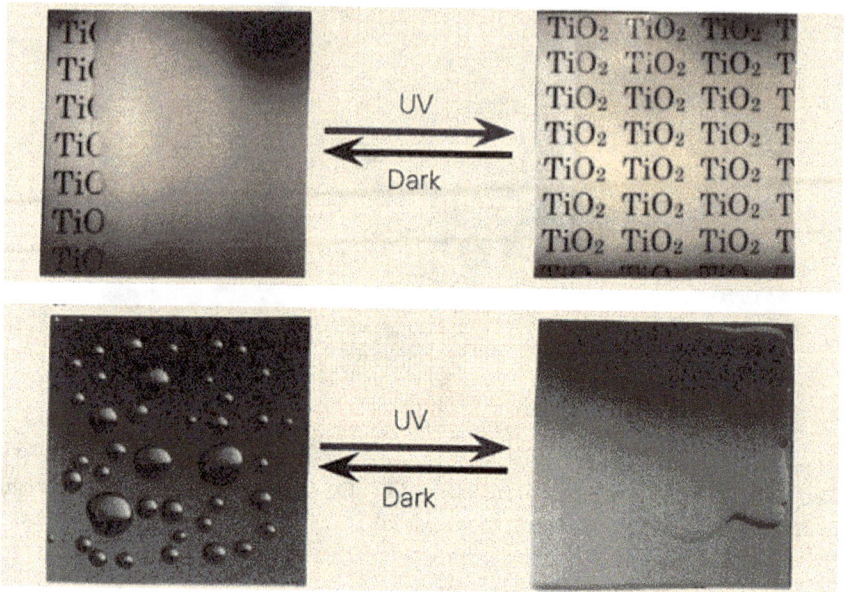

Fig. 5.28 Reversible superamphiphilic surface induced by UV illumination.[46] (Reproduced by permission from Macmillan Publishers Ltd: *Nature*, copyright 1997.)

wettability. The discovery of this magical superamphiphilic surface has remarkably facilitated progress in developing the research and applications of TiO_2.[47-52] The results showed that this kind of surface was composed of hydrophilic and oleophilic regions spanning several tens of nanometers. This formation of a microstructured composite between hydrophilic and oleophilic phases accounts for the unique feature (Fig. 5.29(b)).[53] By illumination of the hydrophobic surface of TiO_2, it results in the conversion of the corresponding Ti^{4+} sites to Ti^{3+} sites, and the formation of hydrophilic domains. These created Ti^{3+} defect sites lead to the conclusion that the nanoscale separation between the hydrophilic and the oleophilic phases accounts for the highly amphiphilic character of the TiO_2 surfaces. This kind of surface can be widely used in antifogging-, antidroplet-, anti-frosting-, self-cleaning, applications, as well as in lubrication to reduce resistance, in the acceleration of drying process, and to increase bio-affinity.[54] This unique feature of amphiphilic surface can be obtained by coating TiO_2 (no matter single crystal or

water spreading   hexadecane spreading

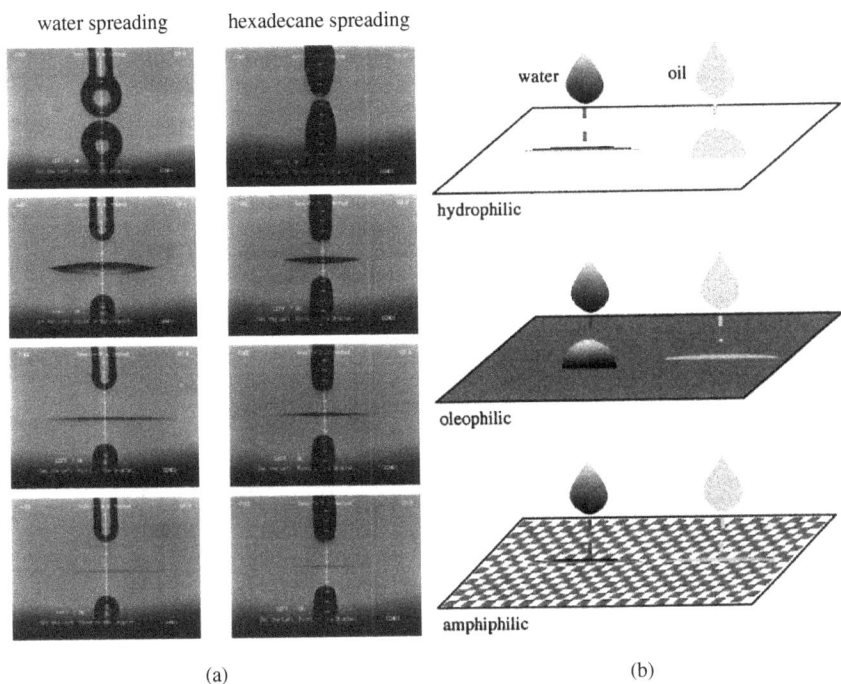

(a)   (b)

Fig. 5.29 (a) Photograph of the spreading of water and hexadecane on a UV-illuminated TiO$_2$(110) single-crystal surface.[47] (b) A nanoscale distribution of hydrophilic and oleophilic regions induced by UV illumination.[53] [Figure (a) — Reprinted with permission from Ref. 47. Copyright 1999, American Chemical Society. Figure (b) — Copyright Wiley–VCH Verlag GmbH & Co. KGaA. Reproduced with permission.]

polycrystal) film on all kinds of substrates such as glass, ceramics, plastic, metal, and polymer. This application has already been applied to many industrial systems, such as car observation mirror, window film, external tile, highway wall board, etc. The relevant development of this research will widely extend the range of applications.

Fujishima's group adopted friction force microscopy (FFM) images and Fourier transform infrared spectroscopy (FTIR) to study the formation mechanism of TiO$_2$ single-crystal highly amphiphilic surfaces. A triangular-shaped Si$_3$N$_4$ cantilever was used for FFM imaging of (TiO$_2$) rutile single crystal. Since the tip of the Si$_3$N$_4$ cantilever is hydrophilic, the interaction between the tip and a hydrophilic surface shows higher friction (bright); in contrast, the friction force between the tip and a hydrophobic surface is lower (dark). Thus, the distribution of hydrophilic and hydrophobic regions could be clearly seen by FFM. The results from Fujishima *et al.* showed that before UV illumination (Fig. 5.30(a)), no difference in contrast was observed for the FFM image or the topographic image, indicating microscopically homogeneous wettability on the surface. However, after UV illumination (Fig. 5.30(b)), a distribution of hydrophilic (bright) and oleophilic (dark) areas was clearly seen on the surface. Figure 5.30(c) shows a medium-scale FFM image, corresponding to the framed area

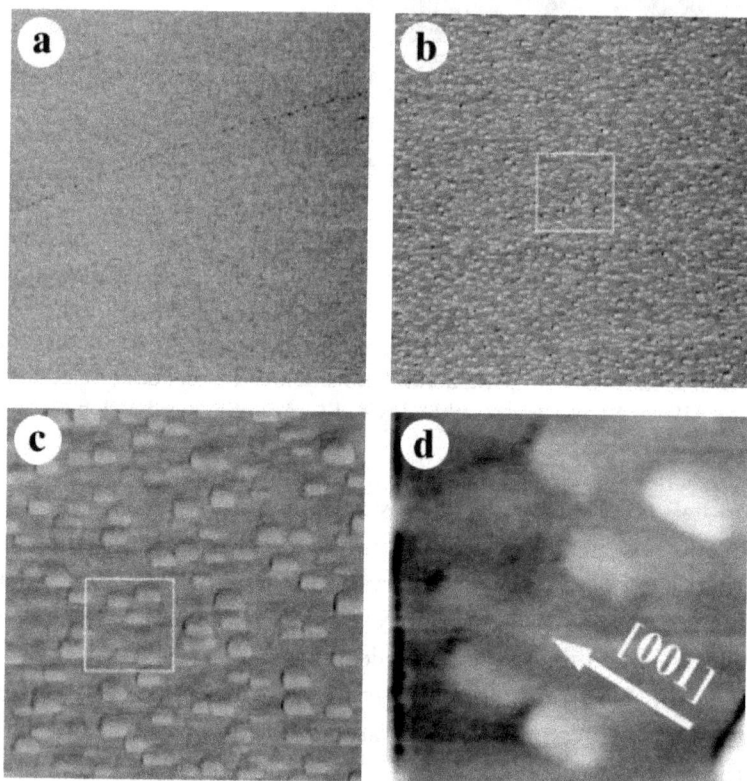

Fig. 5.30 FFM image of the rutile single-crystal TiO_2 (110) surface. (a) Before UV illumination. (b)–(d): After UV illumination.[53] (Copyright Wiley–VCH Verlag GmbH & Co. KGaA. Reproduced with permission.)

in Fig. 5.30(b), which illustrates hydrophilic domains with a regular rectangular shape in the range of 30–80 nm in size. A higher-resolution topographic image (Fig. 5.30(d)) was also acquired, demonstrating that the hydrophilic domains were higher in position than the oleophilic areas. These micro domains were considered to be the flow channels for water and oil, resembling a two-dimensional capillary phenomenon.

FFM observation showed that the micro hydrophilic domains were specially aligned along the direction of oxygen bridging sites (Fig. 5.31), because oxygen vacancies were most likely created by illumination at these sites.[47,55] The atomic coordinations at the TiO_2 surface are different from those in the bulk since the atom arrangements are truncated on the surface. This gives rise to five-coordinated Ti atoms and two-coordinated O atoms, which are more energetically reactive than the six-coordinated Ti and three-coordinated O atoms in the bulk. Via UV illumination, oxygen vacancies are most likely created at

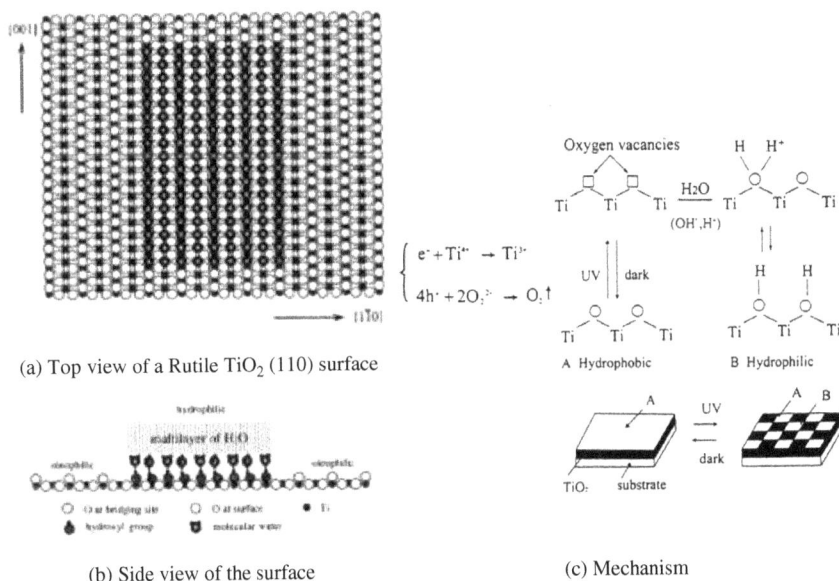

(a) Top view of a Rutile TiO_2 (110) surface

(b) Side view of the surface

(c) Mechanism

Fig. 5.31 A theoretical model for the formation of the rectangular hydrophilic domains on the initially oleophilic TiO_2 surface of a rutile (110) single crystal.[53] (Copyright Wiley–VCH Verlag GmbH & Co. KGaA. Reproduced with permission.)

the two coordinated bridging sites, resulting in the conversion of the corresponding Ti^{4+} sites to Ti^{3+} sites, which are favorable for dissociative water adsorption.[56–58] These defects presumably influence the affinity to chemisorbed water of their surrounding five-coordinated Ti sites, and thereby result in the formation of hydrophilic domains, leaving the others oleophilic. It is reasonable to consider that the rectangular structures, which correspond to the area where the dissociated water is adsorbed onto, are associated with the oxygen vacancy strings along the [001] direction. According to the topographic images, the height of the hydrophilic domains ranges from 0.4 nm to 6.0 nm, which corresponds to a monolayer or multilayer of molecular water adsorption on the chemisorbed hydroxyl groups. But the observed height might be larger than the true height since the capillary force between the hydrophilic tip and the hydrophilic region is much greater than that between the tip and the oleophilic region. The adsorption species were also confirmed by the X-ray photoelectron spectroscopy (XPS) study. It indicated that after UV illumination in the air, the created Ti^{3+} defect sites were healed by the adsorption of the hydroxyls (dissociated water) that coexisted with the additionally adsorbed molecular water; this was also confirmed by the FTIR spectrum. A gradual reversion of these domain structures was observed during storage in the dark.

5.3.3.2 *Changing of wettability of photoresponsive smooth surfaces*

The discovery of UV-induced highly amphiphilic on TiO_2 surface attracted many researchers due to its potential usage in industry.[59–67] People have been paying more attention to the problems of surface stability, light and its intensity, and practicality. There were a series of researches on them.

Common TiO_2 films keep their superhydrophilicity for a short period of time, and it is unstable. After UV illumination, the superhydrophilicity can only be kept within ten-odd hours; a gradual decrease of hydrophilicity of film surface is observed with the passing of time, and finally the superhydrophilicity ability is lost. These were obtained after the treatment of plasmas, heat, or ultrasonic.[68–73] SiO_2 could be added into TiO_2 films, which form the composite SiO_2/TiO_2 films[74–76]; the construction of the nanostructure of the surfaces of TiO_2 films[77–83] can improve the maintenance time of the hydrophilic property, hence enhance the stability.

Usually, the induction of hydrophilicity film is restricted to UV illumination. However, UV radiation had harmful effect on human health. Thus, the application of superhydrophilicity films is limited. In order to realize effective utilization of solar light, the problem of the spectral response range of TiO_2 should be resolved.[84,85] The results show that catalytic and hydrophilic properties of TiO_2 under the visible light can be improved by doping with transition metal ions or nitrogen.[86–94]

Antifogging and self-cleaning properties of TiO_2 films may solve many common interface problems that bother people's living. Recently, great attention has been paid to such superhydrophilic TiO_2 films.[95–98] There have already been many methods to produce TiO_2 films with excellent performances.[99–101] For example, the groups of Fujishima[102] and Cohen[103] reported the multifunctional nanoporous TiO_2 films, which exhibited both antifogging and antireflection properties. Moreover, water droplet contact angle could reach less than 5° within 0.5 s or less.

Since Fujishima's group reported the photogeneration of a highly amphiphilic (both hydrophilic and oleophilic) titanium dioxide surface, the study of wettability of photoresponsive surfaces arouses more interest from researchers. The typical photosensitive materials include semiconductor oxide (such as TiO_2 and ZnO), photoinduced isomerization of azobenzene, spiropyran and its derivatives, etc. The results show that after UV illumination, the water contact angle of the ZnO surface changes from 109° to 5° (Fig. 5.32(a)), while the water contact angle of the TiO_2 surface changes from 54° to 0° (Fig. 5.32(b)).[104]

Ralston's group[105] reported that the thymine dimers attached to solid substrates underwent isomerization when irradiated by UV light with wavelength at 280 nm and 240 nm. That leads to the reversible change of pKa, and thereby the change in contact angle. Firstly, thymine-terminated thiols self-assembled monolayers on gold, and

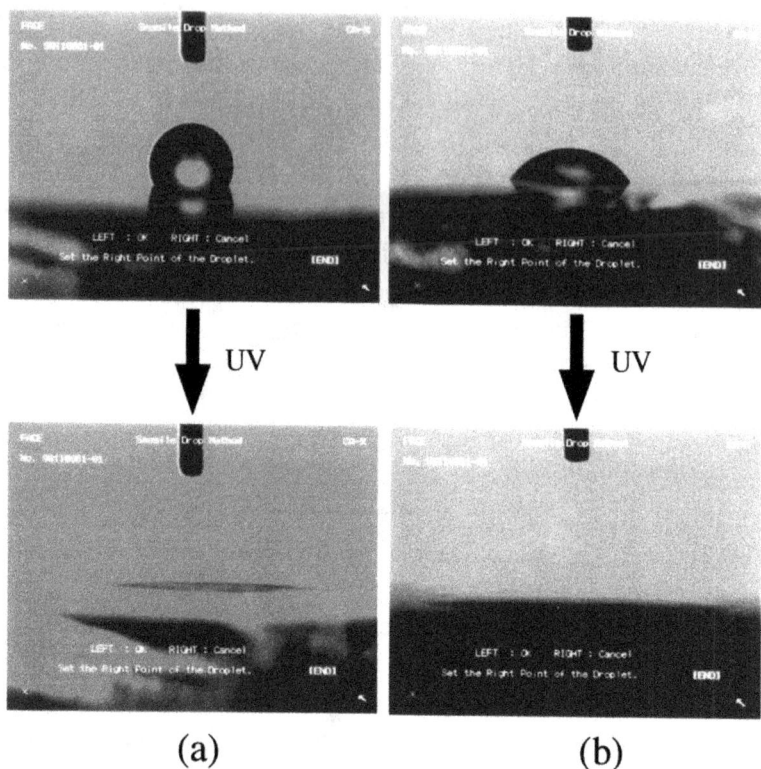

Fig. 5.32 Changes of water contact angles before and after UV illumination: (a) ZnO, and (b) TiO$_2$. (Reprinted with permission from Ref. 104. Copyright 2001, American Chemical Society.)

gave the largest reversible photoinduced contact-angle change (26°) with respect to water at pH 11.1. Moreover, these changes are reversible under illumination. Compared with the spin-cast films, this change can be attributed to the suitable optimum molecular ordering. The thymine monomer underwent dimerization upon irradiation, which induced a change in surface charge (ionized monomer to non-ionized dimer), and gave rise to the increase of contact angle.

However, the extent of photoresponsive wettability on smooth surfaces is small. In order to improve the contact-angle changes before and after illumination, our group[106] reported the wettability changes of azobenzene polymer LB films containing fluorine atom

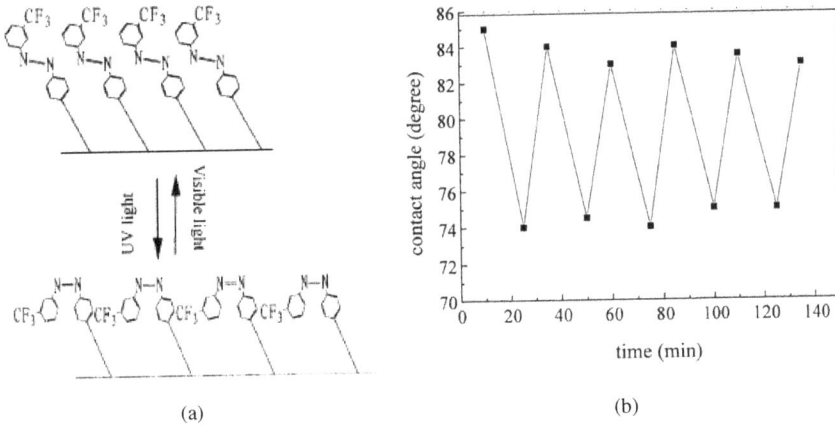

(a)   (b)

Fig. 5.33 (a) Model of the structural change of a single-layer LB film of the polymer accompanied by photoisomerization. (b) Reversible wettability for a 11-layer LB film on glass. The values of the contact angles gradually decrease after several cycles of irradiation. (Reprinted with permission from Ref. 106. Copyright 2001, American Chemical Society.)

with low surface energy before and after illumination. The results demonstrated that the polymer films had reversible change of the wettability under alternate UV (365 nm) and visible-light (436 nm) irradiation. However, the largest photoinduced change of contact angle is only about 10° (Fig. 5.33). It shows that the photoresponsive reversible change of the wettability could greatly improve by using photoresponsive materials to create a rough structure or modify a rough surface.

5.3.3.3 *Rough structure-enhanced change of photoresponsive wettability*

(1) Photoinduced reversible "switch" of photoresponsive semiconductor oxide from superhydrophobicity to superhydrophilicity

Our group has successfully obtained UV photoresponsive reversible superhydrophobicity-to-superhydrophilicity transition "Switch". Firstly, the aligned ZnO nanorods were synthesized by hydrothermal method (Fig. 5.34(a) and (b)).[107] The nanorods grew almost perpendicularly onto the substrates, the surface of the films being the (001) plane of the nanorods. This was confirmed by X-ray diffraction (XRD) with a remarkably enhanced (002) peak. Compared with the random orientation of other ZnO nanocrystal films, these films had the lowest surface free energy. Thus, the superhydrophobicity of the nanorod films was obtained, and it showed a spherical water droplet with a water CA of 161.2°. The water drop rolled off when the surface sloped (Fig. 5.34(a)). Under UV illumination, the wettability of ZnO nanorods film changed from superhydrophobicity to superhydrophilicity. The water droplet spread out on the film, resulting in a CA of about 0°, and penetrated the surface of aligned

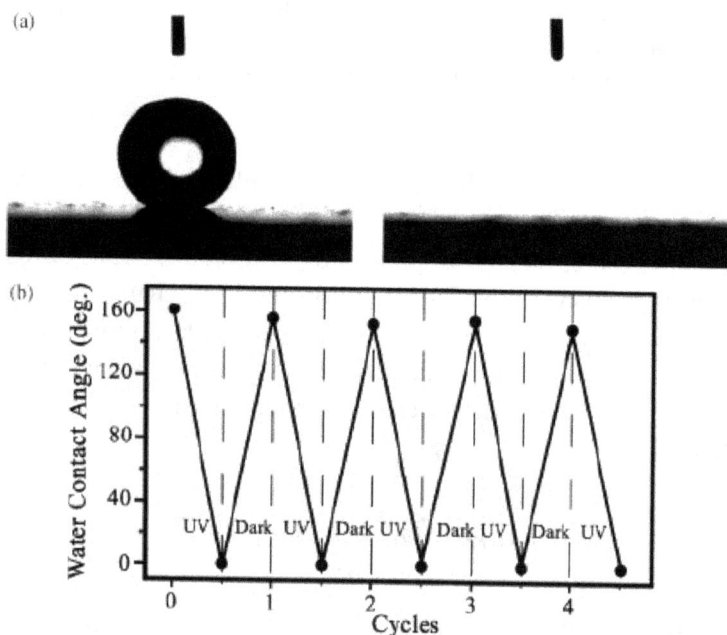

Fig. 5.34 The influence of UV illumination on aligned ZnO film. (a) Photographs of water droplet shape on the aligned ZnO nanorod films before (left) and after (right) UV illumination. (b) Reversible superhydrophobic-superhydrophilic transition of the as-prepared films under the alternation of UV irradiation and dark storage. (Reprinted with permission from Ref. 107. Copyright 2004, American Chemical Society.)

materials (Fig. 5.34(a)). When the UV-irradiated films were placed in the dark, the surface reconverted to its original state of the super-hydrophobicity. The wettability which shows superhydrophobicity and superhydrophilicity at different conditions could be reversibly switched by alternation of ultraviolet (UV) irradiation and dark storage (Fig. 5.34(b)). This kind of interfacial nanomaterial is like a block of carpet, and the diameters of the "surface hair" are only 1/1000 of that of human hairs. The water-repellent and self-cleaning properties of this "carpet" are attributed to its property of superhy-drophobicity. After UV illumination, the superhydrophilicity of the materials was again obtained, and water could be kept in the rough nanostructure. This "switch" process had repeated several times, and good reversibility of the surface wettability was observed.

Our group also reported that the cooperation of the micro/nanoscale hierarchical surface structures of titanium dioxide (TiO_2) film with the lowest surface energy of the crystal face (001) was pre-pared by low-temperature hydrothermal method (Fig. 5.35(b)),[108] and the transparent and conductive SnO_2-aligned nanorod films (Fig. 5.35(c)).[109] A "light switch" of smart surface with wettability that can be switched between superhydrophobicity and superhy-drophilicity has been realized by the alternation of UV-irradiation and dark storage. This property could be applied to new dynamic optical controlled self-cleaning materials.

In addition, we also directly fabricated a superhydrophobic ZnO film with hierarchical surface structure by the chemical vapor deposition (CVD) method. Its wettability switching between super-hydrophobicity and superhydrophilicity was realized by UV illumination.[110] The results showed that different catalysts were pre-sputtered to control the surface morphology and property of the ZnO film. The film catalyzed by the metal Ni exhibited the sub-micrometer-scaled ZnO slices with the sizes from 130 to 360 nm (Fig. 5.36(a)), and the water CA was measured as 110.6°, exhibiting hydrophobicity. When Au was used as the catalyst, the film exhibited hierarchical structure composed of the sub-micro/nanostructures (Fig. 5.36(b)), and many block-shaped protuberances ranging from 320 to 560 nm were distributed on the surface. ZnO nanopapillae

(a)

(b)

(c)

Fig. 5.35 The SEM of semiconductor oxide surface of the photoresponsive reversible superhydrophobicity-to-superhydrophilicity transition "Switch". (a) ZnO film.[107] (b) TiO$_2$ film.[108] (c) SnO$_2$ film.[109] (Reprinted with permission from American Chemical Society (Ref. 107), Wiley–VCH Verlag GmbH & Co. KGaA (Ref. 108), The Royal Society of Chemistry (Ref. 109).)

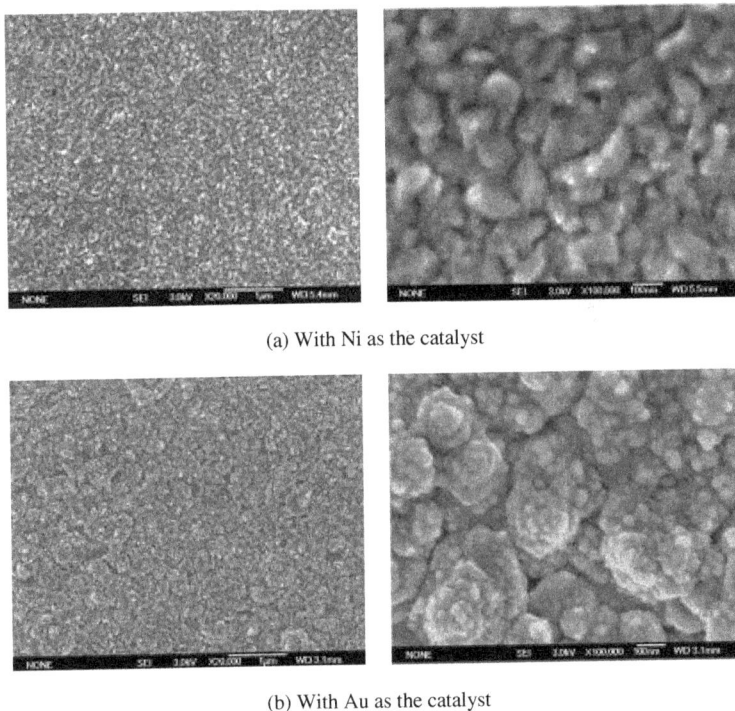

(a) With Ni as the catalyst

(b) With Au as the catalyst

Fig. 5.36   SEM images of the prepared ZnO films by CVD technique. (Reprinted with permission from Ref. 110. Copyright 2004, American Chemical Society.)

with sizes from 21.6 to 41.0 nm were inserted very densely on these protuberances. The surface with these hierarchical structures gave the high water CA of 164.3°, exhibiting superhydrophobicity. After UV illumination, both films became superhydrophilic (CA < 5°). The water droplet spread immediately upon the films. After depositing the UV-irradiated film in dark for several days, the superhydrophobicity could be reversed again. This research result was introduced in detail by *Nature Materials* journal which believed that the microstructure would play a more important role than surface composition in reversible wettability.

Our group also directly fabricated a superhydrophobic tungsten oxide (WO_3) film with nanostructure by the electrochemical deposition method (Fig. 5.37(a)). The prepared inorganic-oxide film showed

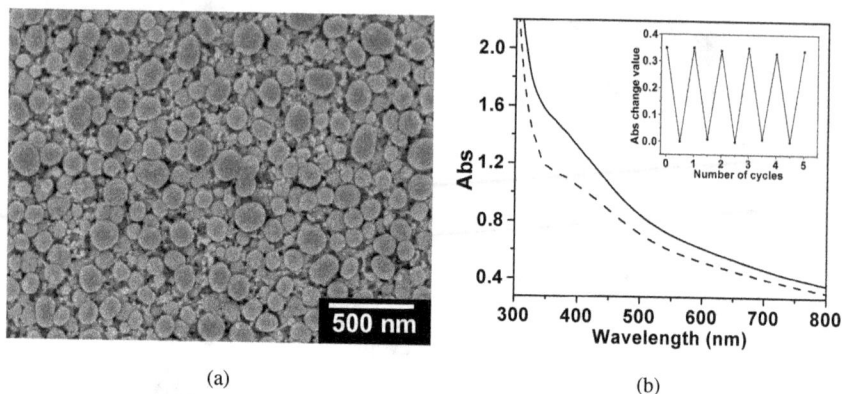

(a) (b)

Fig. 5.37 (a) SEM of WO$_3$ films fabricated by the electrochemical deposition method. (b) Absorption spectra of an electrodeposited tungsten oxide film before (solid line) and after (dashed line) irradiation with UV light. The insert shows the photochromic switching of the absorption change during consecutive cycles of UV irradiation and storage in the dark.[111]

both reversible wettability conversion between superhydrophobic and superhydrophilic states and photochromism upon alternating UV irradiation and storage in dark.[111] In three electrodes systems, a platinum electrode was used as the counter electrode. A piece of ITO glass was used as the cathode, and the reference electrode was an Ag/AgCl electrode. Aqueous Na$_2$WO$_4$ solution was used as the electrolyte for electrodeposition of tungsten oxide films along with oxalic acid to adjust the solution pH. Figure 5.37(a) shows a rough, brown film which exhibits a pebble-beach-like morphology made up of many nanoprotuberances with diameters in the range of 40–350 nm.

The surface wettability of the as-prepared rough film was evaluated by water CA measurements. The CA on the film was 151.3 ± 2.9°, indicating that the surface was superhydrophobic. After UV irradiation, surface wettability changed to a superhydrophilic state. Storage of these irradiated films in the dark for a period of time caused them to become superhydrophobic again, so that a switch of surface wettability from superhydrophobic to superhydrophilic was realized. Since tungsten oxide is an excellent photochromic material, its photochromic behavior was also studied during the wettability conversion

(Fig. 5.37(b)). In addition, a good reversibility was observed for many cycles of coloration and decoloration (Fig. 5.37(b)). Hence, a photo-stimulated dual functional surface switch was realized.

Unlike the above principle, by making use of the photocatalysis of ZnO, Fujishima' group[112] found the wettability change of the composite surface of fluoroalkylsilane (FAS)-modified ZnO film after UV irradiation. Moreover, UV-created micropattern was more stable due to the hydrophilic region originated from the decomposition of FAS. Firstly, they prepared ZnO nano-columnar surface by cathodic electrodeposition. Subsequently, it was modified with a monolayer of FAS, and the CA of water on the surface was measured as 167°. Water CA on the FAS-modified ZnO columnar film showed negligible change after storage over half year, but was found to be reduced to less than 5° after this film was irradiated. It could be explained by the photocatalytic activity of ZnO which cleaved the alkyl chain of FAS. With the cleavage of the fluoroalkyl chain, the hydrophobic FAS monolayer was converted to a layer of hydrophilic silica. Figure 5.38 shows the optical micrograph of water droplets

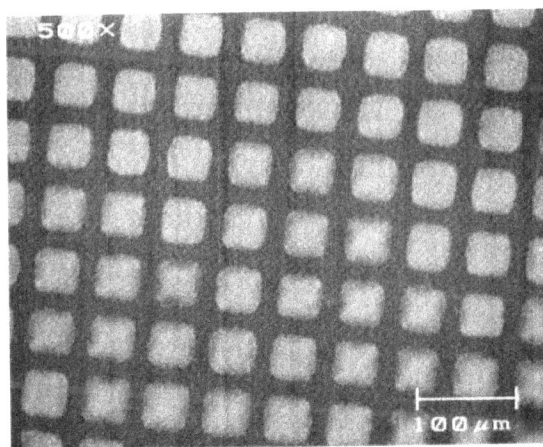

Fig. 5.38 Optical micrograph of an ultrahydrophobic–ultrahydrophilic micropattern prepared by the irradiation of UV light through a photomask. Water selectively dispersed on the ultrahydrophilic area so that sharp contrast could be observed in the picture. (Reprinted with permission from Ref. 112. Copyright 2004, American Chemical Society.)

on an ultrahydrophobic/ultrahydrophilic micropattern fabricated by UV light lithography. The Cu grid acted as a photomask. The bright area was filled with water, corresponding to the ultrahydrophilic area, while the ultrahydrophobic area looked darker by the optical microscopy because the film was highly scattering. This hydrophilic region could act as a microvessel for water confined by the surrounding ultrahydrophobic walls.

Meng et al.[113] reported that the superhydrophobic ZnO nanowire arrays were synthesized by a simple vapor–solid process. By oxygen plasma treatment or annealing at O_2 atmosphere, the surface turned to superhydrophilic due to the O_2 atmosphere which had excess oxygen. The contact angle decreased with increasing time of oxygen plasma treatment. However, with increasing annealing temperature, the contact angle grew larger. After being stored in the dark at ambient atmosphere for 108 h, the wettability gradually reconverted to its original hydrophobic properties. Yin et al.[114] reported that a novel superstructure of the film including nanorod, nanoscrews, and nanodisks, etc. by a solution route under mild reaction conditions with simple methods (Fig. 5.39). In detail, 1 mM of a Zn^{2+}-hexamethylenetetramine mixed solution was kept at 95°C for a desired aging time to get different structure. Low concentrations and long aging-times preferentially formed nanoscrew and nanodisk structure, and different superstructures possessed different hydrophilicity (93°, 165° and 0°). Han et al.[115] reported that zinc oxide (ZnO) surfaces with controllable structures (i.e. microstructure, nanostructure, and micronanobinary structure) had been created by controlling pH at <4 or >10.5 in the Zn (gray) + H_2O_2 reaction. The resulting surface showed superhydrophobicity. It was found that the water contact angle of the surface with micronanobinary structure was greater than that of nanostructure, which in turn was greater than the microstructure. Theoretical analysis completely agreed with the experimental results.

Schmuki et al.[116] reported the wetting behavior of nanotubular TiO_2 layers (π-TiO_2) grown by electrochemical anodization of Ti. The prepared p-TiO_2 layers showed a superhydrophilic behavior, but after being covered by octadecylphosphonic acid monolayer, they showed a superhydrophobic behavior.

(a)

(b)

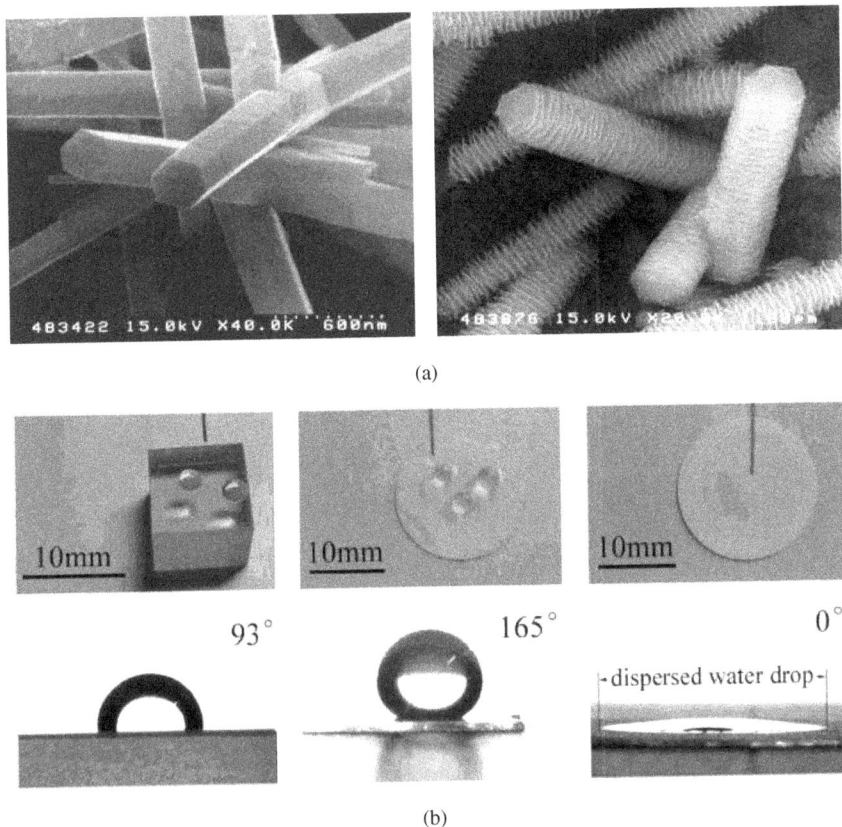

Fig. 5.39   (a) SEM photographs of ZnO films with different structures on glass substrates prepared by heating a Zn^{2+}-HMT aqueous solution at 95°C: (1) Nanorods, (2) Nanoscrews. (b) Photographs of water drops on various surfaces: (1) ZnO single crystal (001) face, (2) Superhydrophobic surface of ZnO film with nanorod superstructure, (3) Superhydrophilic surface of ZnO film with nanoscrew superstructure. (From Ref. 114 — Reproduced by permission of The Royal Society of Chemistry.)

(2) Photoresponsive surface of azobenzene

Azobenzene derivatives exhibit *trans–cis* photoisomerization. With UV irradiation, azobenzene underwent the photoelectrochemical conversion reaction and changed from *trans* (stable) to *cis* (unstable) conformation. When the irradiation was discontinued, azobenzene underwent the reverse reaction i.e. from *cis* (unstable) to *trans* (stable). The process of the reverse reaction could be accelerated by visible light illumination ($\lambda > 400$ nm).[117]

Our group[118] fabricated a photo-switched azobenzene monolayer on rough substrates by a simple electrostatic self-assembly technique, forming an azobenzene monolayer on the laser-etched rough substrate. The transition from being superhydrophobic to being superhydrophilic was obtained by light irradiation and a large reversible CA change of about 66° (Fig. 5.40). The surface of inverse opal was modified by a self-assembled azobenzene monolayer, and the ordered monodisperse air spheres throughout the inverse opal contributed to the control of both structural color and the response of wettability.[119]

(3) Photoresponsive surface of spiropyran

Picraux's group[120] reported that surface roughness could be an effective tool for the amplification of stimulus-induced contact angle switching. A polished silicon wafer bearing random silicon nanowires was modified with the photoinduced isomerization spiropyran-containing

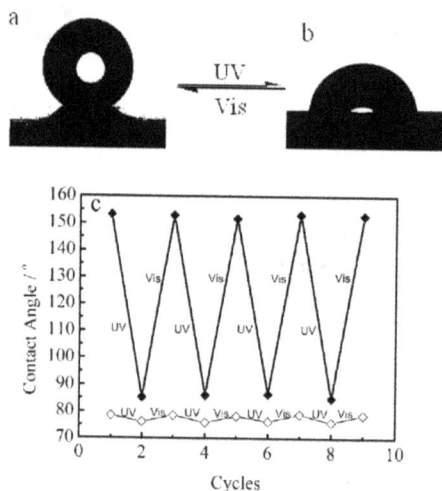

Fig. 5.40   Reversible wettability transition of photoresponsive monolayer by UV and visible light irradiation. (From Ref. 118 — Reproduced by permission of The Royal Society of Chemistry.)

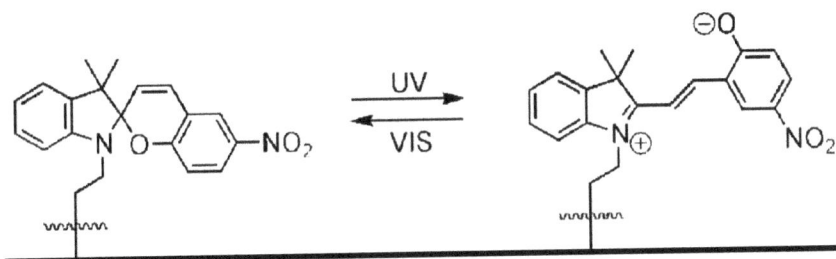

Fig. 5.41   Switching of surface-bound spiropyran between nonpolar (left) and polar (right) forms under visible and UV irradiation, respectively. (Reprinted with permission from Ref. 120. Copyright 2004, American Chemical Society.)

hydrophobic monolayer. Under visible light irradiation, the spiropyran was in a closed, hydrophobic form, whereas UV irradiation converted the spiropyran to a polar, hydrophilic form, reducing the contact angle (Fig. 5.41). Under visible irradiation, the surface of silicon nanowires had the superhydrophobic property.

The average contact angle of the smooth surface under UV irradiation was 12° lower than that under visible irradiation. While on the nanowire surface, this light-induced contact-angle change increased to 23° due to the amplifying effect of the roughness structures (Fig. 5.42). In addition, under visible irradiation of the spiropyran-coated surfaces, the water contact angle hysteresis was measured to be 37° on the smooth surface, whereas on the nanowire surface, a significantly lower value of only 17° was observed. On the smooth spiropyran-coated surface, the advancing water contact angle under UV irradiation (110°) was higher than the receding water contact angle under visible irradiation (85°). This did not fulfill the criterion for liquid motion, and it was found that water drops on the smooth surface could not be driven by light. However, on the spiropyran-coated nanowire surface, the advancing water contact angle under UV irradiation (133°) was lower than the receding water contact angle under visible irradiation (140°). Accordingly, when they applied a UV-visible light gradient across water drops sitting on the nanowire surface, the drops moved toward the UV end of the gradient. Control experiments performed on drops sitting on nanowire

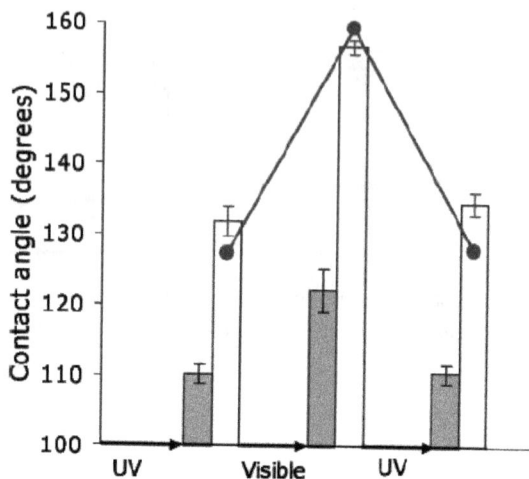

Fig. 5.42 Advancing contact angle changes on the smooth (gray bars) and rough (white bars) photoresponsive surfaces after UV and visible light irradiation. (Reprinted with permission from Ref. 120. Copyright 2004, American Chemical Society.)

surfaces coated with a hydrophobic layer without the spiropyran did not result in any drop motion. Therefore, it can be concluded that the motion of the water droplets on the photoresponsive nanowire surface was due to the roughness-magnified light-induced switching of surface energy of spiropyrans coupled with the lower contact angle hysteresis of the superhydrophobic surface. This research result, for the first time, proved that we can realize water movement on surface by light irradiation, or along the microfluidic channels, to the expected position for analysis. Additionally, according to this research direction, it also points the way to study the effect of electric filed, the solution temperature and acidity for amplifying the change of surface.

The phenomenon of photonic control of water movement will be widely applied to analytical chemistry and drug research. It will also have a profound impact on the microfluidic chip technology. Photonic control of water movement will not destroy electric field. It can move the bubble which can cause the protein denaturation, or move the microscopic mechanical pump. This will significantly improve the development of the microfluid chip device. Another potential application is to reduce the amount of enzyme and protein which is used in the detection of drug development. Usually, the manufacturing and purification of these candidate drugs are time consuming and inefficient. The microfluid chip device can reduce the amount of cell, DNA and protein which is used in the detection of the effects of these candidate drugs. Thus, these trace candidate drugs can bind their targets and be recorded. In this way, the time is reduced, while the detection is performed as much as possible.

5.3.3.4 *The application of photoresponsive surfaces — light-driven motion of liquids*

Ichimura *et al.*[121] reported that a smooth solid surface modified with a crown conformer of ocarboxymethylated calix[4] resorcinarene (CRA-CM, Fig. 5.43(a)) to assemble a photoresponsive monolayer. Asymmetrical photoirradiation caused a gradient in surface free energy due to the photoisomerization of surface azobenzenes, leading to the directional motion of the droplet. The octylazobenzene units in their trans-state would be stretched out to be exposed to the air. UV-vis spectral measurements revealed that photoirradiation of the monolayer with UV (365 nm) light results in the formation of ~90% of cis isomer with a higher dipole moment; hence, the outermost surface of a UV-exposed CRA-CM monolayer is likely terminated by the polar cis-azo groups, leading to an increase in surface free energy. Photoirradiation of the cis-rich surface with blue (436 nm) light causes the cis isomer to reverse into the trans-isomer. Thus, liquids can be driven on the surface by the photoinduced isomerization property.

Figure 5.43(b) shows the lateral photographs of light-driven motion of an olive oil droplet on a silica plate modified with CRA-CM. Controlling spatial photoirradiation caused a gradient in surface free energy due to the photoisomerization of surface. Upon asymmetrical

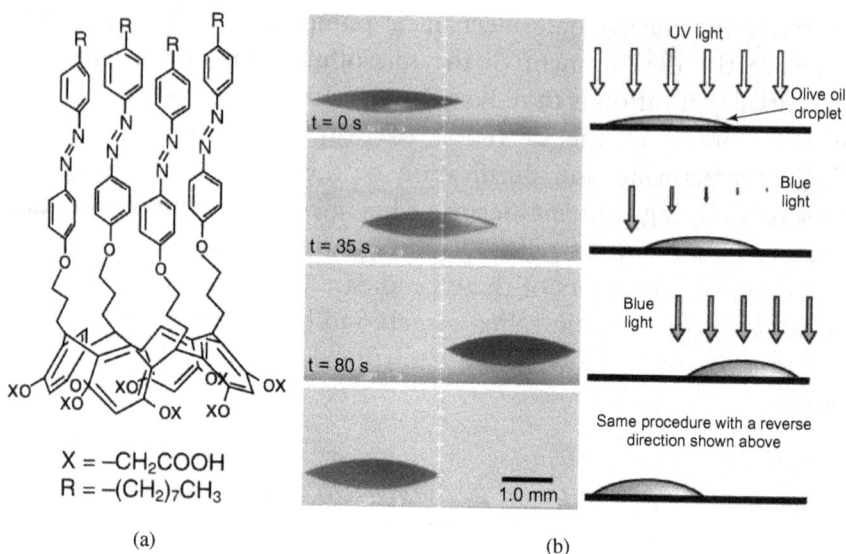

X = –CH₂COOH
R = –(CH₂)₇CH₃

(a)   (b)

Fig. 5.43   Lateral photographs of light-driven motion of an olive oil droplet on a silica plate modified with CRA-CM. (a) Macrocyclic amphiphile tethering photochromic azobenzene units (CRA-CM). (b) Lateral photographs of light-driven motion of an olive oil droplet on a silica plate modified with CRA-CM. (From Ref. 121. Reprinted with permission from AAAS.)

irradiation with blue light perpendicular to the surface, the droplet was moving along the direction of high surface energy. The sessile contact angles were changed from 18° to 25°. The moving of the droplet was constantly maintained by moving the light beam. To stop the movement of the droplet, they irradiated the photoresponsive surface with a homogeneous blue light. Moreover, the moving direction of the droplet was controllable by varying the direction of the photoirradiation, and the velocity of the droplet intrinsically relied on the intensity and gradient of the light.

No deterioration was observed even after more than 100 cycles of the light-driven motions. This photochemical method can position liquids. When an NPC-02 (a binary mixture of 4-propyl-4'-ethoxy- and 4-propyl-4'- butoxy-phenylcyclohexanes) droplet was exposed to alternating irradiation with asymmetrical UV and blue light, the droplet migrated stepwise to capture a glass bead 0.5 mm in diameter (Fig. 5.44). It was possible to convey the bead by the spatially controlled alternating irradiation, and the migration of NPC-02 placed in a glass tube can realize the transportation of micro liquid droplet.

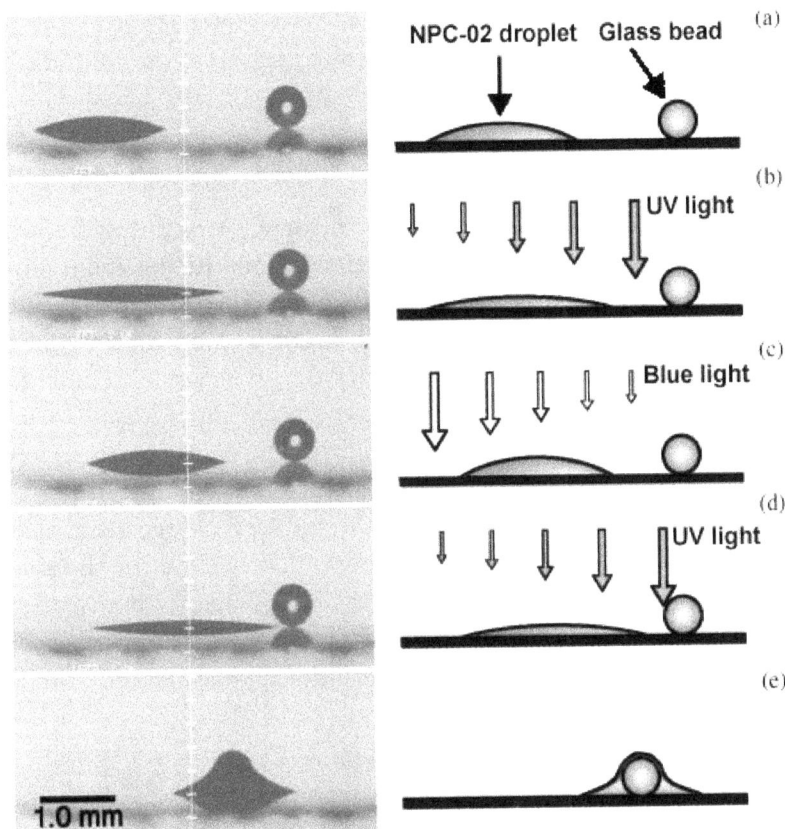

Fig. 5.44 Dynamic process of light-driven motion of an NPC-02 droplet to capture a glass bead. (a) Light-driven motion of an NPC-02 droplet on a CRA-CM-modified plate to capture a glass bead. (b) Followed by UV light irradiation at the right edge of the droplet to cause an asymmetrical spreading. (c) Subsequent irradiation with blue light at the left edge resulted in dewetting, leading to the displacement of the droplet. (d) and (e): The repetition of this stepwise photoirradiation resulted in the approach of the droplet to the bead, which was finally captured by the liquid. (From Ref. 121. Reprinted with permission from AAAS.)

5.3.4 *Electric field-induced wetting*

5.3.4.1 *Electrowetting*

Electrowetting (voltage-induced wetting effect) is that the contact angle of a droplet on a surface can be reduced by applying an electric field between the conducting liquid and a counter electrode underneath the liquid. Based on electrowetting, a droplet on a surface can be manipulated and controlled.

(1) Electrowetting theory

Verheijen and Prins[122] derived a model for electrowetting from the principle of virtual displacement, which includes the possibility that a charge is trapped in or on the wetted surface. They also discussed the experiment by this model. Reversible electrowetting for an aqueous droplet on an insulating layer of 10-μm thickness was experimentally shown in Fig. 5.45. A highly fluorinated layer impregnated with oil was coated on the insulator, providing a contact angle hysteresis lower than 2°. They analyzed the data with their model, and found that until a threshold voltage of 240 V, the induced charge remained in the liquid and was not trapped. For potentials beyond the threshold, the wetting force and the contact angle saturate, in line with the occurrence of trapping of charge in or on the insulating layer. The data were independent of the polarity of the applied electric field, and of the ion type and molarity.

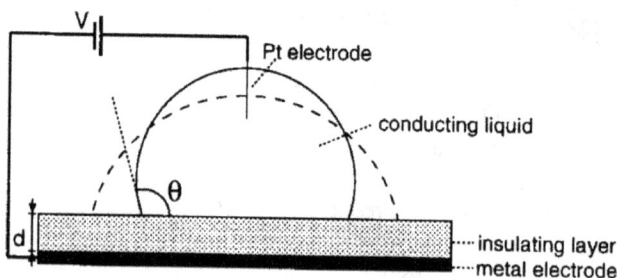

Fig. 5.45 Schematic diagram of an electrowetting experiment. (Reprinted with permission from Ref. 122. Copyright 1999, American Chemical Society.)

Based on the principle of virtual displacement, the two cases that with trapped charge and without trapped charge were discussed respectively, and the corresponding expression of the electrowetting force was obtained.

(a) No trapped charge

A droplet spreads until it has reached a minimum free energy, which is determined by cohesion in the liquid and adhesion between the liquid and the surface. In general, the energy required to create an interface is given by γ, the surface tension (N/m) or surface free energy (J/m^2). In the case of an applied potential, the electric charge distribution change at the liquid/solid interface changes the free energy.

The researchers define their thermodynamic system as the droplet, the insulating layer, the metal electrode, and the voltage source. Throughout the entire derivation, they assumed that the system was in equilibrium at constant potential V. They focused on the change in free energy due to an infinitesimal increase in the base area of the droplet on the solid surface, surrounded by vapor. When a potential V is applied, a charge density σ_L builds up in the liquid phase and induces an image charge density σ_M on the metal electrode. Figure 5.46 showed the edge of a droplet and its virtual displacement.

An infinitesimal increase of the base area dA resulted in a contribution to the free energy from the surface energies as well as an energy contribution due to the additional charge density dσ_L in the liquid and its image charge density dσ_M on the metal electrode. The voltage source performed the work, dW_B. The free energy (F) of the system can be written in differential form:

$$dF = \gamma_{SL}dA - \gamma_{SV}dA + \gamma_{LV}dA\cos\theta + dU - dW_B, \qquad (5.3)$$

where U is the energy stored in the electric field between the liquid and the counter electrode. The parameters γ_{SL}, γ_{SV} and γ_{LV} are the free energies of the solid/liquid, solid/vapor, and liquid/vapor interface, respectively, for the situation in the absence of an electric field. The contact angle, θ, is the angle between the liquid/vapor interface

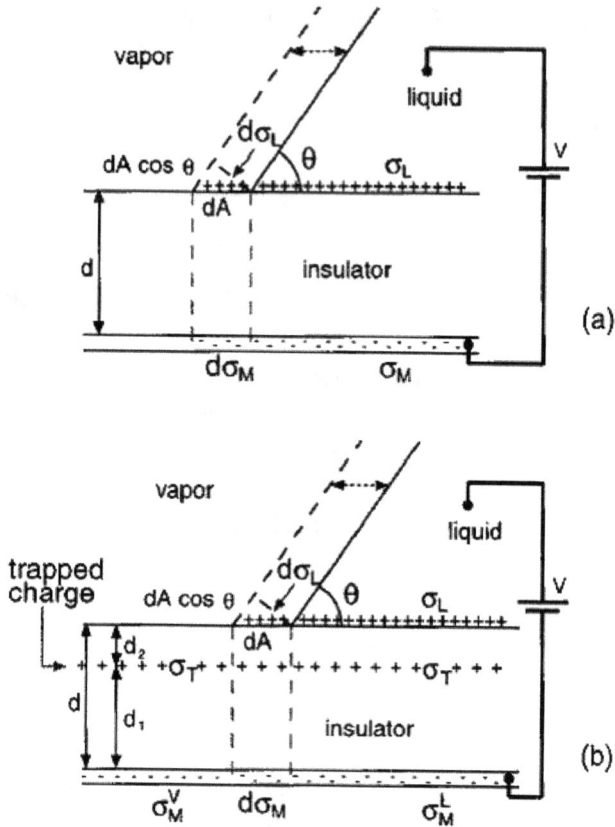

Fig. 5.46   (a) Schematic diagram of the virtual displacement of the contact line in the presence of a potential across the insulator. (b) The virtual displacement of the contact line in the presence of a sheet of trapped charge. (Reprinted with permission from Ref. 122. Copyright 1999, American Chemical Society.)

and the solid/liquid interface at the contact line. Mechanisms for energy dissipation, which may cause contact-angle hysteresis, are not considered.

Without an externally applied voltage, $dU = dW_B = 0$. When $dF/dA = 0$, the minimum free energy was found, relating the surface energies to the contact angle. This equation was obtained by Young:

$$\gamma_{LV}\cos\theta = \gamma_{SV} - \gamma_{SL}. \tag{5.4}$$

For a non-zero potential, the energy of the charge distribution has to be included. In Fig. 5.46(a), the droplet with charge density σ_L was at constant voltage V, while the metal electrode with charge density $\sigma_M = -\sigma_L$ was at ground potential. The electrostatic energy per unit below the liquid was given by

$$\frac{U}{A} = \int_0^d \frac{1}{2} \mathbf{E}\mathbf{D} \ dz, \tag{5.5}$$

where z is the coordinate perpendicular to the surface, d the thickness of the insulating layer, $\mathbf{E}$ the electric field, and $\mathbf{D}$ the charge displacement, with $\mathbf{D} = \varepsilon_0 \varepsilon_r \mathbf{E}$. The increase of free energy owing to the charge distribution in the liquid, upon an infinitesimal increase of droplet base can be described as:

$$\frac{dU}{dA} = \frac{1}{2} d \ \mathbf{E}\mathbf{D} = \frac{1}{2} d \frac{V}{d} \sigma_L = \frac{1}{2} V \sigma_L. \tag{5.6}$$

The electric field originating from the liquid/vapor boundary of the droplet (the so-called fringing or stray field) makes a constant contribution to the free energy: this contribution does not change when the contact line is displaced by dA. Therefore, the stray fields do not contribute to dU. The voltage source performs the work to redistribute the charge; the work per unit area is given by:

$$\frac{dW_B}{dA} = V\sigma_L. \tag{5.7}$$

Young's equation (5.8) with an additional electrowetting term γ_{EW} was achieved by calculating the minimum of free energy and the electrowetting force per unit length due to the applied potential:

$$\gamma_{LV} \cos\theta = \gamma_{SV} - \gamma_{SL} + \gamma_{EW}. \tag{5.8}$$

The electrowetting force is:

$$\gamma_{EW} = \frac{1}{2} \frac{d}{\varepsilon_0 \varepsilon_r} \sigma_L^2, \tag{5.9}$$

where $\sigma_L = \varepsilon_0\varepsilon_r V/d$ (Gauss' law) was used, with ε_r the dielectric constant of the insulating layer and ε_0 the permittivity of vacuum. Equations (5.8) and (5.9) can be rewritten to get the relation (Eq. (5.10)) between θ and V for electrowetting:

$$\gamma_{LV}[\cos\theta(V) - \cos\theta_0] = \frac{1}{2}\frac{\varepsilon_0\varepsilon_r}{d}V^2, \qquad (5.10)$$

where θ_0 is the contact angle at 0 V.

(b) Influence of charge trapping

When a potential difference is applied between the liquid and the metal electrode, electric forces work on the ions in the liquid and pull them toward the insulating layer. When the interaction of the ions with the solid is stronger than that with the liquid, there is a possibility that the charge becomes trapped in or on the insulating layer. Ions are trapped in the three-phase region when the detrapping time is larger than the typical vibration time of the contact line. As a result of excitations of the droplet, e.g. thermal, mechanical, or voltage-induced vibrations, a density of trapped charge arises on both sides of the contact line.

As the precise nature of the trapped charge (e.g. electronic or ionic) is not specified, researchers simply assume that a layer of charge with constant surface charge density σ_T is trapped in the insulating layer at distance d_2 from the top of the insulator as shown in Fig. 5.46(b). To determine the change in electrostatic energy due to the infinitesimal base area increase (dU in Eq. (5.3)), they have to take into account the electrostatic contribution below the liquid dU_L, as well as the contribution below the vapor phase, dU_V:

$$dU = dU_L - dU_V. \qquad (5.11)$$

The sign difference is due to the fact that the virtual displacement dA increases the solid/liquid interface while the solid/vapor interface is decreased. They assumed that the trapped charge was distributed uniformly at constant depth, extending sideways to the left of the contact line to a length scale of at least the insulator thickness. Then, the charge density at the liquid/vapor interface due to fringing fields at the edge of the droplet was kept unchanged by the virtual displacement. Therefore, they omit the electrostatic energy of the fringing field in Eq. (5.9).

The potential as a function of the depth in the insulator is sketched in Fig. 5.47(a). The solid line shows the potential without trapped charge. For the case of charge trapping, the potential beneath the liquid phase is indicated by the long dashed line, while the short dashed line shows the potential beneath the vapor phase. The vertical line indicates the depth where the trapped charge is situated.

Figure 5.47(b) shows a plot of the electrostatic fields with $E = -\nabla V$. It is clear that trapping of charge lowers the electric field at the liquid/solid interface and should consequently reduce the electrowetting force. The charge density of the trapped charge, σ_T, is at potential V_T^L below the liquid and at V_T below the vapor phase. On the metal electrode below the liquid, the charge density was $\sigma_M^V = -(\sigma_L + \sigma_r)$. The charge density on the metal electrode below the vapor phase was $\sigma_M^V = -\sigma_r$. Using the general expression for the energy density

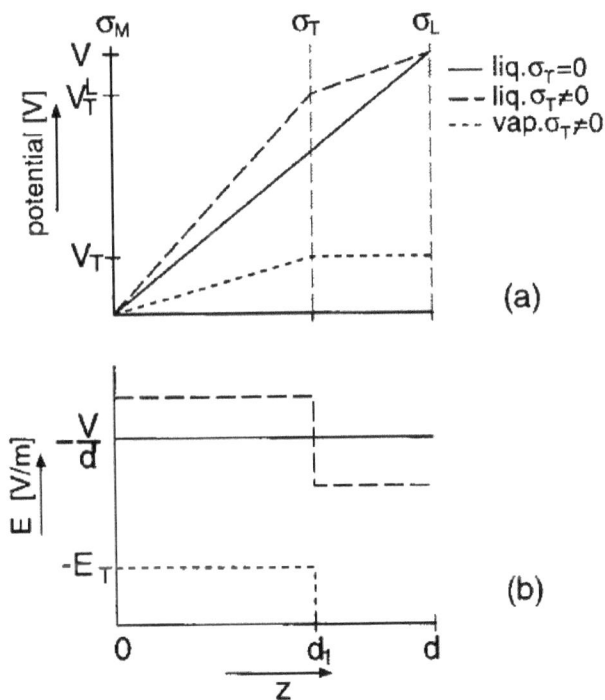

Fig. 5.47   Sketchs of (a) the potential; (b) the electric field in the insulator beneath the liquid and the vapor phase. (Reprinted with permission from Ref. 122. Copyright 1999, American Chemical Society.)

(Eq. (5.3)), they found the electrostatic energy density below the liquid phase:

$$\frac{\mathrm{d}U_L}{\mathrm{d}A} = \frac{1}{2}d_1 E_1 D_1 + \frac{1}{2}d_2 E_2 D_2$$

$$= \frac{1}{2}d_1 \frac{V_T^L}{d_1}(\sigma_T + \sigma_L) + \frac{1}{2}d_2 \frac{V - V_T^L}{d_2}\sigma_L$$

$$= \frac{1}{2}V_T^L \sigma_T + \frac{1}{2}V\sigma_L. \tag{5.12}$$

The energy to create the charge distribution below the vapor phase equals:

$$\frac{\mathrm{d}U_V}{\mathrm{d}A} = \frac{1}{2}d_1 \frac{V_T}{d_1}\sigma_T = \frac{1}{2}V_T \sigma_T. \tag{5.13}$$

The work performed by the voltage source per unit area was given by Eq. (5.5). Using Gauss' law, they found the following relationships between the charge densities and the potentials:

$$\sigma_T = \frac{\varepsilon_0 \varepsilon_r V_T}{d_1}, \tag{5.14}$$

$$\sigma_L = \frac{\varepsilon_0 \varepsilon_r (V - V_T)}{d}, \tag{5.15}$$

$$V_T^L = V_T + \frac{d_1}{d}(V - V_T). \tag{5.16}$$

Using Eqs. (5.1), (5.5), and (5.14), and $\mathrm{d}F/\mathrm{d}A = 0$, they recover Eqs. (5.6) and (5.7), the modified Young's equation. With Eq. (5.13), they found the following relation for the contact angle modulation in the presence of trapped charge:

$$\gamma_{LV}[\cos\theta(V) - \cos\theta_0] = \frac{1}{2}\frac{\varepsilon_0 \varepsilon_r}{d}(V - V_T)^2. \tag{5.17}$$

The electrowetting force was proportional to the square of the applied voltage minus the voltage due to charge trapping.

(2) Electrowetting-based actuation of liquid droplets

Pollack *et al.*[123] have reported rapid actuation of discrete liquid droplets based on direct electrical control of their surface tension. Controlled transfer of droplets at high rates under low voltages offers lots of advantages over current continuous flow and electrokinetic flow technologies. Direct manipulation of discrete droplets enables integrated microfluidic systems to be realized without using conventional pumps, valves, or channels. Such systems are flexible, efficient, and capable of performing complex and highly parallel microfluidic processing.

The electrowetting microactuator is illuminated in Fig. 5.48. A droplet of polarizable and conductive liquid is sandwiched between two sets of planar electrodes. The upper plate consists of a single continuous ground electrode, while the bottom plate consists of an array of independently addressable control electrodes. Both of the two electrode surfaces are covered by a thin layer of hydrophobic insulation. The system geometry and droplet volume are controlled such that the footprint of the droplet overlaps at least two adjacent control electrodes while also making contact to the upper ground electrode. The environmental medium may be either air or an immiscible fluid such as silicone oil to prevent evaporation of the droplet.

A droplet was initially placed on the center of a grounded electrode, and the potential on the adjacent electrode was increased

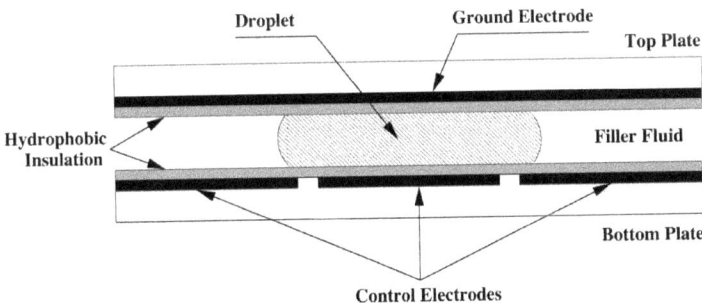

Fig. 5.48   Schematic diagram of the cross-section of the electrowetting microactuator. (Reprinted with permission from Ref. 123. Copyright 2000, American Institute of Physics.)

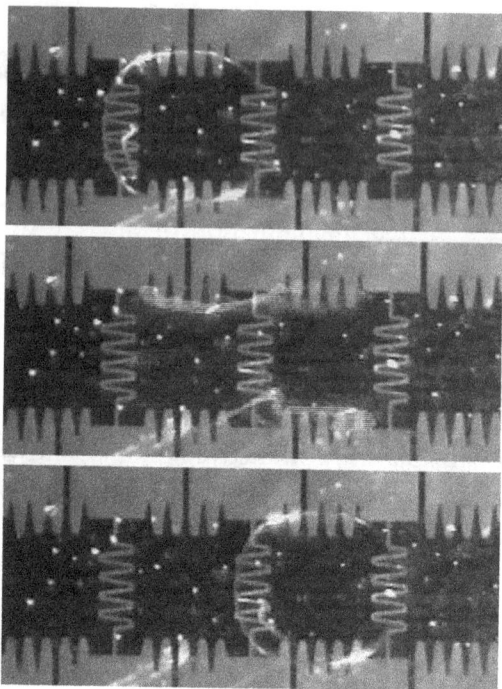

Fig. 5.49 Video frames of a moving droplet. (Reprinted with permission from Ref. 123. Copyright 2000, American Institute of Physics.)

until motion was observed. A voltage of 30–40 V were typically required to initiate movement of the droplet. Once this threshold was exceeded, movement was both rapid and repeatable. Pollack *et al.* believe that contact angle hysteresis is the mechanism responsible for this threshold effect. Three successive video frames (~30 frames/s) of a moving droplet at 80 V of applied potential are shown in Fig. 5.49.

(3) Electrowetting-based liquid droplets complete wetting

Krupenkin *et al.*[124] have reported a novel dynamic electrical control of the wetting behavior of liquids on nanostructured surfaces, which spans the entire possible range from the superhydrophobic behavior to nearly complete wetting. Moreover, the dynamic control was obtained at voltages as low as 22 V. They have demonstrated that the liquid droplet on a nanostructured surface exhibits the contact angle which is primarily depended on the surface tension of the liquid. They studied high-surface-tension liquids such as water (surface tension 72 mN/m) and molten salt (1-ethyl-3-methyl-1H-imidazolium tetrafluoroborate, 62 mN/m), liquids with lower surface tensions (such as cyclopentanol, 33 mN/m, and octanol, 28 mN/m), and liquids with even lower surface tensions (such as 2-propanol, 24 mN/m, and methanol, 23 mN/m). Typical results are illustrated in Fig. 5.50. With no voltage applied, a droplet of molten salt formed a highly mobile ball on the 4-μm pitch substrate (Fig. 5.50(a)), with the contact angle 0–180°; A droplet of cyclopentanol formed an immobile droplet on the 1.75-μm pitch

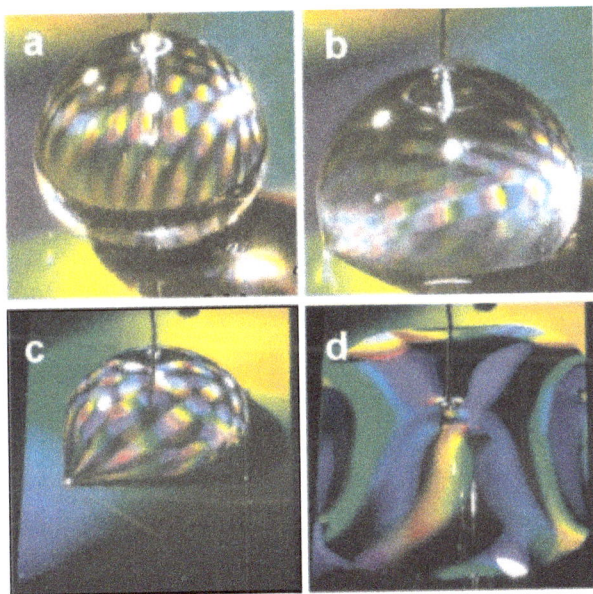

Fig. 5.50   Different wetting states of a liquid droplet on the nanostructured substrate. (a) With no voltage applied, a droplet of molten salt formed a highly mobile ball on the 4-μm pitch substrate. (b) With the application of about 22 V, the droplet of molten salt underwent a sharp transition to the immobile droplet state. (c) With no voltage applied, a droplet of cyclopentanol formed an immobile droplet on the 1.75-μm pitch substrate. The unusual "square" shape of the base of the droplet reflects the underlying symmetry of the nanopost array. (d) With the application of about 50 V, the contact angle of the cyclopentanol droplet dramatically decreased and the droplet filled most of the substrate area. (Reprinted with permission from Ref. 124. Copyright 2004, American Chemical Society.)

substrate (Fig. 5.50(c)). The unusual "square" shape of the base of the droplet reflects the underlying symmetry of the nanopost array. At even lower surface tensions, the droplet appeared to completely wet the substrate. Four frames from the video recording demonstrated electrically induced transitions between different wetting states of a liquid droplet on the nanostructured substrate. Applying the voltage between the droplet (contacted through the Pt wire) and the substrate, when the voltage was about 22 V, the droplet of molten salt underwent a sharp transition to the immobile droplet state (Fig. 5.50(b)). With the voltage being about 50 V, the contact angle of the cyclopentanol droplet dramatically decreased and the droplet filled most of the substrate area (Fig. 5.50(d)).

To understand the microscopic nature of these transitions and, in particular, the degree of the liquid penetration in the nanostructured layer, a UV polymerizable monomeric liquid (NA72, Norland, Inc., surface tension 40 mN/m) was employed. After dispensing on the substrate, the NA72 droplet was polymerized *in situ* by UV irradiation. The silicon substrate was then carefully broken, and the bottom of the solidified droplet as well as the delaminated complementary piece of the substrate were investigated by scanning electron microscopy. For the rolling ball state (Fig. 5.51(a)), there was no liquid penetration in the nanostructured layer. The bottom of the droplet showed clear imprints of the tips of the nanoposts, but there were no signs of nanoposts embedding into the polymer. It was a completely different case for the immobile droplet (Fig. 5.51(b)), where the liquid clearly penetrated all the way to the bottom of the nanostructured layer, dramatically increasing the area of nanostructured layer. A similar investigation indicated that the electrowetting-induced transition between the states of the droplet was also accompanied by the liquid penetrating through the nanostructured layer, as seen in Fig. 5.51(c).

The ability to dynamically change the interaction between the liquid and the nanostructured substrate potentially opens a wide range of exciting new applications. The particular areas of interest include thermal management of microelectronics, optical communications, microfluidics, lab-on-a-chip devices, chemical microreactors, drag reduction systems, etc. In addition, technology based on this can also be applied in the range from optical network and advanced microcell to self-cleaning the windshield and low resistance hull, and so on.

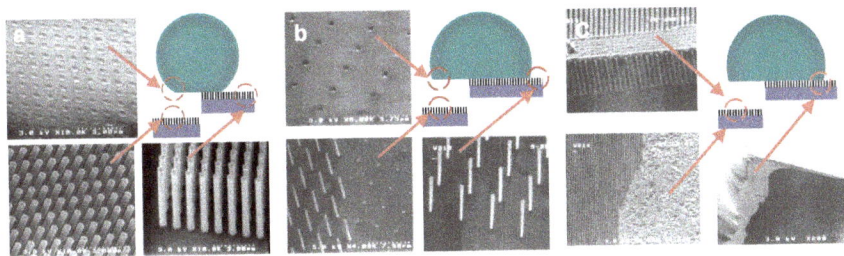

Fig. 5.51   Scanning electron micrographs of solidified *in situ* droplets of NA72. (a) The rolling ball state showed no penetration of liquid through the nanostructured layer. The bottom of the droplet showed clear imprints of the nanopost tips, but no penetration into the liquid. All the nanoposts on the delaminated piece of the substrate were undamaged. (b) Immobile droplets exhibited complete penetration of the liquid through the nanopost array. All the nanoposts underneath the droplet were embedded in the liquid and broken at their bases upon delamination from the substrate. No damage to the nanoposts located outside the droplet was observed. (c) The droplet solidified during the electrowetting process. The resulting immobile droplet exhibited complete penetration of the liquid through the nanopost array. The droplet was preceded by a precursor layer, which was formed by the liquid wetting the nanopost array. (Reprinted with permission from Ref. 124. Copyright 2004, American Chemical Society.)

(4) Electrowetting applications

(a) Liquid microlens

Krupenkin and co-workers[125,126] have reported a tunable liquid microlens capable of adjusting its focal length and lateral position. The microlens consists of a droplet of a transparent conductive liquid placed on a dielectric substrate with a low surface-energy coating.

As shown in Fig. 5.52(a), the electrode design was etched on ITO glass. Applied voltages are indicated as V_0 (voltage applied to the ground electrode) through V_4. Lightly shaded area represents an approximate droplet position. Figure 5.52(b) shows the device cross section.

The researchers took an alternative approach, and did not impregnate the substrate surface with any lubricant. Instead, they attempted to separate the droplet from the substrate surface by creating conditions under which a "lubricating" liquid would penetrate spontaneously under the droplet. This lubricating liquid should

(a)

(b)

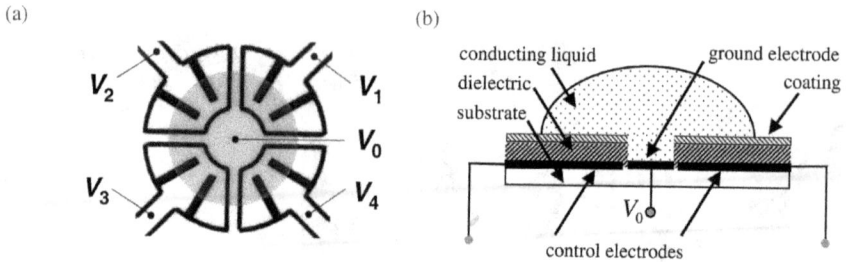

Fig. 5.52 Schematic representation of the tunable liquid microlens. (a) Electrode's design. (b) Device's cross section. (Reprinted with permission from Ref. 125. Copyright 2003, American Institute of Physics.)

be immiscible with the primary liquid that forms the droplet. A thin layer of such "lubricant" would prevent the droplet from directly interacting with the contaminants and inhomogeneities of the solid surface, thus completely eliminating the contact angle hysteresis and the stick-slip behavior. The focal length f of the resulting microlens is determined by the contact angle θ, the droplet volume Ω, and the refractive indexes of the liquid n_L and surrounding media n_V as

$$f^3 = \frac{3\Omega}{\pi(1 - \cos\theta)(2 - \cos^2\theta - \cos\theta)(n_L - n_V)^3}.$$

By varying the voltage applied to a set of electrodes positioned underneath the dielectric substrate, both the position and curvature of the microlens can be changed reversibly. The lens consisted of a 6-µL droplet of the aqueous solution of potassium sulfate (0.01 M concentration). Two cases were investigated, one without any lubricating liquid present and the other with about 0.1 ml of silicone oil (T11, Gelest Inc.) used to form a lubricating layer. Both the microlens curvature adjustment and the lateral motion were demonstrated using a constant as well as an alternating (2 kHz frequency) driving voltage.

Figure 5.53 shows the microlens profiles. The electrodes are transparent and not visible. A 200-mm grating is visible under the substrate. Figure 5.53(a) is the microlens in its unperturbed state when no voltage is applied. Figure 5.53(b) is the microlens in its spread state when identical voltage of 80 V is applied to all four electrodes. One can

(a)   (b)

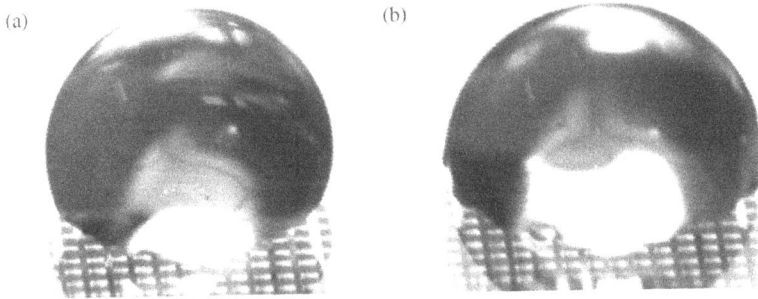

Fig. 5.53 Microlens profiles. (a) Microlens in its unperturbed state when no voltage is applied. (b) Microlens in its spread state when identical voltage of 80 V is applied to all four electrodes. (Reprinted with permission from Ref. 125. Copyright 2003, American Institute of Physics.)

notice some distortion of the droplet circumference due to the gaps between the electrode segments. From the observed droplet profiles, the microlens curvature and, thus, its focal length could be calculated.

(b) Video-speed electronic paper based on electrowetting

Hayes and Feenstra have developed a video-speed electronic paper based on electrowetting, which is realized by the movement of a confined water-oil interface (Fig. 5.54).[127] In equilibrium (Fig. 5.54(a), no voltage applied), a colored oil film lies naturally between the water and the hydrophobic insulator coating of an electrode, because

$$\gamma_{o,w} + \gamma_{o,i} < \gamma_{w,i},$$

where γ is the interfacial tension, and the subscripts o, w and i denote the oil, water and insulator, respectively. Owing to the dominance of interfacial over gravitational forces in small systems (<2 mm), such an oil film is continuous and stable in all orientations. However, when a voltage V is applied between the substrate electrode and the water, an electrostatic term (~$0.5CV^2$, where C is the parallel-plate capacitance) is added to the energy balance, and the stacked state is no longer energetically favorable (Fig. 5.54(b)). The system can lower its energy by moving the water into contact with the insulator, then displacing the oil.

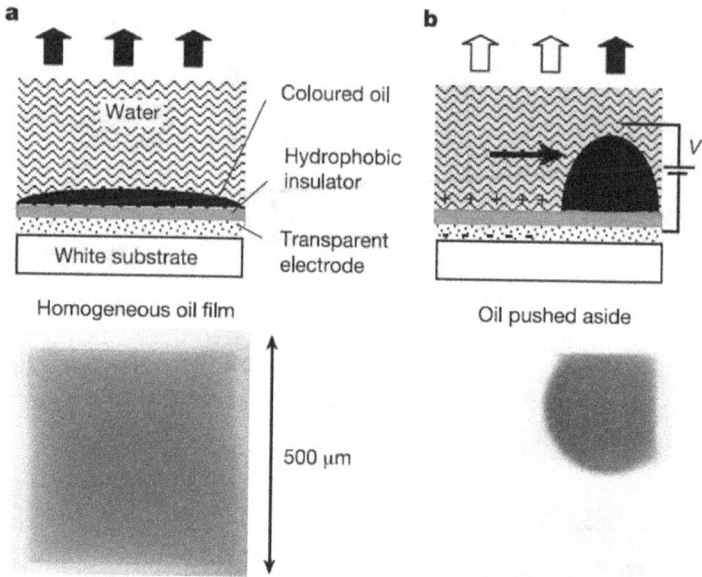

Fig. 5.54   Electrowetting display principle.[127] (Reprinted by permission from Macmillan Publishers Ltd: *Nature*, copyright 2003.)

One of the major drawbacks of current reflective display technologies is their low color-reflectivity, mostly because of the use of RGB (red-green-blue) color segmentation. Figure 5.55 shows a color concept for an electrowetting display that has a reflectivity that is four times higher than that of a reflective color liquid crystal display (LCD) and twice as high as other emerging technologies. In this case, a single display pixel would consist of three subpixels. In each of the subpixels a second oil layer is added, adjacent to the top plate. With a hydrophilic pixel wall, the sandwich structure with two oil layers is the stable equilibrium state (Fig. 5.55(a)). Each oil layer can be independently and reversibly switched by applying a voltage difference between the water and the top or bottom electrode.

A demonstration of the independent switching of two layers in a single cell is given in Fig. 5.55(b). The electrodes cover only half of the top and bottom plate, orthogonally oriented with respect to one

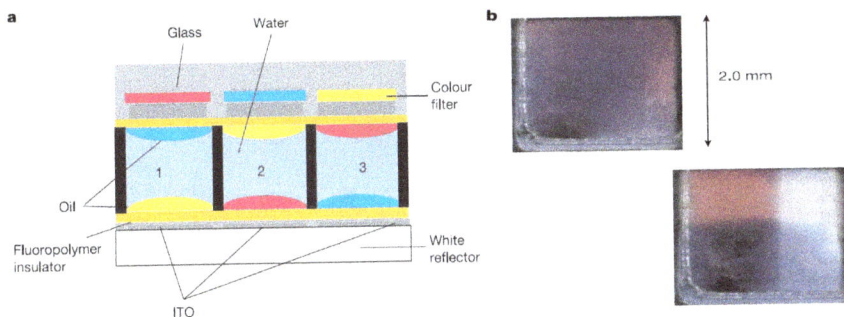

Fig. 5.55 High-brightness color electrowetting display principle.[127] (a) Illustration showing a single subpixelated display with two active layers. (b) Top view of a working cell containing magenta (top) and cyan (bottom) oil films separated by water. (Reprinted by permission from Macmillan Publishers Ltd: *Nature*, copyright 2003.)

another. The top layer is magenta and the bottom layer is cyan. As a result, a different optical response is obtained in the bottom left (two layers present), the upper left and bottom right (one layer) and the upper right corner (both layers absent).

This reflective display based on electrowetting has very attractive electro-optical characteristics, with high switching speed and a straightforward path to a high-brightness full-color display, so that video content can be displayed. In addition, the display has a low power consumption and low voltage technology, and can be made very thin, so it can be called "electronic paper".

5.3.4.2 *Electric field-induced wetting transition*

Lahann *et al.*[128] have reported the design of surfaces that exhibit dynamic changes in interfacial properties, such as wettability, in response to an electrical potential. The change in wetting behavior was caused by surface-confined, single-layered molecules undergoing conformational transitions between a hydrophilic and a moderately hydrophobic state.

They chose (16-Mercapto) hexadecanoic acid (MHA) as a model molecule because it (i) self-assembles on Au (111) into a monolayer, and (ii) has a hydrophobic chain capped by a hydrophilic

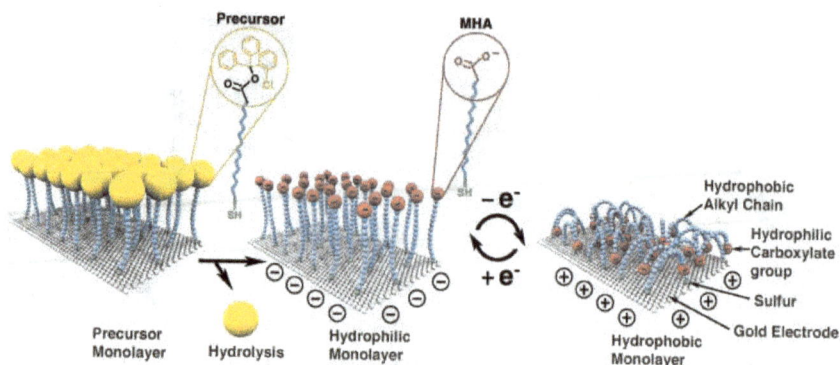

Fig. 5.56   Idealized representation of the transition between straight (hydrophilic) and bent (hydrophobic) molecular conformations (ions and solvent molecules are not shown). (From Ref. 128. Reprinted with permission from AAAS.)

carboxylate group, thus potentially facilitating changes in the overall surface properties. To create a monolayer with sufficient spacing between the individual MHA molecules, they used a strategy that exploited synthesis and self-assembly of a MHA derivative with a globular end group, which resulted in a SAM that was densely packed with respect to the space-filling end groups but showed low-density packing with respect to the hydrophobic chains. Subsequent cleavage of the space-filling end groups established a low-density SAM of MHA (Fig. 5.56).

Their study demonstrates reversible control of surface switching for a low-density monolayer MHA. Because of synergistic molecular reorientations, amplification into macroscopic changes in surface properties is observed. Figure 5.57 shows the advancing (open symbols) and receding (solid symbols) contact angles for the low-density (Fig. 5.57(a)) and the dense (Fig. 5.57(b)), SAMs, respectively, determined while applying voltage to the underlying gold electrode. Four switch cycles were conducted. For the low-density SAM, the switching of the receding contact angles occurs as the surface polarization was changed alternately. Although the advancing contact angle was independent from the applied potential,

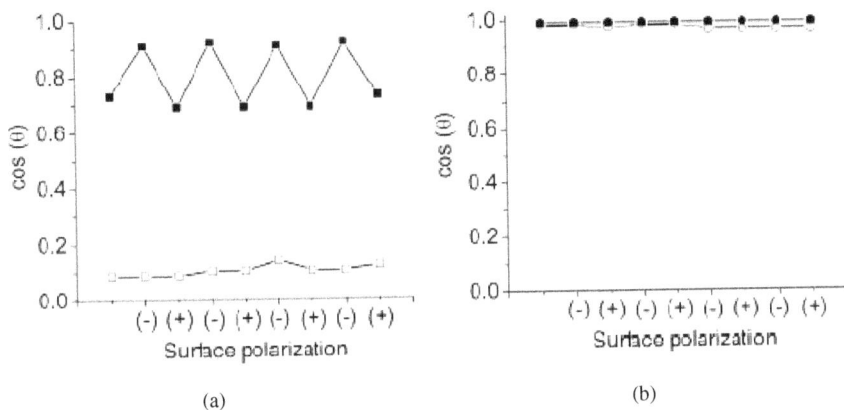

Fig. 5.57   The advancing (open symbols) and receding (solid symbols) contact angles for (a) the low-density; (b) the dense SAMs of MHA were determined while applying voltage to the underlying gold electrode. (From Ref. 128. Reprinted with permission from AAAS.)

the receding contact angle showed a sharp step whenever the polarization of the surface was changed (Fig. 5.57(a), solid squares). The large contact angle hysteresis can be caused by surface roughness or chemical heterogeneity of the surface.[129] Because scanning force microscopy did not reveal notable differences in surface roughness between systems configuration of molecules in bent and straight states, the assumption of a chemical discontinuity along the solid–air–liquid contact line best explains the large hysteresis in contact angles and the pronounced sensitivity of the receding contact angle. The system reflects the behavior of a smooth but chemically heterogeneous system that is composed of a low-energy area (exposed to air) and a high-energy area (exposed to solution). The observed changes in receding contact angle then signify molecular transitions at the high-energy area. The contact angle drop was a reversible phenomenon, because the assembly was switched several cycles between its hydrophilic (straight molecules) and hydrophobic (bent molecules) states. In contrast, switching could not be observed for the dense SAM (Fig. 5.57(b)).

5.3.4.3 *Jump of water droplet from a superhydrophobic film by vertical electric field*

Hashimoto and co-workers[130] have investigated a vertical electric field that induces a water droplet on a superhydrophobic to jump up from the surface without splitting itself. Figure 5.58 shows the schematic illustration of the experimental system and electrode arrangement. The superhydrophobic film-coated glass was laid down horizontally and sandwiched vertically by a pair of parallel metal (aluminium tape) electrodes (size, 80 mm × 50 mm; distance, 10 mm) connected to a DC power supply.

The film used in this study was transparent, with an average contact angle of 159.1° and sliding angle of 1°. Figure 5.59 shows sequence photographs of the motion of a 5-mg water droplet induced by 9.0 kV DC on the superhydrophobic film. The shape of the water droplet did not change remarkably as the applied voltage increased, but the droplet jumped off from the superhydrophobic film to stick on the upper electrode. When the vertical electrode distance was decreased to less than 10 mm under 9.0 kV DC, an electric discharge was exchanged between upper electrode and the top portion of the spherical water droplet on the superhydrophobic film. Water droplet did not jump under 9.0 kV DC when the droplet size was increased.

This trend was different from that for the droplet motion on the superhydrophobic film under the horizontal electric field. Under a

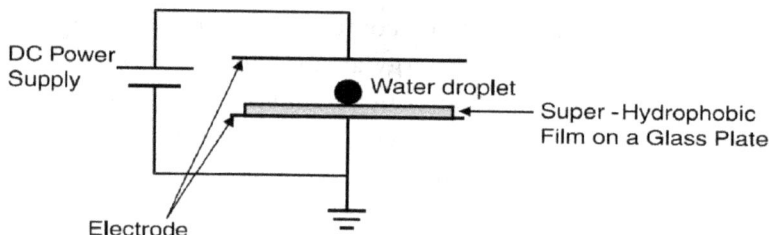

Fig. 5.58   Schematic illustration of the experimental system and electrode arrangement. (Reprinted from Ref. 130. Copyright 2002, with permission from Elsevier.)

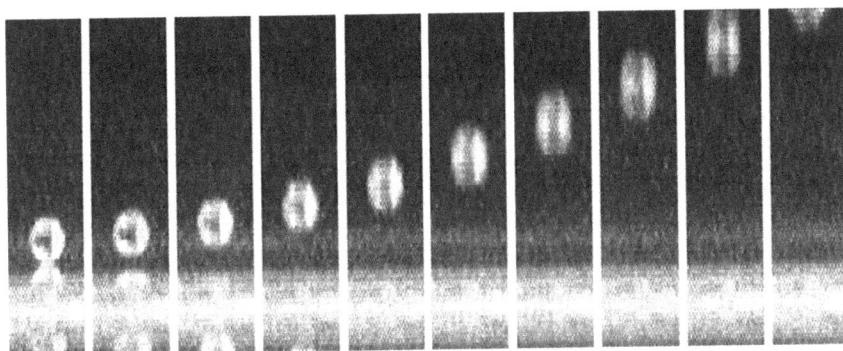

Fig. 5.59 Droplet jumped by static electric field on a superhydrophobic film. (Reprinted from Ref. 130. Copyright 2002, with permission from Elsevier.)

horizontal electric field, Coulomb force has to overcome the moving resistance on the superhydrophobic surface for the droplet motion. Thus, a larger droplet size increases the Coulomb force, and also increases the moving resistance because of the increase of the three-phase (solid–liquid–air) contact lines. Droplet size has more effect on Coulomb force than on the moving resistance, as the former depends on the contact area while the latter depends on the contact line. Therefore, the electric field required for droplet movement decreases with increasing droplet size. Under a vertical electric field, Coulomb force on the droplet should overcome the sum of the adhesion force and gravity. Both adhesion force and gravity increase as the droplet size increases. The adhesion force is originally the same as the moving resistance mentioned above and depends on the three-phase contact lines. However, gravity of a water droplet depends on the droplet mass, which depends on the droplet volume. The droplet size has more effects on gravity than on Coulomb force. Thus, the electric field required for droplet jump increases as the droplet size increases.

To attain the jump of a water droplet without splitting, precise control of Coulomb force is required. Although the voltage required for the droplet jump is still high, further investigation and detailed electrode design may reduce the voltage and provide new types of intelligent devices to control small water droplets.

5.3.5 *Magnetic-responsive surface*

Willner and co-workers[131] developed a two-phase system consisting of an aqueous buffer solution and a toluene phase that includes the suspended undecanoate-capped magnetic nanoparticles (average diameter about 4.5 nm; saturated magnetization, Ms, 38.5 emu g^{-1}). This system is used to control the interfacial properties of the electrode surface. The magnetic attraction of the functionalized particles to the electrode by means of an external magnet yields a hydrophobic interface that acts as an insulating layer to prohibit interfacial electron transfer. Meanwhile, the retraction of the magnetic particles from the electrode to the upper toluene phase by means of the external magnet generates a hydrophilic electrode that reveals effective interfacial electron transfer. Faradaic impedance spectroscopy was used to probe the electron-transfer resistance and double-layer capacitance of the electrode surface upon the attraction and retraction of the functionalized magnetic particles to and from the electrode, respectively, by means of the external magnet (Data: Ret = 170 Ω, Cdl = 40 μF sm-2 in the hydrophilic state of the electrode and Ret = 22 kΩ, Cdl = 0.5 μF sm-2 in the hydrophobic state of the interface). The reversible magnetic-control of the interface enables magnetic switching of the surface wetting behavior and bioelectrocatalytic oxidation of glucose in the presence of glucose oxidase and ferrocene dicarboxylic acid to "ON" and "OFF" states (Fig. 5.60).

Our group recently reported an intelligent system that can be used in reversible no-lost transportation of superparamagnetic microdroplet by using superhydrophobic surface with high adhesion. As shown in Fig. 5.61, a superparamagnetic microdroplet was placed on an ordinary superhydrophobic surface, above which was positioned the polystyrene (PS) nanotube layer with the distance of 2 mm in height. Magnets A and B were assembled outside (Fig. 5.61(a)). When magnet A was switched on, the microdroplet was magnetized and attracted to fly upward (Fig. 5.61(b)). Due to the strong adhesive effect of the nanotube layer, the superparamagnetic microdroplet was stuck to the PS nanotube surface (Fig. 5.61(c)).

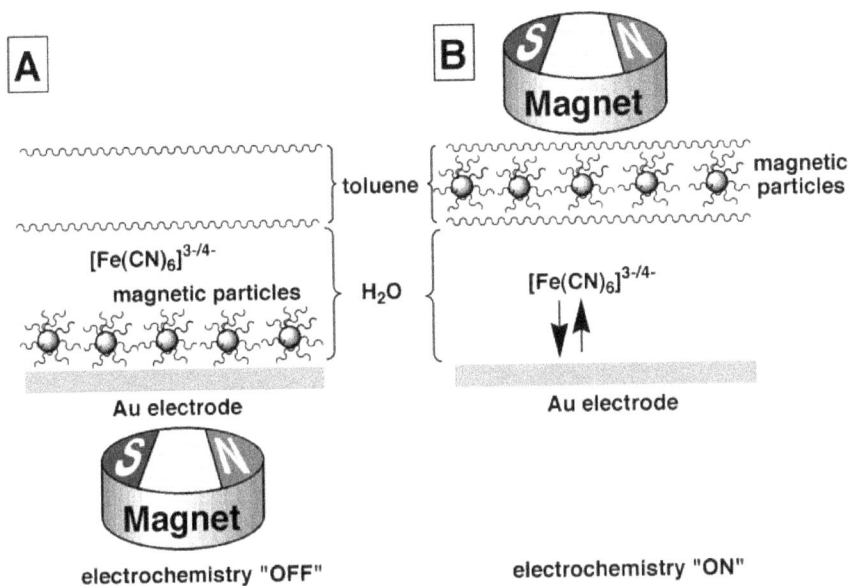

Fig. 5.60 Magnetically controlled reversible translocation of the functionalized magnetic particles between the organic phase above the aqueous electrolyte and the electrode surface. (Reprinted with permission from Ref. 131. Copyright 2004, American Chemical Society.)

When magnet A was switched off and magnet B was switched on, the direction of magnetic force was reversed and the microdroplet fell back to the initial surface (Fig. 5.61(d)). The transportation was completed when both magnets were removed (Fig. 5.61(e)). Such transportation can be repeated by alternating magnetic fields (Fig. 5.61(f)). Notably, no loss of the liquid was observed during the transporting process. It was also demonstrated that the intensity of external magnetic fields decreased with the number of superparamagnetic nanoparticles per microdroplet increased. Compared to other transportation systems, this smart system is more effective, agile, and can be potentially used in a wide range of applications, such as localized chemical or biological reactions, traced analysis, and *in situ* detection.

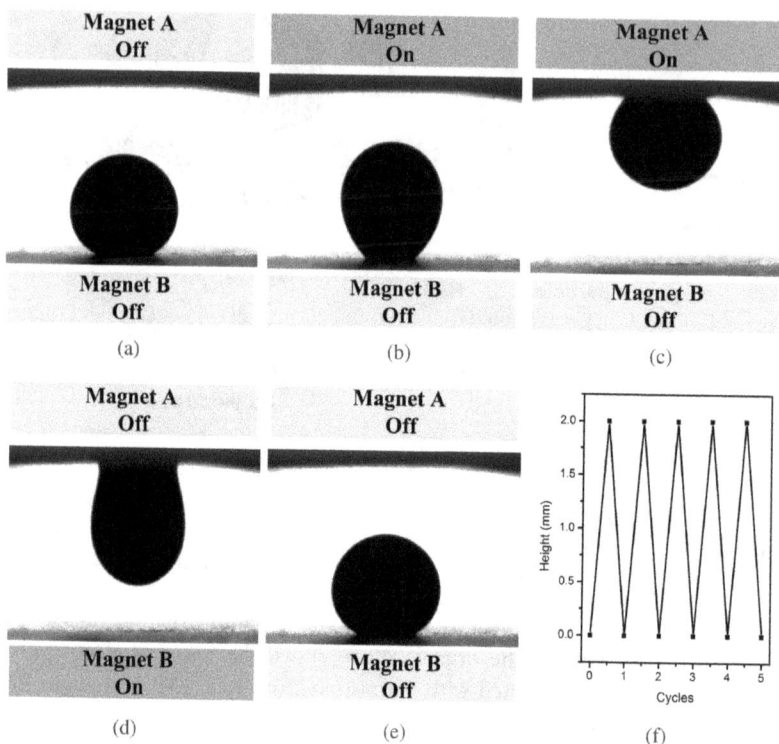

Fig. 5.61 No-lost transport processes of a superparamagnetic microdroplet in alternating magnetic fields. (Reprinted with permission from Ref. 132. Copyright 2007, American Chemical Society.)

5.3.6 *Mechanical force induced wetting changing behavior*

Han and co-workers[133] from Changchun Institute of Applied Chemistry, Chinese Academy of Sciences, found that reversible switching between superhydrophobicity and superhydrophilicity of a polyamide film with a triangular net-like structure could be achieved by biaxially extending and unloading the elastic film. Figure 5.62(a) shows that before biaxial extension, the average side length of the triangle is around 200 μm. The average size of the polyamide elastic fiber is 20 μm, and contact angle is 151.2° (inset). Figure 5.62(b) shows that with an extension ratio near 120%, the average side length

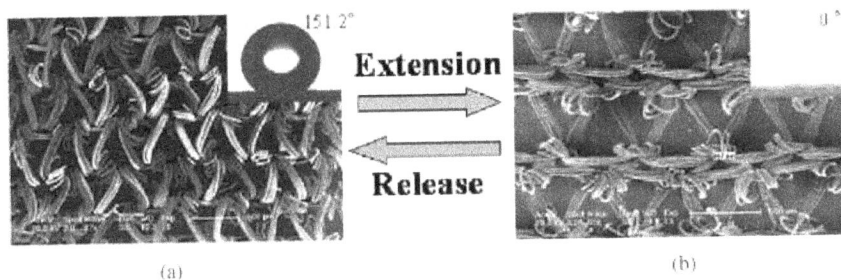

Fig. 5.62   Change in structure and wetting behavior of the triangular net-like polyamide film upon extension and unloading.[133] (Copyright Wiley–VCH Verlag GmbH & Co. KGaA. Reproduced with permission.)

of the triangle is around 450 µm, with contact angle of about 0° (inset). Both the change of the average side-length of the triangular net-like structure upon biaxial extension and unloading, and the surface tension of the water droplet, are believed to be responsible for the reversible switching between superhydrophobicity and superhydrophilicity.

5.3.7 *pH-responsive surfaces*

pH-responsive materials are volume changeable polymer gels in response to pH value and ionic concentration. Such macromolecular gel network has dissociated ion groups, the network structure and charge density of which can change with medium pH changing and have an impact on the osmotic pressure of the gel; on the other hand, changes in the ionic concentration will lead to changes in the volume of the gel. In recent years, pH responsive gel has attracted a lot of attention as a responsive intelligent polymer gel. For example, Jennings and co-workers[134] have prepared pH-responsive carboxyl-modified polymethylene films on the gold surface. Film resistances decrease by up to five orders of magnitude over the critical pH range, which can be tuned from five to ten. Stroeve and co-workers[135] have demonstrated the modification of the surface properties of pore walls in polycarbonate track-etched (PCTE) membranes by formation of self-assembled monolayers (SAMs) from ω-substituted thiols onto electroless gold deposited on the pore walls of the membrane. The surface-modification procedure described in their work may provide an alternative route to prepare porous ion-selective membranes.

Bergbreiter and co-workers[136] have reported composite membranes with pH-switchable permselectivity for both cationic and anionic redox-active probe molecules. The pH-dependent changes in charge of the dendrimer amine groups and the carboxylic groups derived from the random coil polyanhydride chains serve as independent ion gate. These gates pass cations only at high pH, anions only at low pH, and both cations and anions at intermediate pH (Fig. 5.63(a)).

Whiteside and co-workers[137,138] examined the wetting by water of low-density polyethylene film modified at the polymer–water (air) interface by introduction of polar organic functional groups, for example carboxyl and amino, in 1988. They performed a contact angle titration on a single type of surface. On the other hand, they prepared carboxyl acids at the surface of monolayer on gold. Minko and co-workers[139] reported on a thin polyelectrolyte film (mixed polyelectrolyte brush) with a gradual change of the composition (ratio between two different oppositely charged surface-grafted weak polyelectrolytes) across the sample. The gradient of surface composition

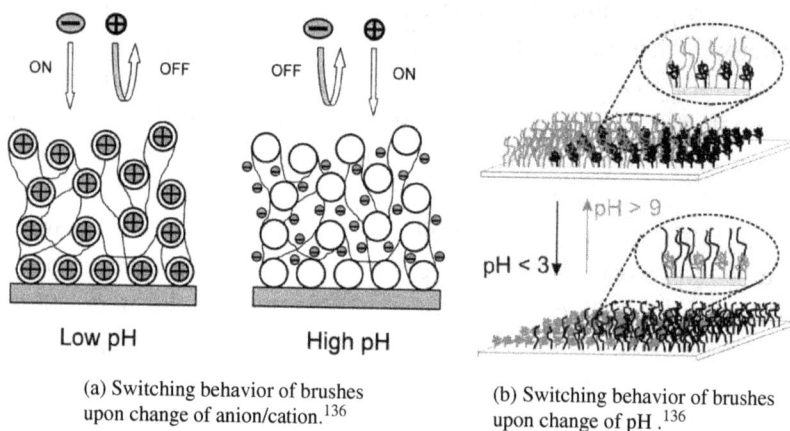

(a) Switching behavior of brushes upon change of anion/cation.[136]

(b) Switching behavior of brushes upon change of pH.[136]

Fig. 5.63   Schematic illustration of the pH-switchable "On/Off" function of the composite film. (Reprinted with permission from American Chemical Society.)

creates a gradient in surface charge density and, consequently, a gradient of the wetting behavior. The gradient film is sensitive to a pH signal and can be reversibly switched via pH change.

Zhang and co-workers[140,141] used pH-responsive materials on a self-assembled monolayer (SAM) of dendron thiol for the pH-responsive surface, which can be superhydrophobic in an acidic and even neutral aqueous environment, but superhydrophilic under basic conditions (Fig. 5.64). It is made in the following way. Firstly, gold electrodeposition was employed on a "full-of-defect" matrix of a dendritic thiol SAM to obtain a rough gold surface, which exhibited micro/nanoscale roughness. Then the above rough gold substrate was immersed in a mixed solution of HS $(CH_2)_9CH_3$ HS

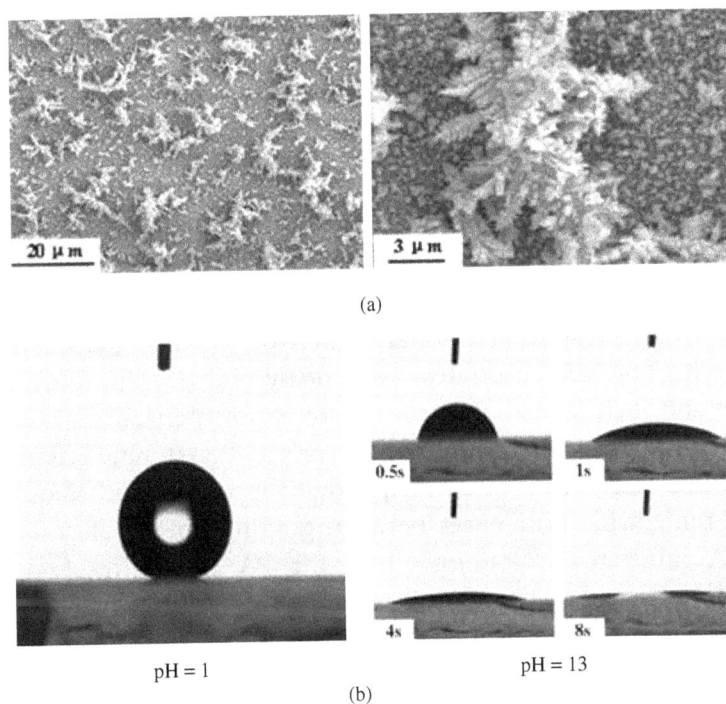

(a)

(b)

pH = 1   pH = 13

Fig. 5.64 pH-responsive superhydrophilic/superhydrophobic switch. (a) SEM pictures of rough surface. (b) CA of different pH values.[141] (Copyright Wiley–VCH Verlag GmbH & Co. KGaA. Reproduced with permission.)

$(CH_2)_{10}COOH$ overnight. In this way, a mixed monolayer containing both alkyl and carboxylic acid groups can form on the surface of the gold. Moreover, substrates modified with this kind of monolayer exhibit a pH-responsive wetting behavior due to the deprotonation of the surface carboxylic acid groups. It is expected that on a rough surface, this effect can be amplified. The wetting properties of the substrate are examined by means of contact angle measurements. For an acid droplet (pH = 1) the contact angle can reach as high as 154° showing superhydrophobicity, while a base droplet (pH = 13) will spread out on this surface in less than ten seconds with contact angle 0° showing superhydrophilicity. After being rinsed with distilled water and dried with an N_2 stream, the base-exposed surface can recover its pH-responsive property, thus the pH surface sensitivity is preserved. Moreover, the surface is mechanically and chemically stable, as its responsive property does not change, even over one month without any special protection.

5.3.8 *Solvent-responsive polymer surfaces*

5.3.8.1 *Changes of the wetting behavior on solvent-responsive polymer surfaces*

Minko *et al.*[142] developed a route to fabricate mixed brushlike layers on polyamide substrates (PA-6, PA-6I, PA-66) by the "grafting from" approach. The PA substrates were firstly functionalized by NH_3 plasma and then the azo initiator of radical polymerization was covalently bound to the functionalized PA surface. Polystyrene was grafted in the first step and poly(2-vinylpyridine) in the second step. They found remarkable differences between grafting on Si wafers and on the PA substrates. "Grafting" from the PA surface resulted in a dramatic increase of the surface roughness of the film which could be explained by grafting in a swollen surface layer of PA. Because of this effect, they found a substantial amount of grafted polymers even on not-functionalized PA substrates, which was explained by grafting via chain transfer reaction. They also performed the same grafting procedure on the surface of PA fabric. In this case the switching behavior was amplified by the texture of the material and the wetting

(a)

(b)

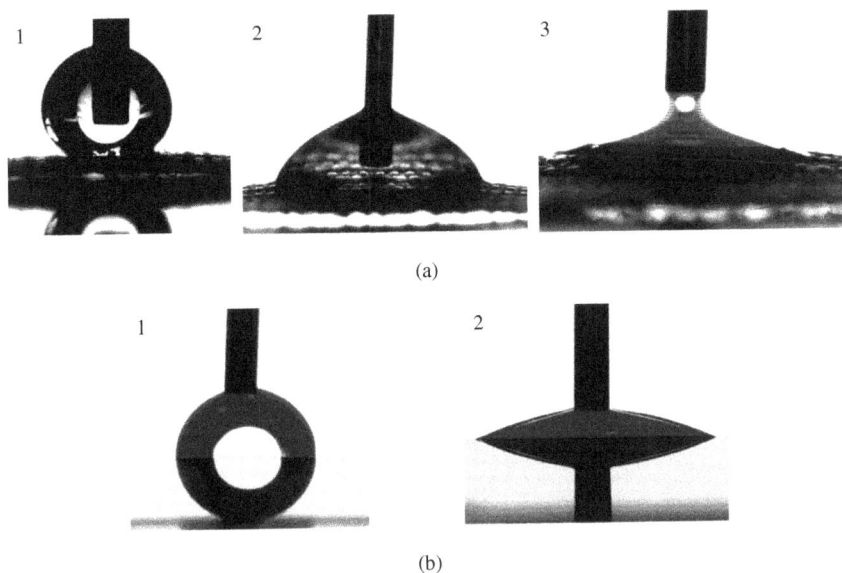

Fig. 5.65   Contact angle photographs of drops on substrates with grafted PS-PVP brush from PA-6 textile. (a) After exposure to (1) toluene, CA = 150°; (2) ethanol, CA = 50°; (3) water, pH = 3, wicking regime. (b) From PA-6 plates after exposure to toluene CA = 90° (1) and water, pH = 3, CA = 20° (2). (Reprinted with permission from Ref. 143. Copyright 2003, American Chemical Society.)

behavior of the fabric with the mixed brush was switched from complete wetting to highly hydrophobic state (150° water contact angle, see Fig. 5.65).

Based on the above research, they reported a route to fabricate two-level structured self-adaptive surfaces (SAS) of polymer materials. The first level of structure was built by a rough polymer film consisting of needle-like structures of micrometer size (Fig. 5.66(a) and (b)). The second level of structure was formed by the nanoscopic self-assembled domains of a demixed polymer brush irreversibly grafted onto the needles. The brush consisted of hydrophobic and hydrophilic polymers and the morphology resulted from interplay between lateral and vertical phase segregation of the polymers, which switched the morphology and surface properties upon exposure to different solvents. In selective solvents the preferred polymers preferentially occupied the top of the surface (Fig. 5.66(c) and (e)), while

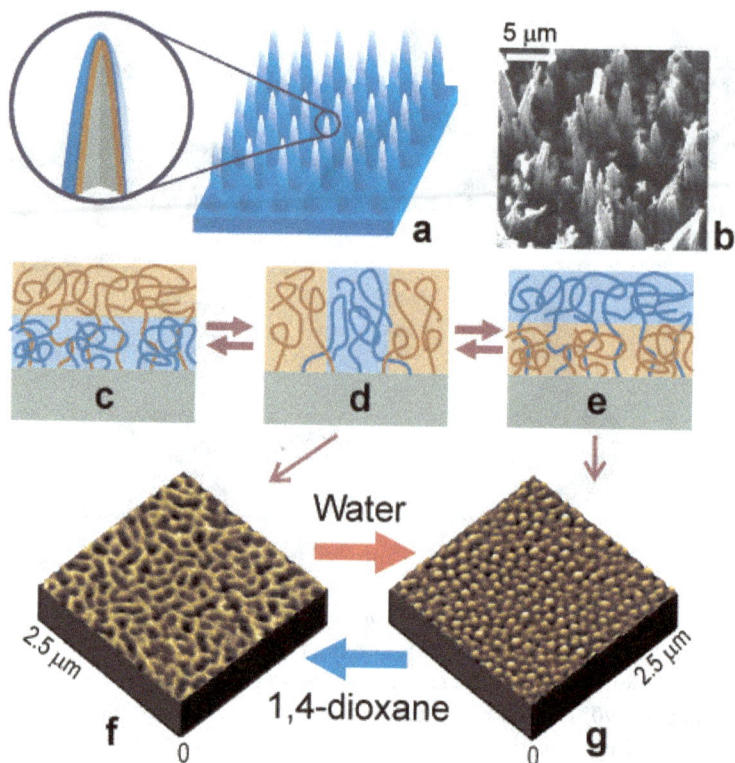

Fig. 5.66 Two-level structure of self-adaptive surfaces (SAS). (a) Schematic representation of needle-like surface morphology of the PTFE surface (first level). (b) SEM image of the PTFE film after 600 seconds of plasma etching. (c)–(e): The second level structures. (f) and (g): AFM images (model smooth substrate) of the different morphologies after exposure to different solvents. (Reprinted with permission from Ref. 144. Copyright 2003, American Chemical Society.)

in nonselective solvents, both polymers were present in the top layer (Fig. 5.66(d)). Therefore, by exposing the surface to different solvents that were selective to one of the components of the brush, reversible transition between the superhydrophobicity and the superhydrophilicity could be controlled. After the surface was exposed to toluene, it was superhydrophobic with a contact angle of 160°, and a drop of water could roll easily on the surface, indicating also a very small hysteresis of the contact angle (Fig. 5.67(a)). After the same sample was immersed in an acid (pH = 3) water bath for several

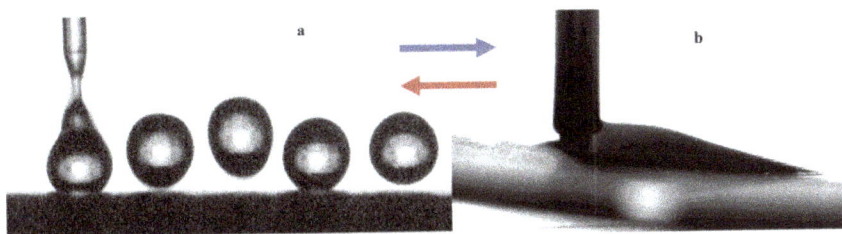

Fig. 5.67 Photograph of a water drop deposited onto the SAS. (a) The stroboscopic image shows that a water drop jumps and rolls on the ultrahydrophobic surface obtained after exposure of the sample to toluene. (b) Exposure to acidic water switches the sample to a hydrophilic state and the water drop spreads on the substrate. (Reprinted with permission from Ref. 144. Copyright 2003, American Chemical Society.)

minutes and dried, however, a drop of water spread on the surface because of the wicking effect (Fig. 5.67(b)). It was noteworthy that after the treatment with toluene, the wetting hysteresis depended on the time of the water drop contact with the substrate. After 30 seconds, the difference between advancing and receding contact angles was within the error of the measurement (2°), while in several minutes the difference increased dramatically due to the slow switching of the surface composition to a more hydrophilic state.

5.3.8.2 *Applications of solvent responsive polymer surfaces*

Ionov *et al.*[145] reported a novel type of chemical patterning based on thin stimuli-responsive polymer films (switch from hydrophilic to hydrophobic, or from hydrophobic to hydrophilic), which was fabricated from the mixed brush of poly(2-vinylpyridine) and polyisoprene (Fig. 5.68). In their work, cross-linking of polyisoprene via illumination through a photomask results in formation of patterns with suppressed switching, which formed "unchanging background" of the pattern. The mixed brush demonstrates switching behavior upon exposure to different solvents, and shows three main characters: (1) the contact angle of this film changed from 89° to 69° after immersion in toluene and then exposed to UV illumination; in other words, light illumination caused switching from hydrophobic to hydrophilic of the polymer brushes; (2) the contact angle remained 69° via UV illumination after

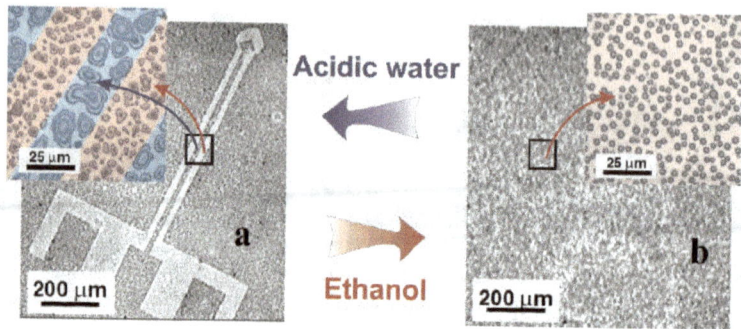

Fig. 5.68 Chemical patterning based on thin stimuli-responsive polymer films. (Reprinted with permission from Ref. 145. Copyright 2003, American Chemical Society.)

immersing in ethanol, indicating no changes of wetting behavior; (3) the contact angle of this film changed from 42° to 69° after immersion in acid solution (pH = 2) and then exposed to UV illumination, showing switching from hydrophilic to hydrophobic. It could be concluded that this polymer brush could change from a hydrophobic film (toluene, no illumination) to a hydrophilic film (pH = 2, no illumination). However, this ability of the film was lost when exposed to UV illumination and became an inert material.

Based on the above, the researchers fabricated an "inert" substrate to the environment via UV illumination, then immersed the polymer brushes into different solvents. As the non-illumination part could change its wetting behavior in response to different solvents, the substrate formed the very pattern as existed on the template, as shown in Fig. 5.69. When exposed to UV illumination (through a template with patterns), rinsed by ethanol, blown dry, and then treated by water vapor, the pattern did not show up. When treated with acid solution (pH = 2) instead of ethanol, the water vapor formed a pattern as on the template. The water droplet did not wet the light-illuminating part, while spread on the non-illumination part easily.

In recent years, the solvent-responsive wetting behavior of surface has been promoted a lot.[146] Anastasiadis *et al.*[147] developed a methodology for creating responsive polymer surfaces that could alter the

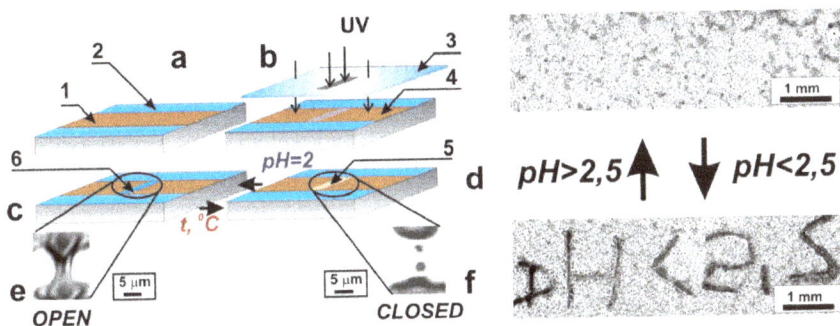

Fig. 5.69 The "Open" and "Close" effect of solvent responsive polymer surface. Left: Fabrication of a switchable channel employing environment responsive lithography (for explanation see the main text). The images (e) and (f) show open and closed states of the switchable channel, respectively, as they appear in optical microscopy. Right: Example of a smart sensor from the mixed brush grafted to Si wafer which displays the result of the analysis of acidic aqueous solution: the wafer was exposed to neutral water (top) and to water with pH 2.3 (bottom). The image appeared upon exposure to water vapor only if the sample had been treated with acidic water solution of pH < 2.5. (Reprinted with permission from Ref. 145. Copyright 2003, American Chemical Society.)

wetting characteristics when exposed to different humid environment. They used the surface partitioning of block copolymers at the polymer/air interface and utilized a hydrophilic group at the end of the surface-active block. It is demonstrated that when exposed to water vapor, the end group, hidden below the surface when in contact with dry air, presents itself to the surface thus significantly reducing the water contact angle. This function can be reversed and repeated again and again by successive exposures of the surface to dry and wet environment, respectively.

Granville and Brittain[148] reported a $Si/SiO_2//PMA\text{-}b\text{-}PPFA$ semi-fluorinated diblock copolymer brush synthesized on a porous silica substrate. Upon solvent and thermal treatment, the surfaces exhibit surface rearrangement properties. The solvent treatment resulted in rearrangement from an ultrahydrophobic polymer surface (CA = 135°) to a less hydrophobic surface (CA = 90°). Furthermore, the surface was shown to extend and generate an ultrahydrophobic surface upon annealing at 60°C after six minutes.

Han and co-workers[149] reported a responsive polymer composite film, which was generated by using reversibly switchable surface morphology of polystyrene-block-poly (methyl methacrylate) (PS-b-PMMA) films in response to different block selective solvents on the rough isotactic poly(propylene) (i-PP) substrate. The composite films had a take-off water contact angle (CA) of 137.4° and after they were exposed to CS_2 vapor (a selective solvent for PS) for 12 h, the average CA increased to 159.6°. Following treatment with acetone vapor (a selective solvent of PMMA) for 30 h, the hydrophobicity of the composite films decreased greatly, i.e. the average CA decreased to 117.0° (Fig. 5.70(b)). For comparison, on the flat Si substrate, the water contact angles (CAs) of polystyrene-block-poly(methyl methacrylate) (PS-b-PMMA) film could be reversibly switchable between 71° and 90° upon exposure to different block-selective solvents (Fig. 5.70(a)). Our group considered the surface composition, microstructures provided by the rough i-PP substrate, and nanostructures formed by the microphase separation of PS-b-PMMA diblock copolymer being responsible for the superhydrophobicity of the composite film after CS_2 swelling.

Our group and researchers from Sichuan University have designed and synthesized a novel segmented poly-(carbonate urethane)s containing both hydrophobic fluorinated alkyl group and hydrophilic phosphatidylcholine polar head groups on the side chain.[150]

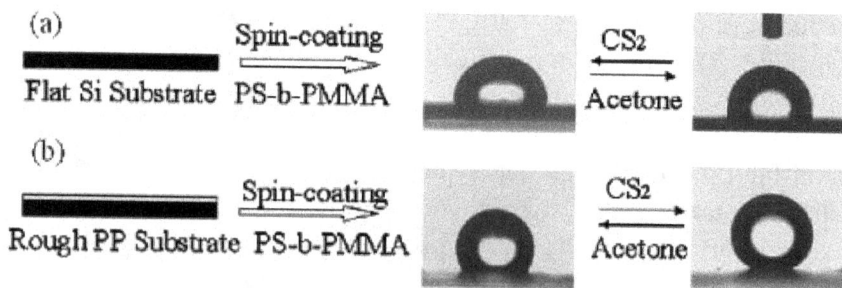

Fig. 5.70 Schematic illustration of how to fabricate a wettability-responsive composite film: (a) on the flat Si substrate; (b) on the rough PP substrate.[149] (Copyright Wiley–VCH Verlag GmbH & Co. KGaA. Reproduced with permission.)

Investigation indicated a reversible overturn of the phosphatidyl-choline groups with the movement of the hydrophobic fluorinated alkyl groups when the samples were treated in dry air or water. The change in environment from air to water induced a reorganization of the surface in order to minimize the interfacial free energy, resulting in a macroscopic change of surface wettability. The good environmental responsiveness of such biomembrane-mimicking films may find successful applications as biomaterials.

5.3.9 *Changes of the wetting behavior by electro-chemical control*

The wetting behavior on material surface depends on the surface properties, which is often adjusted by adsorption of certain molecules. Sondag-Huethorst and Fokkink[151] reported the electrochemical and electrowetting behavior of ferrocene (Fcbtenninated alkanethiol $(FcCO_2C_{11}H_{22}\text{-}SH)$ monolayers adsorbed on gold in 1 M $HClO_4$. It is found that upon oxidation, the wettability of the surface increases: the advancing contact angle of the $FcCO_2C_{11}H_{22}\text{-}SH$ monolayer with the electrolyte (θ_a^{red}) decreases from 74° to 49°, the latter being the advancing contact angle of the oxidized Fc^+ cation, θ_a^{ox}. The meniscus recedes and the receding contact angle, θ_r^{red}, becomes 56° and can be continuously stepped between θ_r^{red} and θ_a^{ox} upon renewed reduction. Due to contact angle hysteresis, this potential-dependent change in contact angle is only partly reversible. A decrease in θ_r^{red} is observed within the number of applied steps due to disordering of the mono-layer on continuous oxidation/reduction. However, this disruptive effect is reduced for a mixed $FcCO_2C_{11}H_{22}SH/C_{12}H_{25}SH$ monolayer. The $C_{12}H_{25}SH$ thiol increases the distance between the bulky Fc groups, and consequently the monolayer becomes more stable. For a mixed monolayer with 60% $FcCO_2C_{11}H_{22}SH$ and 40% $C_{12}H_{25}SH$, the contact angle can reversibly be switched between 49° and 59°. They suggested that the change in wetting behavior after oxidation/reduction was due to specific interactions of the charged Fc^+ ion with anions from the electrolyte, and due to the hydrophilic nature of the ClO_4^- anion, the wettability of the surface increased.

Whitesides and co-workers[152] used applied electrical potentials, in combination with self-assembled monolayers (SAMs) of alkanethiolates supported on gold films, to control the wettability of a surface over wide ranges. This surface can be transformed from non-wetting to wetting, or the reverse, with time constants of seconds. The method is based on a competition between reductive electrochemical desorption of a hydrophobic SAM and its re-formation from alkanethiol in solution. Self-assembled monolayers formed from either $CH_3(CH_2)_{15}SH$ or $CH_3(CH_2)_2SH$ were hydrophobic ($80° < \theta < 110°$, θ = contact angle) toward aqueous solutions of electrolyte at neutral potentials but became hydrophilic ($\theta \sim 10°$) at reducing potentials ($E < -1.3$ V vs. Ag wire). After the reducing potentials were removed, contact angles of aqueous solutions containing $CH_3(CH_2)_2SH$ returned to their initial values ($\theta \sim 80°$). Because the change in wettability was dependent on the structure of the organic molecule in the monolayer, it was possible to prepare patterned SAMs in which certain regions were transformed from hydrophobic to hydrophilic by changing potential, while other regions were inert.

Berggren and co-workers[153] reported an electronic wettability switch (EWS), based on a combination of a solid polymer electrolyte and an electrochemically active conjugated polymer (Fig. 5.71). The degree of oxidation or reduction of the polymer at the surface actively controls the wettability. They have primarily investigated this phenomenon for polyaniline (PANI) doped with dodecylbenzenesulfonic acid (DBSA), as the active layer. For the first kind of EWS (EWS-1), the active areas were approximately 1 cm². The most conductive, protonated form of polyaniline is the green-colored emeraldine salt (or protonated emeraldine form). PANI in this state can be switched electrochemically to the oxidized blue protonated perni-graniline form, or reduced to the transparent leucoemeraldine form. Bi-stable switching is achieved within the two adjacent films when a voltage, typically 2 V, is applied. Water contact angle measurements show clear contrast between the three forms of polyaniline in the EWS-1 structure, as measured with the EWS-1 device. The green emeraldine salt yielded a static water contact angle of $\theta = 28° \pm 2°$. When PANI was

(a)

Water droplets

(b)

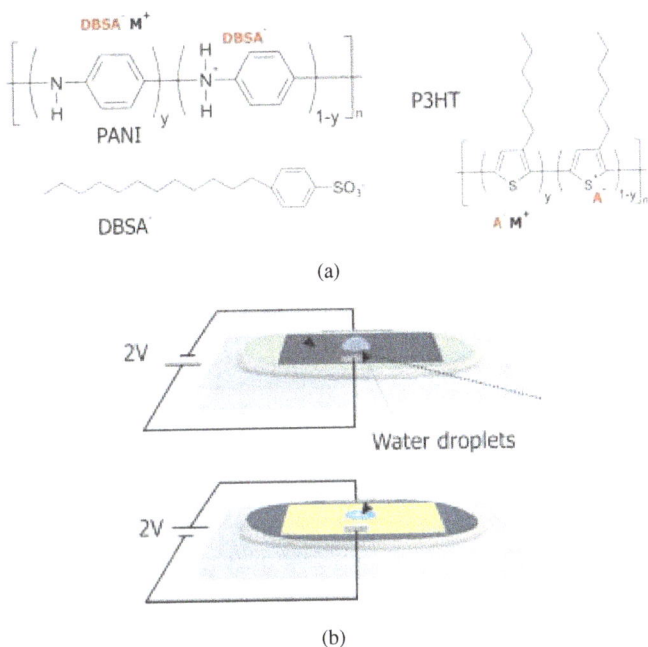

Fig. 5.71 (a) Electro-chemical oxidation and reduction of polyaniline and polythiophene. (b) Dodecylbenzenesulfonic acid doped polyaniline bi-stable electro-wetting switcher.[153] (Copyright Wiley–VCH Verlag GmbH & Co. KGaA. Reproduced with permission.)

oxidized, the water contact angle of droplets added to the surface increased to $\theta = 37° \pm 2°$, while if the polymer was reduced, the angle decreased to $\theta = 9° \pm 2°$. Electronically addressable wettability on planar surfaces would be useful for a wide range of applications, such as those found in the graphical printing, patterning industry, microfluidic devices, growth control of living cell clusters and adsorption on biomaterials.

Yan and co-workers[154] have demonstrated a simple facile electropolymerization of superhydrophobic PPy films and the reversible switching of the PPy films between superhydrophobicity and superhydrophilicity by changing the different oxidative and reductive state of PPy porous films.

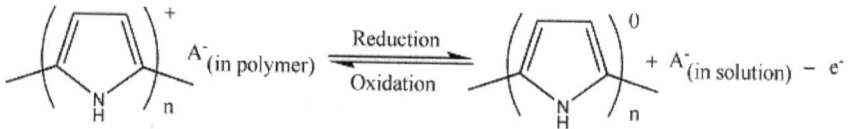

The reported PPy films were synthesized on a conducting surface such as Au-coated glass by electrochemical polymerization. The electroplating solution contained 0.1-M pyrrole, 0.05-M tetraethylammonium perfluorooctanesulfonate (TEAPFOS, $Et_4N^+CF_3(CF_2)_7SO_3^-$), and 2.0×10^{-4} M $FeCl_3$ in acetonitrile was used, and the electropolymerization was carried out galvanostatically. The SEM image reveals an extended porous structure of the PPy film. The pore size ranges from 10 to 50 μm. A magnified image of a single pore shows that the pore wall is also highly porous. The smaller pores are on the order of 1–4 μm. The particle assemblies surrounding the smaller pores are composed of submicron PPy particles with diameters of 0.5–1 μm. When low-surface-energy perfluorooctanesulfonate (PFOS) was used as a dopant in the polymerization, the PPy films showed superhydrophobicity with a static water contact angle as high as 152° ± 2°.

In their works, Fe^{3+} acts as catalyst in the synthesis of PPy. The need to regenerate Fe^{3+} by the electrode is believed to confine the chemical polymerization to the vicinity of the electrode surface, leading to closely coupled electropolymerization and chemical polymerization. This combined polymerization process promotes the growth of porous and rough PPy film. This porous structure with roughness in both coarse and fine scales is most effective in achieving superhydrophobicity and superhydrophilicity.

(1) Electro-chemical polymerization of the PPy porous films:

(2) Catalytic chemistry polymerization of the PPy porous films:

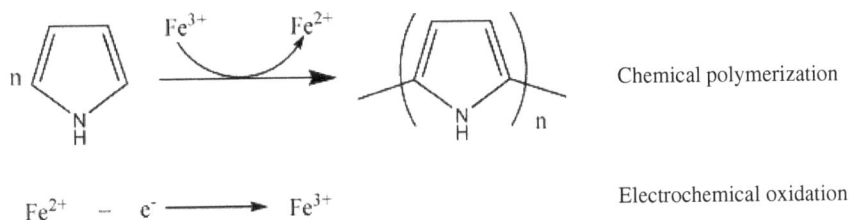

Chemical polymerization

$$Fe^{2+} - e^- \longrightarrow Fe^{3+}$$

Electrochemical oxidation

When the PFOS-doped PPy was held in a 0.05-M solution of TEAPFOS in acetonitrile at a negative potential (-0.6 V vs. Ag/AgCl reference electrode) for 20 minutes, the superhydrophobic PPy film was converted to neutral PPy film, which exhibited a water contact angle of about $0°$, indicating that a superhydrophilic PPy surface was achieved; when the superhydrophilic (neutral PPy) film was held in a 0.05-M solution of TEAPFOS in acetonitrile at a positive potential (1.0 V vs. Ag/AgCl reference electrode) for 20 minutes, the super-hydrophobic (oxidized PPy) surface was regenerated.

5.3.10 *Dual/multiple-responsive polymer films*

As shown above, with the development of the combination of responsive materials and surface roughness, several photo-/thermo-/pH-responsive smart interfacial materials can switch between super-hydrophilicity and superhydrophobicity. However, all of these surfaces are responsive to only one kind of external stimuli, which has restricted the applications of responsive smart interfacial materials in many multiple-responsive fields. For example, certain drug needs to be released at specific part of body, where the temperature and pH is different from elsewhere. Thus the material of this drug has to be responsive to the environmental temperature and pH at the same time. That is why the research on dual-responsive or multiple-responsive smart wetting material is of great importance.

Recently, we developed a dual-stimuli–responsive surface with tunable wettability, reversible switching between superhydrophilicity

Thermo-responsive   **pH-responsive**

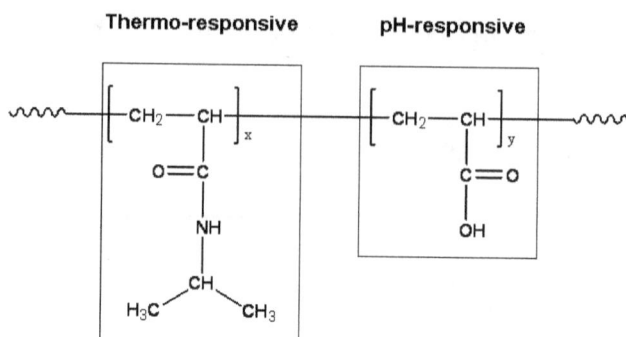

Fig. 5.72 Molecular structure of dual-responsive smart copolymer.[155]

and superhydrophobicity, in response to both temperature (T) and pH.[155] Such surfaces are obtained by simply fabricating a poly(N-iso-propyl acrylamide-*co*-acrylic acid) (P(NIPAAm-*co*-AAc)) copolymer (Fig. 5.72) thin film on a roughly etched silicon substrate. Reversible switching between superhydrophilicity and superhydrophobicity can be realized over both a narrow temperature range of about 10°C and over a relatively wide pH range of about 10.

Figure 5.73(a) shows how the dual-responsive surfaces work under the external stimuli (temperature and pH). The x-axis represents the pH value and the y-axis the temperature. The general trend in the change of wettability is that half of the CAs of the water profiles are almost all larger than 130° (the background color is red) and the others are almost all smaller than 20° (the background color is blue). The film is hydrophilic at low temperatures and hydrophobic at high temperatures (when the pH is fixed), which is similar to a single, thermally responsive film. Interestingly, the film simultaneously shows a very strong dependence on the pH value. That is, the film is hydrophobic at low pH, and hydrophilic at high pH.

To determine the effect of pH on the CAs for the dual-responsive films, further studies have been developed in detail and are summarized in Fig. 5.73(b). At pH 7 (the third column in Fig. 5.73(a)), the water profiles show that the film changes from being superhydrophilic to superhydrophobic at 32°C, which is the same LCST as

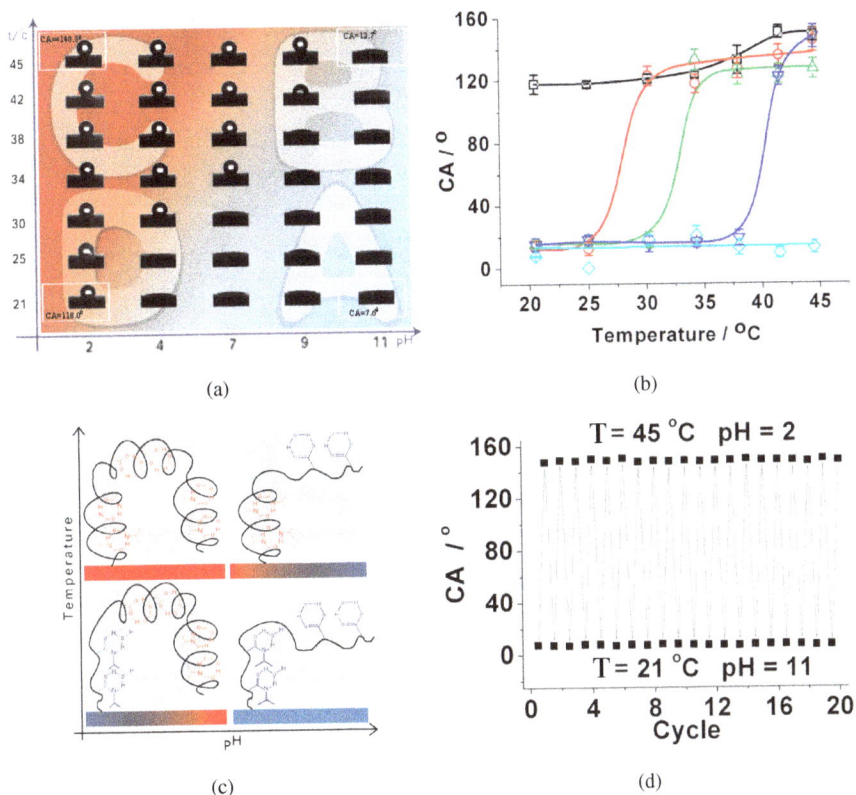

Fig. 5.73 Surface contact angle of dual-responsive smart material varies at different temperature and pH.[155]

for homo-PNIPAAm (the LCST of the copolymer here can be defined as the transition point of the CAs against temperature). This is probably due to the low content of AAc in the copolymer, which has no effect on the LCST in the neutral environment. However, at pH 4, the LCST of the copolymer is about 24°C, while at pH 9, the LCST of the copolymer is about 36°C. More interestingly, at pH values of 2 (the first column in Fig. 5.73(a)) and 11 (the fifth column in Fig. 5.73(a)), the water profiles show no obvious variation with the change of temperature.

Hypothetical conformations of hydrogen bonding between the copolymers and water in the regions A, B, C, and D are suggested in

Fig. 5.73(c), and they correspond with those in Fig. 5.73(a). In region A, both the PAAc and PNIPAAm components are at their most hydrophilic state (high pH and low temperature). The loosely coiled conformation of the P (NIPAAm-co-AAc) chains and intermolecular hydrogen bonding with water molecules lead to a high surface free energy and a small water CA. In region C, the AAc chain will become uncharged. Both the PAAc and PNIPAAm components are at their most hydrophobic state (low pH and high temperature). The compact and collapsed conformation of the P (NIPAAm-co-AAc) chains, which is induced by intramolecular hydrogen bonding between C=O, N–H, and the acrylic acid groups in the copolymer chains, leads to a low surface free energy and large water CAs. The film is hydrophobic and the rough surface increases the air/water interface. Regions B and D are the transition phases. The wettability of the film is decided by the competition of the intermolecular and intramolecular hydrogen bonding. Also the measurements of water CAs at high and low temperatures and pH are conducted on two sample stages, one set at T = 21°C and pH 11 (region A in Fig. 5.73(a)) and the other at T = 45°C and pH 2 (region C in Fig. 5.73(a)), indicates a excellent reversibility for 20 cycles (Fig. 5.73(d)) and a quick transformation between superhydrophilicity and superhydrophobicity have been recorded, as a single cycle lasts only several minutes. This reversibility remains after the samples have been laid aside without special protection for at least two months, which shows that the copolymer film is stable.

The dual-responsive smart material shown above can switch from superhydrophilicity to superhydrophobicity according to different temperature and pH. However, there is another important factor for drug delivery in human body — the concentration of glucose. For example, one of the serious problems in the treatment of insulin-dependent diabetes mellitus is the difficulty in controlling the insulin dosage. An overdose of insulin leads to an excessively decreased blood sugar level, which may result in serious hypoglycemia. Considering the complex conditions in the human body (with variations in temperature, pH, and glucose concentration), multi-responsive surfaces may prove to be suitable candidates for carrying the drug.[156]

We introduced a NIPAAM-*co*-PBA copolymer film onto a rough silicon substrate by a typical surface-initiated atom transfer radical polymerization. Thus, through the introduction of a pH/glucose-sensitive component, PBA, into a temperature-sensitive component, PNIPAAm, a NIPAAM-*co*-PBA block polymer that exhibited sensitivity towards pH, glucose, and temperature was obtained. The multi-stimulus–responsive surface can reversibly switch between super-hydrophilicity and superhydrophobicity in response to glucose, temperature, and pH changes, which may find applications in various fields including drug delivery, diagnostics, separations, and cell culture.

Figure 5.74 shows the reversible surface wettability changes between superhydrophilicity and superhydrophobicity. We demonstrated that: (1) At the pH and concentration of given glucose, the

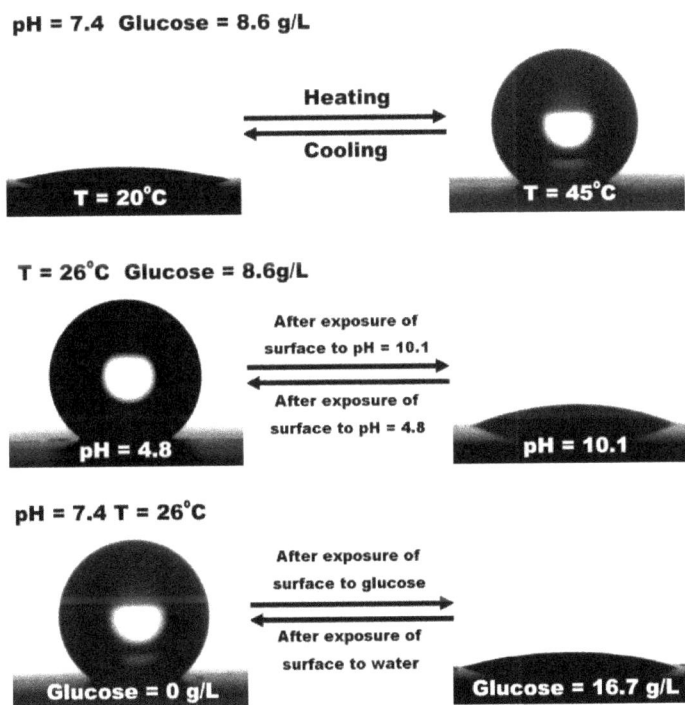

pH = 7.4 Glucose = 8.6 g/L

Heating
Cooling

T = 20°C   T = 45°C

T = 26°C Glucose = 8.6g/L

After exposure of surface to pH = 10.1
After exposure of surface to pH = 4.8

pH = 4.8   pH = 10.1

pH = 7.4 T = 26°C

After exposure of surface to glucose
After exposure of surface to water

Glucose = 0 g/L   Glucose = 16.7 g/L

Fig. 5.74 Responsive wetting behavior to (a) temperature; (b) pH; (c) glucose concentration.[156]

wetting behavior changes at different temperature. It is thought to be caused by the reversible formation of intermolecular hydrogen bonding between PNIPAAm chains and water molecules and intramolecular hydrogen bonding between $C = O$ and N–H groups in PNIPAAm chains. (2) At given temperature and concentration of glucose, the pH induced change of wetting behavior is caused by the changing of PBA molecule. (3)At given temperature and pH, the change of wetting behavior occurs because charged borates can form a stable complex with glucose, and the latter is very hydrophilic.

We could fabricate many kinds of smart interfacial materials by combining the existing materials and methods, and inducing multiple-function. They are promising "living" materials compared to the novel natural materials, which could be stimuli-sensing, self-driving and self-controlled. The composites of smart interfacial materials include sensitive materials and functional materials. The sensitive materials include thermal sensitive materials, photoinduced isomer-ization materials, magnetostrictive materials and pH-sensitive materials, etc. The Functional materials include conductive materials, semi-conductive materials, magnetic materials, etc. The smart interfa-cial materials possessing specific wetting behavior also include the multiple-combining of superhydrophobicity, superhydrophilicity, superolephobicity, superolephilicity. The smart interfacial materials system has a close relationship with many advanced sciences and technologies, such as material sciences, physics, chemistry, mechanics, biology and bio-mimic studies; it is very important and has various benefits to our industry and society. The consistent development of smart interfacial materials will promote the development of numer-ous technologies, help us explore a lot of new research fields, and also change dramatically the designing concept of new materials and structures.

References

1. Feng X J, Jiang L. *Adv. Mater.*, 2006, **18**: 3063.
2. Coulson S R, Woodward I, Badyal J P S, Brewer S A, Willis C. *J. Phys. Chem. B.*, 2000, **104**: 8836.

3. Coulson S R, Woodward I S, Badyal J P S, Brewer S A, Willis C. *Chem. Mater.*, 2000, **12**: 2031.
4. Kiuru M, Alakoski E. *Mater. Lett.*, 2004, **58**: 2213.
5. Nicolas M, Guittard F, Géribaldi S. *Langmuir*, 2006, **22**: 3081.
6. Ma M, Mao Y, Gupta M, Gleason K K, Rutledge G C. *Macromolecules*, 2005, **38**: 9742.
7. Tsujii K, Yamamoto T, Onda T, Shibuichi S. *Angew. Chem. Int. Ed. Engl.*, 1997, **36**: 1011.
8. Shibuichi S, Yamamoto T, Onda T, Tsujii K. *J. Colloid Interf. Sci.*, 1998, **208**: 287.
9. Chen W, Fadeev A Y, Hsieh M C, Öner D, Youngblood J, McCarthy T J. *Langmuir*, 1999, **15**: 3395.
10. Wang C F, Chiou S F, Ko F H, Chou C T, Lin H C, Huang C F, Chang F C. *Macromol. Rapid Commun.*, 2006, **27**: 333.
11. Jiang L, Wang R, Yang B, Li T J, Trigk D A, Fujishima A, Hashimoto K, Zhu D B. *Pure Appl. Chem.*, 2000, **72**: 73.
12. Li H J, Wang X B, Song Y L, Liu Y Q, Li Q S, Jiang L, Zhu D B. *Angew. Chem. Int. Ed.*, 2001, **40**: 1743.
13. Li H J, Wang X B, Song Y L, Liu Y Q, Li Q S, Jiang L, Zhu D B. *Chem. J. Chinese Universities*, 2001, **22**: 759.
14. Xie Q D, Xu J, Feng L, Jiang L, Tang W H, Luo X, Han C C. *Adv. Mater.*, 2004, **16**: 302.
15. Feng L, Zhang Z Y, Mai Z H, Ma Y M, Liu B Q, Jiang L, Zhu D B. *Angew. Chem. Int. Ed.*, 2004, **43**: 2012.
16. Russell T P. *Science*, 2002, **297**: 964.
17. Liu Y, Mu L, Liu B, Kong J. *Chem. Eur. J.*, 2005, **11**: 2622.
18. Takeshita N, Paradis L A, Oner D, McCarthy T J, Chen W. *Langmuir*, 2004, **20**: 8131.
19. Wu X F, Shi G Q. *Nanotechnology*, 2005, **16**: 2056.
20. Feng L, Li S H, Li H J, Zhai J, Song Y L, Jiang L, Zhu D B. *Angew. Chem. Int. Ed.*, 2002, **41**: 1221.
21. Wang J X, Wen Y Q, Feng X J, Song Y L, Jiang L. *Macromol Rapid Commun.*, 2006, **27**: 188.
22. Zhang J, Li J, Han Y. *Macromol. Rapid Commun.*, 2004, **25**: 1105.
23. Shiu J, Kuo C, Chen P, Mou C. *Chem. Mater.*, 2004, **16**: 561.
24. Nakajima A, Abe K, Hashimoto K, Watanabe T. *Thin Solid Films*, 2000, **376**: 140.
25. Li Y, Cai W P, Cao B Q, Duan G T, Sun F Q, Li C C, Jia L C. *Nanotechnology*, 2006, **17**: 238.

26. Wang S T, Feng L, Liu H, Sun T L, Zhang X, Jiang L, Zhu D B. *Chem. Phys. Chem.*, 2005, **6**: 1475.
27. Chaudhury M K, Whitesides G M. *Science*, 1992, **256**: 1539.
28. Zhang J, Xue L, Han Y. *Langmuir*, 2005, **21**: 5.
29. Lu X, Zhang J, Zhang C, Han Y. *Macromol. Rapid Commun.*, 2005, **26**: 637.
30. Yu X, Wang Z, Zhang X. *Langmuir*, 2006, **22**: 4483.
31. Daniel S, Chaudhury M K, Chen J C. *Science*, 2001, **291**: 633.
32. Crevoisier D, Fabre P, Corpart J, Liebler L. *Science*, 1999, **285**: 1246.
33. Wischerhoff E, Zacher T, Laschewsky A, Rekaï E D. *Angew. Chem. Int. Ed.*, 2000, **39**: 4602.
34. Liang L, Feng X, Liu J, Rieke P C, Fryxell G E. *Macromolecules*, 1998, **31**: 7845.
35. Liang L, Rieke P C, Fryxell G E, Liu J, Engelhard M H, Alford K L. *J. Phys. Chem. B.*, 2000, **104**: 11667.
36. Liang L, Rieke P C, Liu J, Fryxell G E, Young J S, Engelhard M H, Alford K L. *Langmuir*, 2000, **16**: 8016.
37. Kuckling D, Harmon M E, Frank C W. *Macromolecules*, 2002, **35**: 6377.
38. Yakushiji T, Sakai K, Kikuchi A, Aoyagi T, Sakurai Y, Okano T. *Langmuir*, 1998, **14**: 4657.
39. Lin S, Chen K, Liang R. *Polymer*, 1999, **40**: 2619.
40. Sun T L, Wang G J, Feng L, Liu B Q, Ma Y M, Jiang L, Zhu D B. *Angew. Chem. Int. Ed.*, 2004, **43**: 357.
41. Fu Q, Rao G V R, Basame S B, Keller D J, Artyushkova K, Fulghum J E, López G P. *J. Am. Chem. Soc.*, 2004, **126**: 8904.
42. Sun T L, Liu H, Song W L, Wang X, Jiang L, Li L, Zhu D B. *Angew. Chem. Int. Ed.*, 2004, **43**: 4663.
43. Sun T L, Song W L, Jiang L. *Chem. Commun.*, 2005, 1723.
44. Hu Z, Chen Y, Wang C, Zheng Y, Li Y. *Nature*, 1998, **393**: 149.
45. Shirtcliffe N J, McHale G, Newton M I, Perry C C, Roach P. *Chem. Commun.*, 2005, 3135.
46. Wang R, Hashimoto K, Fujishima A, Chikuni M, Kojima E, Kitamura A, Shimohigoshi M, Watanabe T. *Nature*, 1997, **388**: 431.
47. Wang R, Sakai N, Fujishima A, Watanabe T, Hashimoto K. *J. Phys. Chem. B.*, 1999, **103**: 2188.
48. Sakai, N, Wang R, Fujishima A, Watanabe T, Hashimoto K. *Langmuir*, 1998, **14**: 5918.

49. Machida M, Norirmoto K, Watanabe T, Hashimoto K, Fujishima A. *J. Mater. Sci.*, 1999, **34**: 2569.

50. Miyauchi M, Nakajima A, Hashimoto K, Watanabe T. *Adv. Mater.*, 2000, **12**: 1923.

51. Sakai N, Fujishima A, Watanabe T, Hashimoto K. *J. Phys. Chem. B*, 2001, **105**: 3023.

52. Nakajima A, Koizumi S, Watanabe T, Hashimoto K. *Langmuir*, 2000, **16**: 7048.

53. Wang R, Hashimoto K, Fujishima A, Chikuni M, Kojima E, Kitamura A, Shimohigoshi M, Watanabe T. *Adv. Mater.*, 1998, **10**: 135.

54. Highfield J G, Grätzel M. *J. Phys. Chem.*, 1988, **92**: 464.

55. Hugenschmidt M B, Gamble L, Campbell C T. *Surf. Sci.*, 1994, **302**: 329.

56. Henderson M A. *Langmuir*, 1996, **12**: 5093.

57. Henderson M A. *Surf. Sci.*, 1996, **355**: 151.

58. Fujishima A, Hashimoto K, Watanabe T. *TiO$_2$ Photocatalyst, Fundamentals and Applications*. BKC Inc.: Tokyo, 1999, p. 66.

59. Sakai N, Fujishima A, Watanabe T, Hashimoto K. *J. Phys. Chem. B.*, 2003, **107**: 1028.

60. Stevens N, Priest C I, Sedev R, Ralston J. *Langmuir*, 2003, **19**: 3272.

61. Miyauchi M, Kieda N, Hishita S, Mitsuhashi T, Nakajima A, Watanabe T, Hashimoto K. *Surf. Sci.*, 2002, **511**: 401.

62. Nakajima A, Koizumi S, Watanabe T, Hashimoto K. *J. Photochem. Photobio. A: Chem.*, 2001, **146**: 129.

63. Miyauchi M, Nakajima A, Fujishima A, Hashimoto K, Watanabe T. *Chem. Mater.*, 2000, **12**: 3.

64. Nakajima A, Koizumi S, Watanabe T, Hashimoto K. *Langmuir*, 2000, **16**: 7048.

65. Langlet A, Permpoon S, Riassetto D, Berthomé G, Pernot E, Joud J. *J. Photochem. Photobio. A: Chem.*, 2006, **181**: 203.

66. Takeuchi M, Sakamoto K, Martra G, Coluccia S, Anpo M. *J. Phys. Chem. B.*, 2005, **109**: 15422.

67. Jiang F Z. Work report of postdoctoral research (Institute of Chemistry, Chinese Academy of Sciences), 2001.

68. Setzer W N, Liou S, Easterling G E, Simmons R C, Gullion L M, Meehan E J. *Heteroatom Chemistry*, 1998, **9**: 123.

69. Zhang Y R, Asahina S, Suzuki M, Yoshihara S, Shirakashi T. *Surface & Coatings Technology*, 2005, **196**: 303.

70. Gao Y F, Masuda Y, Koumoto K. *Langmuir*, 2004, **20**: 3188.
71. Shibata T, Irie H, Hashimoto K. *J. Phys. Chem. B.*, 2003, **107**: 10696.
72. Pinzari F, Ascarelli P, Cappelli E, Giorgi R. *Langmuir*, 2002, **18**: 5457.
73. Du Y K, Gan Y Q, Yang P, Zhao F, Hua N P, Jiang L. *Thin Solid Films*, 2005, **491**: 133.
74. Irie H, Mori H, Hashimoto K. *Vacuum.*, 2004, **74**: 625.
75. Guan K S, Lu B J, Yin Y S. *Surf. Coat. Technol.*, 2003, **173**: 219.
76. Liu Q J, Wu X H, Wang B L, Liu Q. *Mater. Res. Bull.*, 2002, **37**: 2255.
77. Yu J G, Zhao X J, Zhao Q G, Wang G. *Mater. Chem. Phys.*, 2001, **68**: 253.
78. Yu J C, Yu J G, Tang H Y, Zhang L Z. *J. Mater. Chem.*, 2002, **12**: 81.
79. Yu J G, Zhao X J. *J. Mater. Sci. Lett.*, 2001, **20**: 671.
80. Yu J C, Yu J G, Ho W K, Zhao J C. *J. Photochem. Photobio. A: Chem.*, 2002, **148**: 331.
81. Yu J G, Yu J C, Ho W K, Jiang Z T. *New J. Chem.*, 2002, **26**: 607.
82. Censo D D, Exnar I, Graetzel M. *Electrochem. Commun.*, 2005, **7**: 1066.
83. Kommireddy D S, Patel A A, Shutava T G, Mills D K, Lvov Y M. *J. Nanosci. Nanotech.*, 2005, **5**: 1081.
84. Miyauchi M, Nakajima A, Hashimoto K, Watanabe T. *Adv. Mater.*, 2000, **12**: 1923.
85. Cai R, Van G M, Aw P K, Itoh K. *Comptes Rendus Chimie.*, 2006, **9**: 829.
86. Asahi R, Morikawa T, Aoki K, Taga. *Science*, 2001, **293**: 269.
87. Dohshi S, Takeuchi M, Anpo M. *Catalysis Today*, 2003, **85**: 199.
88. Kang M S. *Mater. Lett.*, 2005, **59**: 3122.
89. Ren D S, Bei Z M, Shen J, Cui X L, Yang X L, Zhang Z J. *J. Sol-gel Sci. Technol.*, 2005, **34**: 123.
90. Bei Z M, Ren D S, Cui X L, *et al. J. Mater. Res.*, 2004, **19**: 3189.
91. Ren D S, Bei Z M, Huang L, Shen J, Cui X L, Yang X L, Zhang Z J. *Acta Physico-Chimica Sinica.*, 2004, **20**: 414.
92. Yu J G, Zhou M H, Yu H G, Zhang Q J, Yu Y. *Mater. Chem. Phys.*, 2006, **95**: 193.
93. Premkumar J. *Chem. Mater.*, 2004, **16**: 3980.
94. Weng W J, Ma M, Du P Y, Zhao G L, Shen G, Wang J X, Han G R. *Surf. Coat. Technol.*, 2005, **198**: 340.
95. Kuo C S, Tseng Y H, Li Y Y. *Chem. Lett.*, 2006, **35**: 356.
96. Cebeci F C, Wu Z Z, Zhai L, Robert E. Cohen R E, Rubner M F. *Langmuir*, 2006, **22**: 2856.

97. Karuppuchamy S, Jeong J M, Amalnerkar D P, *et al. Vacuum*, 2006, **80**: 494.

98. Karuppuchamy S, Jeong J M. *Mater. Chem. Phys.*, 2005, **93**: 251.

99. Yamashita H, Nishio S, Katayama I, Nishiyama N, Fujii H. *Catalysis Today*, 2006, **111**: 254.

100. Gu Z Z, Fujishima A, Sato O. *Appl. Phys. Lett.*, 2004, **85**: 5067.

101. Asthana Y, Kim D P. *Mater. Sci. Forum*, 2006, **510–511**: 42.

102. Zhang X T, Sato O, Taguchi M, Einaga Y, Murakami T, Fujishima A. *Chem. Mater.*, 2005, **17**: 696.

103. Cebeci F C, Wu, Z Z, Zhai L, Cohen R E, Rubner M F. *Langmuir*, 2006, **22**: 2856.

104. Sun R D, Nakajima A, Fujishima A, Watanabe T, Hashimoto K. *J. Phys. Chem. B.*, 2001, **105**: 1984.

105. Abbott S, Ralston J, Reynolds G, Hayes R. *Langmuir*, 1999, **15**: 8923.

106. Feng C, Zhang Y J, Jin J, Song Y L, Xie L Y, Qu G R, Jiang L, Zhu D B. *Langmuir*, 2001, **17**: 4593.

107. Feng X J, Feng L, Jin M H, Zhai J, Jiang L, Zhu D B. *J. Am. Chem. Soc.*, 2004, **126**: 62.

108. Feng X J, Zhai J, Jiang L. *Angew. Chem. Int. Ed.*, 2005, **44**: 5115.

109. Zhu W Q, Feng X J, Feng L, Jiang L. *Chem. Commun.*, 2006, 2753.

110. Liu H, Feng L, Zhai J, Jiang L, Zhu D B. *Langmuir*, 2004, **20**: 5659.

111. Wang S T, Feng X J, Yao J N, Jiang L. *Angew. Chem. Int. Ed.*, 2006, **45**: 1264.

112. Zhang X, Sato O, Fujishima A. *Langmuir*, 2004, **20**: 6065.

113. Meng X Q, Zhao D X, Zhang J Y, Shen D Z, Lu Y M, Dong L, Xiao Z Y, Liu Y C, Fan X W. *Chem. Phys. Lett.*, 2005, **413**: 450.

114. Yin S, Sato T. *J. Mater. Chem.*, 2005, **15**: 4584.

115. Zhang J L, Huang W H, Han Y C. *Langmuir*, 2006, **22**: 2946.

116. Balaur E, Macak J M, Tsuchiya H, Schmuki P. *J. Mater. Chem.*, 2005, **15**: 4488.

117. Wang R, Jiang L, Iyoda T, Tryk D A, Hashimoto K, Fujishima A. *Langmuir*, 1996, **12**: 2052.

118. Jiang W H, Wang G J, He Y N, Wang X G, An Y L, Song Y L, Jiang L. *Chem. Commun.*, 2005, 3550.

119. Ge H L, Wang G J, He Y N, Wang X G, Song Y L, Jiang L, Zhu D B. *Chem. Phys. Chem.*, 2006, **7**: 575.

120. Rosario R, Gust D, Garcia A A, Hayes M, Taraci J L, Clement T, Dailey J W, Picraux S T. *J. Phys. Chem. B.*, 2004, **108**: 12640.

121. Ichimura K, Oh S, Nakagawa M. *Science*, 2000, **288**: 1624.
122. Verheijen H J J, Prins M W J. *Langmuir*, 1999, **15**: 6616.
123. Pollack M G, Fair R B, Shenderov A D. *Appl. Phys. Lett.*, 2000, **77**: 1725.
124. Krupenkin T N, Taylor J A, Schneider T M, Yang S. *Langmuir*, 2004, **20**: 3824.
125. Krupenkin T, Yang S, Mach P. *Appl. Phys. Lett.*, 2003, **82**: 316.
126. Yang S, Krupenkin T, Mach P, Chandross E. *Adv. Mater.*, 2003, **15**: 940.
127. Hayes R A, Feenstra B J. *Nature*, 2003, **425**: 383.
128. Lahann J, Mitragotri S, Tran T-N, Kaido H, Sundaram J, Choi I S, Hoffer S, Somorjai G, Langer R. *Science*, 2003, **299**: 371.
129. Chen W, Fadeev A Y, Hsieh M C, Öner D, Youngblood J, McCarthy T J. *Langmuir*, 1999, **15**: 3395.
130. Takeda K, Nakajima A, Hashimoto K, Watanabe T. *Surf. Sci.*, 2002, **519**: L589.
131. Katz E, Sheeney-Haj-Ichia L, Basnar B, Felner I, Willner I. *Langmuir*, 2004, **20**: 9714.
132. Hong X, Jiang L. *J. Am. Chem. Soc.*, 2007, **129**: 1478.
133. Zhang J, Lu X, Huang W, Han Y. *Macromol. Rapid Commun.*, 2005, **26**: 477.
134. Bai D S, Habersberger B M, Jennings G K. *J. Am. Chem. Soc.*, 2005, **127**: 16486.
135. Hou Z, Abbott N L, Stroeve P. *Langmuir*, 2000, **16**: 2401.
136. Liu Y, Zhao M, Bergbreiter D E, Crooks R M. *J. Am. Chem. Soc.*, 1997, **119**: 8720.
137. Holmes-Farley S R, Bain C D, Whitesides G M. *Langmuir*, 1988, **4**: 921.
138. Bain C D, Whitesides G M. *Langmuir*, 1989, **5**: 1370.
139. Ionov L, Houbenov N, Sidorenko A, Stamm M, Luzinov I, Minko S. *Langmuir*, 2004, **20**: 9916.
140. Jiang Y G, Wang Z Q, Yu X, Shi F, Xu H P, Zhang X. *Langmuir*, 2005, **21**: 1986.
141. Yu X, Wang Z Q, Jiang Y G, Shi F, Zhang X. *Adv. Mater.*, 2005, **17**: 1289.
142. Minko S, Patil S, Datsyuk V, Simon F, Eichhorn K, Motornov M, Usov D, Tokarev I, Stamm M. *Langmuir*, 2002, **18**: 289.
143. Motornov M, Minko S, Eichhorn K, Nitschke M, Simon F, Stamm M. *Langmuir*, 2003, **19**: 8077.

144. Minko S, Müller M, Motornov M, Nitschke M, Grundke K, Stamm M. *J. Am. Chem. Soc.*, 2003, **125**: 3896.

145. Ionov L, Minko S, Stamm M, Gohy J, Jérôme R, Scholl A. *J. Am. Chem. Soc.*, 2003, **125**: 8302.

146. Uhlmann P, Ionov L, Houbenov N, Nitschke M, Grundke K, Motornov M, Minko S, Stamm M. *Prog. in Org. Coatings*, 2006, **55**: 168.

147. Anastasiadis S H, Retsos H, Pispas S, Hadjichristidis N, Neophytides S. *Macromolecules*, 2003, **36**: 1994.

148. Granville A M, Brittain W J. *Macromol. Rapid Commun.*, 2004, **25**: 1298.

149. Lu X Y, Peng J A, Li B Y, Zhang C C, Han Y C. *Macromol. Rapid Commun.*, 2006, **27**: 136.

150. Tan H, Sun T, Li J, Guo M, Xie X, Zhong Y, Fu Q, Jiang L. *Macromol. Rapid Commun.*, 2005, **26**: 1418.

151. S-Huethorst J A M, Fokkink L G J. *Langmuir*, 1994, **10**: 4380.

152. Abbott N L, Gorman C B, Whitesides G M. *Langmuir*, 1995, **11**: 16.

153. Isaksson J, Tengstedt C, Fahlman M, Robinson N, Berggren M. *Adv. Mater.*, 2004, **16**: 319.

154. Xu L, Chen W, Mulchandani A, Yan Y. *Angew. Chem. Int. Ed.*, 2005, **44**: 6009.

155. Xia F, Feng L, Wang S T, Sun T L, Song W L, Jiang W H, Jiang L. *Adv. Mater.*, 2006, **18**: 432.

156. Xia F, Ge H, Hou Y, Sun T L, Zhang G Z, Jiang L. *Adv. Mater.*, 2007, **19**: 2520.

Chapter 6

Conclusion and Prospect

Bioinspired smart material has been attracting more and more interest ever since it came into being. Research on bioinspired smart materials has great value, for it combines the processes of learning from nature, mimicking nature and surpassing nature. As the source of knowledge innovation, it provides new ideas, new theories and new methods for innovation of science and technology. During the process of evolution in several millions of years, plants and animals have already developed to be adaptive to the environment, with certain structural-functional characteristics approaching perfection. Mimicking the structures and properties of nature will build a bridge between biology and technology. This will help us solve technological problems. Through reproducing the biological basic principles, we can figure out the technological solutions. To observe new phenomenon, realize new laws, propose new concepts, and set up new theories in the system of bioinspired smart materials will lay a solid foundation for the bioinspired science system, and greatly enrich the connotation of biological science, material science, information science, mathematics, mechanics, technology, and systems science. Nowadays, the hot topics in the research of smart materials include unique hydrophilic/hydrophobic surfaces with controllable wettability, fresh-water harvesting materials, anti-agglomeration materials, molecular recognition materials, structural-color controllable materials, energy-saving materials, energy-transition materials, responsive structure materials, etc. This gives the definite destination of applications in human health, environment, resource and energy.

Breakthrough in designing new smart materials and bioinspired smart devices is being conceived by utilizing new methods, new principles, and new physical properties. The application of bioinspired smart materials will greatly impact on the allocation adjustment of pillar industries in national economy, the design of new products, the formation of new industry, and reconstruction of traditional industries.

Nowadays, many countries have elaborately prepared their long-term plans on smart bionics. In USA there is a long-term program related to bionic technology, giving priority to the development of advanced manufacturing process (simulation, biotechnology), advanced materials, and advanced military equipments. In Germany, the federal ministry of education and research has already devoted a great mounts of manpower and materials to "Technology in the 21st century" focusing on adaptive electronic technique, bioinspired materials, biosensors, etc. Besides, many other countries such as Great Britain, Japan, and Russia, have drawn up programs on bionics in order to compete at the source of innovation. Bioinspired smart materials have been a key research direction in the "Research key points of industry technology" of Japan as early as 1994. The research achievements made by these countries have already been applied in economy, military affairs, human health and have created great economic results. American Materials Research Society held a meeting in Boston on the theme of "BioInspired Nanoscale Hybrid Systems" in December 2002. It was pointed out that the combination of nanostructure materials and bioinspired research is a new highlight of the scientific research today. Bioinspired smart material is one of the ten research fields of material science which are pushed vigorously by the European Union. In 2004, American Materials Research Society Fall Meeting at Boston set up a symposium on "Mechanical Properties of BioInspired and Biological Materials". In November 2005, *Science* magazine organized a special section titled as "Design for Living".[1] It reviewed the latest progress of design and application of bioinspired functional materials. The guiding role of the inner-relationship research between living structure and function has been emphasized so as to show the importance, as a new cutting-edge field, of the design and fabrication of bioinspired function materials. In February

2006, China announced its "National Program for Medium-to-Long-Term Scientific and Technological Development" and made the "Technology of bioinspired materials and structure" the starting point of the development of new material technology.

The organic combination of smart materials and interface materials, i.e. providing interface materials with smart properties, is going to be a new field of research. The research on bioinspired smart materials is an important fundamental field containing interface effect, multiscale structure effect, cooperation of physical properties, weak interaction effect and bioinspired intelligence. The rush of research on "bioinspired smart materials" is expected to progress rapidly in the 21st century.

As early as 1980s, scientists had been discovering fine crystalline particles along with the development of research on nanoscale materials. These materials are of great importance not because of their "beauty" but the bright future for practical applications. So, the research work at nanoscale has attracted more and more attention from scientists and engineers. These little groups have names such as nanoparticle, nanocrystal, atomic cluster and quantum dot. Currently, a great number of research projects are focused on fabricating nanostructures with both uniform size and unique optical, electrical, magnetic and mechanical properties.[2]

Now, these efforts have achieved some "return": the efficiency of chemical catalysis has improved and ceramics of highly specific surface area can be fabricated. Besides, the application of semiconductor nanocluster in the commercial field brings us confidence about this technique as well. However, in order to acquire a new type of high-level functional material, scientists should not only consider the properties of a single nanoparticle (e.g. high specific surface, quantum confinement effect), but also consider the cooperation between the nanostructures, especially the macroscopic properties after "assembly" as well.

The concept of "binary cooperative complementary nanoscale material" proposed by us has been taken as a new sight by scientists in the design and fabrication of novel functional materials.[3] This concept is based on the fabrication of a unique solid surface with nanostructures that have cooperative complementary properties. Under some equivalent conditions, this kind of special surfaces will exhibit unexpected properties which have great value in theory and practical application.

Actually, the yin and the yang represent the two opposing fundamental properties of nature and the universe in ancient Taiji Chinese philosophy. These ancient philosophy principles still play an important role in classical and modern science. Along with the development of research work, people begin to learn that the binary cooperative complementary properties exhibit universality. For example, in the construction of chemical elements of the universe, the physical and chemical properties of materials are combined into the binary cooperative concept. As material science progresses, people have been able to utilize binary cooperative complementary effect to fabricate unique functional materials. Up to now, binary atomic materials (binary alloys) and binary molecular materials have been identified to have great potential in basic research and practical applications. The most famous example of binary atomic materials is the combination of Fe and Cr resulting in a new kind of steel (introducing Co into Fe to be the hardest alloy up to date). Furthermore, many other binary alloys have displayed lots of unique properties.[4] Amphiphilic molecules, as an example of binary molecules, have been applied in the fields like surfactant. By combining both electron donor and acceptor in the same molecule, scientists have successfully fabricated nonlinear optical materials. Besides, through the synthesis of donor-acceptor complex, organic conductor, even superconductor has been realized.[5] The most typical high-level binary molecules are DNA and RNA. Still, there is plenty room waiting for us to explore. The atomic and molecular systems of binary cooperative complementary structures have displayed the special language of materials for us.

The expansion of the concept of "binary cooperative complementary effect" from molecular level to nanostructure induces a novel material system, i.e. binary nanoscale material. Before reviewing this topic in great detail, let us start with binary interface material which has priority over designing high-level functional materials. The purpose of developing binary cooperative complementary interface materials is to fabricate macroscopic objects with unique properties.

Through applying the concept of binary cooperative complementary effect on solid surface at nanoscale, we hope to achieve nanostructures with cooperative complementary properties such as: hydrophilicity/hydrophobicity, conductor/insulator, bulge/hollow, P-type/N-type semiconductor, oxidation/reduction state, ferromagnetism/antiferromagnetism and so on. No matter what couple of cooperative complementary nanostructures they are, as long as they are under some equivalent conditions, they will display characteristic interface properties. As described in the previous chapters, the unique surface properties of being superamphiphilic/superamphiphobic have displayed the great importance of the concept of binary cooperative complementarity in designing novel interface materials. Next, we are going to show several examples in order to explain the unique properties of the binary cooperative complementarity nanoscale materials.

6.1 Super-Lattice Surface Structures (Stable and Metastable Binary Cooperative Complementary Structure)

Photoinduced surface-phase transitions (surface reconstruction) generate binary (stable cubic phase and metastable hexagonal phase) nanostructures showing several different types of nanopatterns on the surfaces of C_{60} single crystals.[6,7] C_{60} solid may condense into both face-centered cubic (fcc) and hexagonal close-packed (hcp) phase at room temperature. The hcp phase is a metastable one, and cohesive energy difference between the hcp and fcc phases is as small as 0.9 kcal mol^{-1}. This provides a possibility for the coexistence of the two phases on the same surface. Upon photon absorption in the 1.62–2.2 eV range, Frenkel excitons are generated from a strong exciton-lattice coupling in solid C_{60}. These excitons are found hopping between nearest neighbor molecules, that is, anisotropic hopping along the three equivalent [110] directions on the (111) surface or two equivalent [011] directions on (100) face. Under higher light intensity, collective molecular displacement is induced on the surface of the crystal, which results in a large scale surface molecular rearrangement. On the (111) surface, the collective displacement is along the three [110] directions, forming a triangular pattern. On the (100) surface, the displacement is along two equivalent [011] directions forming a cubic pattern. Photoinduced surface reconstruction can be understood as a general phenomenon, and may occur on a wide variety of molecular crystals.

As mentioned in Chapter 1, the assembly of nanoparticles of two different materials into a binary nanoparticle superlattice (BNSL) can also be a kind of new materials. As reported by Shevchenko et al.,[8] they demonstrated formation of more than 15 different BNSL structures, using combinations of semiconducting, metallic and magnetic nanoparticle building blocks. The electrical charges on sterically stablilized nanoparticles determine BNSL stoichiometry and additional contributions from entropic, van der Waals, steric and dipolar forces that stabilize the variety of BNSL structures. Kalsin et al.[9] reported that self-assembly of charged, equally sized metal nanoparticles of two types (gold and silver) led to the formation of large, sphalerite (diamond-like) crystals, in which each nanoparticle has four oppositely charged neighbors (Fig. 6.1). Formation of these non-closed-packed structure is a consequence of elecstatic effects specific to the nanoscale, where the thickness of the screening layer is commensurate with the dimensions of the assembling objects. Because of electrostatic stabilization of larger crystallizing particles by smaller ones, better-quality crystals can be obtained from more polydisperse nanoparticle solutions.

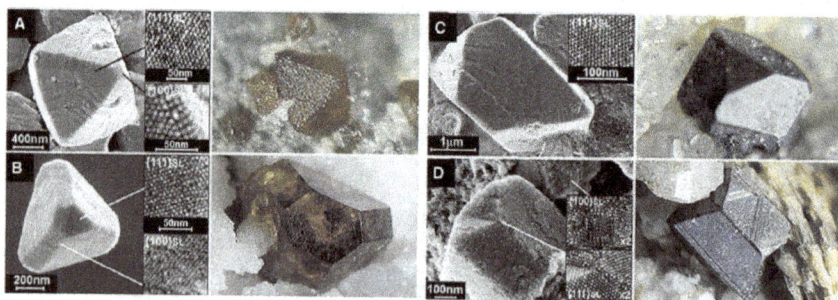

Fig. 6.1 Morphologies of the AuMUA-AgTMA crystals (left of A–D) and their corresponding macroscopic sphalerite images (right of A–D). (A) Octahedron. (B) Cut tetrahedron. (C) Octahedron with two triangular faces cut. (D) Twinned octahedron. (From Ref. 9. Reprinted with permission from AAAS.)

6.2 Optically Controllable Superconducting System (Superconducting/Normal-conducting Phase Binary Cooperative Complementary Structure)

Binary nanostructures formed from superconducting and normal-conducting phases provide an optically controllable superconducting system.[10] The photoinduced modification of near-surface region of K_3C_{60} samples generates very large enhancement of the diamagnetic shielding fraction value. Before illumination, only ~10% by volume of K_3C_{60}-quenched sample exhibited diamagnetism. However, after illumination, values close to 100% of shielding fraction value were obtained. Interestingly, the diamagnetic shielding fraction value was found to return to its original value upon heating, demonstrating that the transformation was reversible. By using Raman spectroscopy, this effect was addressed as photopolymerization and thermal depolymerization of C_{60} molecules in the normal-conducting phase at grain boundaries of the superconducting phase. The nanodomain structures were directly observed by AFM. The domains appear as spherical surface and are about 30–60 nm in size. Upon illumination in a wavelength range of 300–500 nm, an obvious shrinkage occurred near the domain boundaries, and a clear outline of the domain structures was produced, as shown in Fig. 6.3(b). The obvious change upon illumination was in the domain boundaries. These results indicate that the electronic connection between the superconducting nanodomain structure may greatly influence the physical properties of the superconducting system.

6.3 Chiroptical Switch (Chiral/Achiral Binary Cooperative Complementary Structure)

Liu and co-workers[11] (Institute of Chemistry, Chinese Academy of Sciences) reported their work on chiroptical switch based on an achiral molecule. They found that an achiral 5-(octadecyloxy)-2-(2-thiazolylazo)phenol (TARC18) could form a chiral Langmuir-Schaefer (LS) film through organization of the molecules at the air/water interface (Fig. 6.2). The results suggest that TARC18

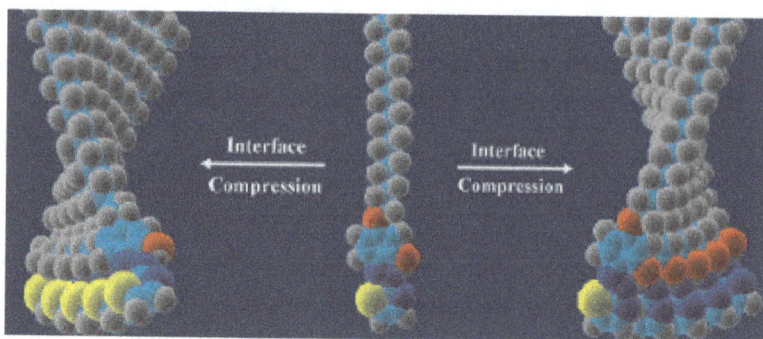

Fig. 6.2 The possible helical stacking of the achiral TARC18 molecule in LS films.[11] (Copyright Wiley–VCH Verlag GmbH & Co. KGaA. Reproduced with permission.)

molecules could be aligned in two different ways in the LS films. These two kinds of assemblies give different UV-vis absorption spectra and circular dichroism (CD) spectra. It was more interesting that the chirality of the TARC18 LS films could be destroyed by exposing the film to HCl gas and could be recovered in air. The number of repeat cycles can be more than ten. In short, the supramolecular chiroptical switch can be achieved from totally achiral molecules and this will bring more opportunities in the design and fabrication of novel chiral materials.

6.4 Novel Mesoporous Structure (Crystalline/Amorphous Phase Binary Cooperative Complementary Structure)

Zhao *et al.*[12] (Fudan University, China) reported their work on fabricating nanoporous metal oxide crystals during the crystalline phase transition process where mesoporous silica was utilized. These metal oxides include Co_3O_4, In_2O_3, Cr_2O_3, NiO, Fe_2O_3, MnxOy, CeO_2, and WO_3. Figure 6.3 is a schematic diagram to help visualize the possible structural correlation between the 2D hexagonal and 3D bicontinuous cubic mesostructures. The rudimentary and final channel structures viewed from $[100]_h$ and $[211]_c$ directions are shown in parts (A) and

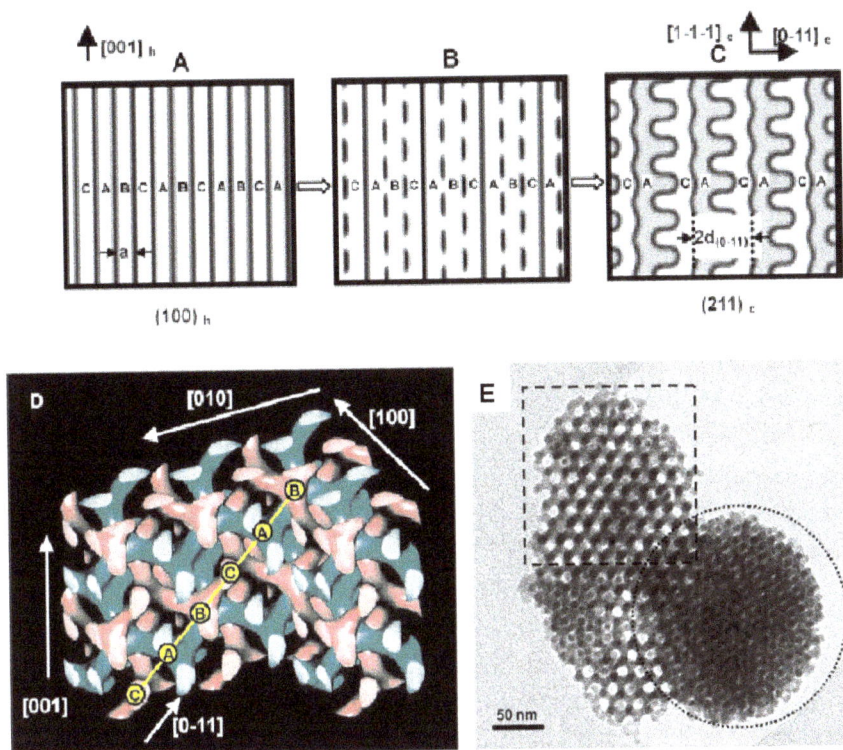

Fig. 6.3 (A) (100) plane of 2D hexagonal phase. (B) An intermediate phase. (C) (211) plane of cubic gyroidal phase. (D) 3D model of double gyroidal channel structures. (E) TEM image of a small fragment of In_2O_3 prepared from the strategy. (Reprinted with permission from Ref. 12. Copyright 2004 American Chemical Society.)

(C), respectively, considering the fact that the cylinder axis (parallel to the (100) plane) of the hexagonal phase is equivalent to the body diagonal (parallel to the (211) plane) of the gyroidal cubic phase. Each channel is labeled as A, B or C. For the 2D hexagonal mesostructure, the channels A, B and C are equal with respect to each other and align side by side (the interdistance is the cell parameter, a, of the 2D hexagonal phase). An intermediate state (Fig. 6.3), caused by the thermal-induced framework rearrangement, assumes that channel B begins contacting channel A and C at equidistant sites (note the adjacent A and C channels are always separate). A final form, i.e. 3D bicontinuous cubic mesostructure, is achieved by uniformly dividing channel B and merging totally with channels A and C at the contacting sites, accompanying the curvature change. The bicontinuous channels (A and C) are aligned along the [0–11] direction, and the single channel is extended along the [1-1-1] direction. An interval of $2d$ (0-11) is calculated between adjacent equal (A or C) channels on this projection plane. It should be noted that the same channels (A or C) connect as a whole at periodic positions above and beneath this hypothetic plane, producing gigantic bicontinuous networks. The simulated perspective model of interwoven channels gives a further 3D insight into the planar structures. The double channels are marked by light green and pink, respectively. At the A or C sites and parallel with the [1-1-1] axis (namely, the body diagonal direction, not shown here), only a light green channel or pink channel is observed, indicating the extension of a single continuous channel along this direction, whereas at B sites and along the same direction, pink and light green channels appear alternately and link with adjacent channels (A or C). Figure 6.3(E) shows the TEM image of a small fragment of In_2O_3 prepared from the strategy mentioned above, which is viewed from [111] zone axis. The square and circular domains are indicative of single uncoupled subframework and coupled frameworks, respectively.

6.5 Interface of the Engineered Magnetism (Ferromagnetic/Antiferromagnetic Binary Cooperative Complementary Structure)

Interface-selective probing of magnetism is a key issue for the design and realization of spin-electronic junction devices. For example, Yamada *et al.*[13] reported that magnetization-induced second-harmonic generation was used to probe the local magnetic properties at the interface of the perovskite ferromagnet $La_{0.6}Sr_{0.4}MnO_3$ (LSMO) with nonmagnetic insulating layers, as used in spin-tunnel junctions. By grading the doping profile on an atomic scale at the interface, robust ferromagnetism can be achieved around room temperature. Figure 6.4 shows the atomic stacking sequences for perovskite-type magnetic oxide interfaces. On the left is a LSMO film capped with a 5-unit-cells (2 nm) layer of $LaAlO_3$ (LAO) or $SrTiO_3$ (STO). On the

A STO or LAO / LSMO   STO / LAO / LSMO

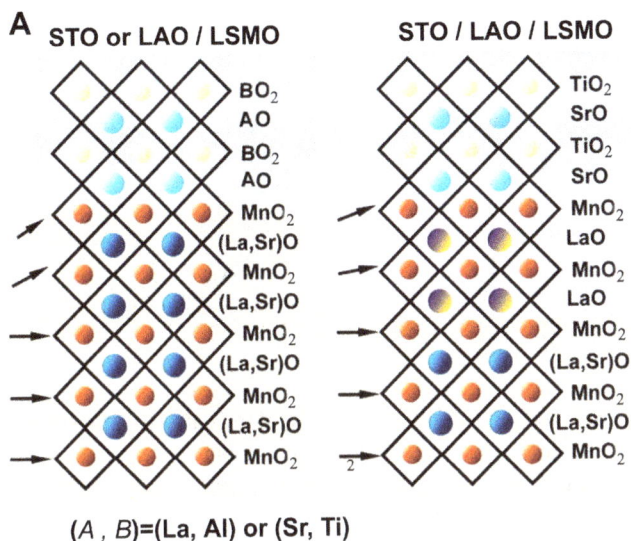

(A, B) = (La, Al) or (Sr, Ti)

Fig. 6.4 Schematic diagram of the atomic stacking sequence for perovskite-type magnetic oxide interfaces. (From Ref. 13. Reprinted with permission from AAAS.)

right is a 2-unit-cells layer of undoped LMO which is inserted between STO and LSMO layers to grade the hole concentration in the manganite layer near the interface. In these films, the top surface is terminated by the MnO_2 or BO_2 (B = Ti or Al) layer. The arrows depict possible arrangements of magnetic moments at the respective MnO_2 layers, showing considerable spin canting near the interface (left) and robust ferromagnetism (right). The results should lead to improvements in the performance of spin-tunnel junction.

Ohtomo and Hwang[14] also reported that polarity discontinuities at the interfaces between different crystalline materials (heterointerfaces) could lead to nontrivial local atomic and electronic structure, owing to the presence of dangling bonds and incomplete atomic coordinations (Fig. 6.5). The $(LaO)^+/(TiO_2)^0$ interface was grown by direct deposition on this surface using a KrF excimer laser at a repetition rate of 2 Hz and a laser fluence at the target surface of ~1Jcm^{-2}. The $(AlO_2)^-/(SrO)^0$ interface was grown by first depositing a monolayer of SrO from a single crystal target, then switching to LaAlO$_3$ *in situ*. After

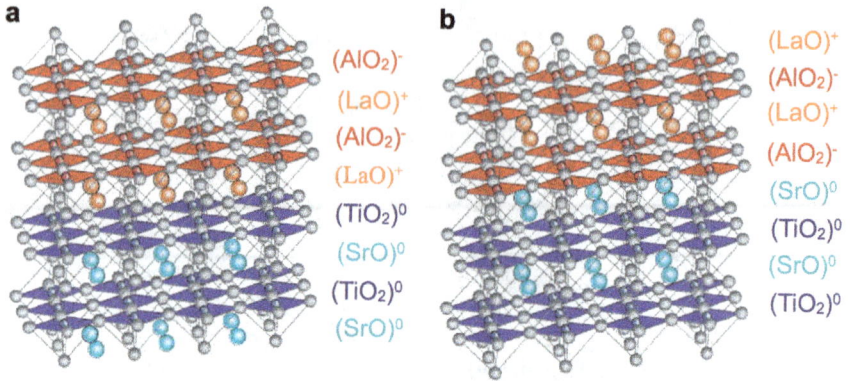

Fig. 6.5 (a) Schematic illustration of the resulting $(LaO)^+/(TiO_2)^0$ interface, indicating the composition of each layer and the ionic charge state of each layer. (b) Schematic illustration of the resulting $(AlO_2)^-/(SrO)^0$ interface.[14] (Reprinted by permission from Macmillan Publishers Ltd: *Nature*, copyright 2004.)

growth of $LaAlO_3$, atomic force microscopy revealed that the original step and terrance structure of the substrate surface were preserved.

Liu *et al.*[15] (Institute of Metal Research, Chinese Academy of Sciences) reported the exchange coupling and remanence enhancement in nanocomposite $(Nd, Dy)(Fe, Co, Nb, B)_{5.5}/\alpha$-Fe thin films prepared by sputtering and heating treatments. Figure 6.6 shows the TEM bright-field image of the as-deposited Ti (10 nm)/ [NdDyFeCoNbB (16 nm) Fe(6 nm)] × 20/Ti (10 nm)/(glass-ceramics substrate) multilayer thin film. The cross-sectional view shows that the distribution of the soft- and hard-phase layers is very uniform. Selected-area electron diffraction analysis revealed that the wide NdDyFeCoNbB layer was amorphous, and the narrow one corresponded to Fe. Clearly, the ratio and distribution of the soft and hard phases could be controlled well in this way.

6.6 Ionic/Nonionic Conductor Binary Cooperative Complementary Structure

Ionic conductivity in solid electrolytes can be improved by dissolving appropriate impurities into the structure or by introducing interfaces

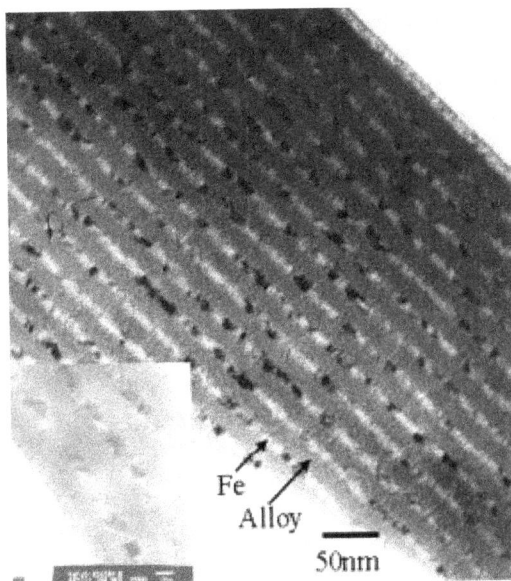

Fig. 6.6 TEM bright-field image of the as-deposited Ti (10 nm)/[NdDyFeCoNbB (16 nm) Fe(6 nm)] × 20/Ti (10 nm)/(glass-ceramics substrate) multilayer thin film.[15] (Copyright Wiley–VCH Verlag GmbH & Co. KGaA. Reproduced with permission.)

that cause the redistribution of ions in the space-charge regions. Heterojunctions in two-phase systems should be particularly efficient at improving ionic conduction, and a qualitatively different conductivity behaviour is expected when interface spacing is comparable to or smaller than the width of the space-charge regions in comparatively large crystals. Eberman and co-workers[16] reported the preparation of defined heterolayered films composed of CaF_2 and BaF_2 by molecular-beam epitaxy. The prepared film exhibits ionic conductivity (parallel to the interfaces) which increases proportionally with interface density for interfacial spacing greater than 50 nm. The results are in excellent agreement with semi-infinite space-charge calculations assuming a redistribution of fluoride ions at the interfaces. If the spacing is reduced further, the boundary zones overlap and the predicted mesoscopic size effect is observed. As a result, the single layers lose their individuality and an artificial ionically conducting material with

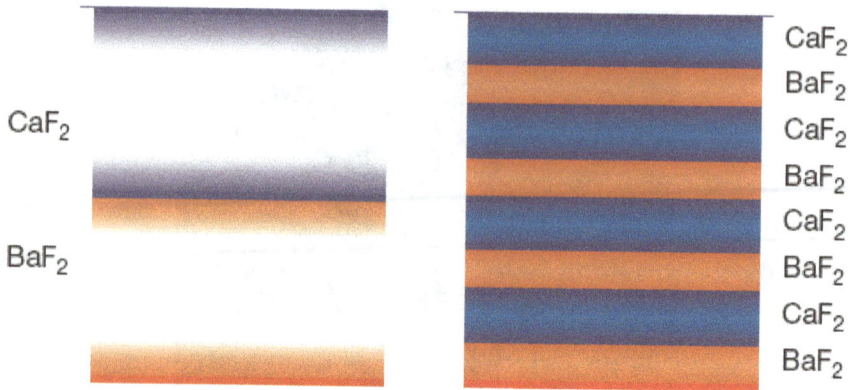

Fig. 6.7   Comparison of conductivity profiles in the simi-infinite space-charge and mesoscale situations. Left: the concentration or (parallel) conductivity profiles are sketched for the semi-infinite space-charge situation (period > 8λ). Right: mesoscale situation (period < 8λ) in which the layers have lost their individuality and an ensemble has formed with qualitatively new conductivity properties.[16] (Reprinted by permission from Macmillan Publishers Ltd: *Nature*, copyright 2000.)

anomalous transport properties is generated (Fig. 6.7). The reported results should lead to fundamental insight into ionic contact processes and to tailored ionic conductors of potential relevance for medium temperature applications.

6.7   Concave/Convex Periodic Binary Cooperative Complementary Structure

The state of a two-particle system is said to be entangled when its quantum-mechanical wave function cannot be factorized into two single-particle wave functions. This leads to one of the strongest counter-intuitive features of quantum mechanics, as is called non-locality. Experimental realization of quantum entanglement is relatively easy for photons; a starting photon can spontaneously split into a pair of entangled photons inside a nonlinear crystal. Altewischer *et al.*[17] reported the investigation of the effects of nanostructured metal optical elements on the properties of entangled photons. They placed optically thick metal films perforated with a periodic array of subwavelength holes in the paths of the two

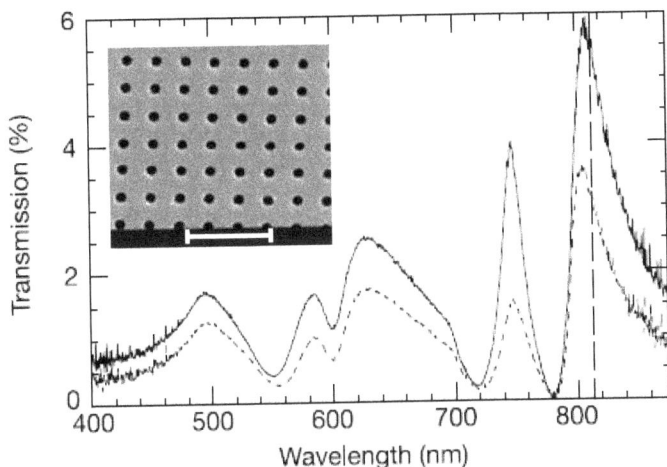

Fig. 6.8 Wavelength-dependent transmission of the arrays used in the experiment. The dashed vertical line shows the wavelength of 813 nm used in the entanglement experiment. Inset is a scanning electron microscope picture of part of a typical hole array.[17] (Reprinted by permission from Macmillan Publishers Ltd: *Nature*, copyright 2002.)

entangled photons as in Fig. 6.8. Such arrays convert photons into surface-plasmon waves — optically excited compressive charge density waves — which tunnel through the holes before reradiating as photons at the far side. Their coincidence-counting measurements showed that entanglement did survive, demonstrating that the surface plasmons have a true quantum nature. Focusing one of the photon beams on its array reduces the quality of the entanglement. The propagation of the surface plasmons makes the array effectively act as a "which way" detector.

The fabrication of high-performance integrated circuits requires the availability of materials with low or ultralow dielectric constant (low-k: $k \leq 2.5$; ultralow-k: $k \leq 2.0$) because such dielectrics not only lower line-to-line noise in interconnect conductors, but also minimize power dissipation by reducing the capacitance between the interconnects. Ree and co-workers[18] reported the preparation of low- and ultralow-k nanoporous organosilicate dielectrics from blends of polymethylsilsesquioxane (PMSSQ) precursor with globular ethyl acrylate-terminated polypropylenimine dendrimers, which act as porogens, as shown in Fig. 6.9. These dendrimers were found to mix well with the PMSSQ precursor and after their sacrificial thermal decompositions resulted in closed, spherical pores of < 2.0 nm radius with a very narrow distribution even at high loading. This pore size and distribution were the smallest and the narrowest respectively ever achieved in porous spin-on dielectrics. The method therefore successfully delivers low- and ultralow-*k* PMSSQ dielectric films that should prove very useful in advanced integrated circuits.

6.8 Organic/Inorganic Binary Cooperative Complementary Structure

Besides the reports mentioned above, the organic–inorganic hybrid material also represents a novel kind of cooperative complementary composite material. This kind of material has the characteristics of both organic and inorganic materials. Through combining the material function, it realizes the property complementation and optimization. In the organic–inorganic material, the area of the interface between organic and inorganic phases is large, which makes the interactions between the two phases strong. As a result, the clear interface becomes unsharp. The micro-domain dimension is usually at nanoscale and sometimes even at molecular level. According to this structure characteristic, the hybrid material is hardly the same as the traditional one, and have lots of unique properties. There are a large number of methods in fabricating organic–inorganic material, and these methods cover a variety of materials which have already been reported, such as electronic materials, biomaterials, luminescent materials, magnetic materials and so on.[19]

Binary cooperative complementary nanointerface material, as a new concept, has been introduced into the field of designing novel functional materials. It is based on the cooperative complementary properties that occur on the same surface. In some circumstances, these interfaces exhibit unexpected properties. This has great impact not only on theory but also on practical applications. Of course, this concept can be applied in bulk materials instead of constraining

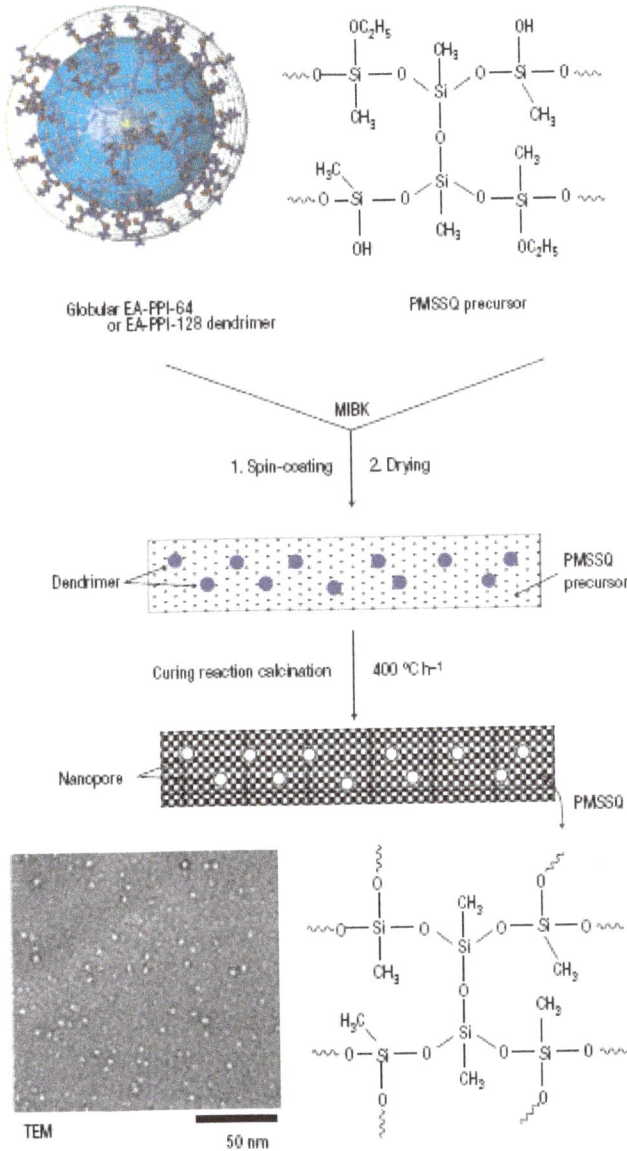

Fig. 6.9 Preparation procedure of a nanoporous organosilicate dielectric thin film and its TEM image.[18] (Reprinted by permission from Macmillan Publishers Ltd: *Nature Materials*, copyright 2005.)

only on surfaces, e.g. binary cooperative complementary nanoscale materials.

Langmuir has published a special issue on the subject of stimuli-responsive materials focusing on an area of intensive current study that has emerged from the application of bioinspired mechanisms to create functional system using traditional materials. The editors hoped to initiate new endeavors and discussions among scientists working with continued inspiration and enthusiasm toward the design of the next generation of stimuli-responsive materials. The special issue contains 50 reports on Polymers, Colloids and Multicomponent Systems, eight of which are about wettability. They center on diverse materials that have one common feature: they respond to an external stimulus such as light, electric or magnetic fields, or a change in temperature, pH, ionic strength or a mechanical deformation, in a controlled and reversible way. The publication of this special issue provides a gauge of the importance of each theme.

The articles presented in this special issue of *Langmuir* demonstrate that the field has evolved well and a lot of novel and intriguing, materials have been developed. The focus of these articles varies, including fundamental studies of the interactions driving the adaptation of material properties, and synthetic approaches that are successful in creating adaptive materials. The response of materials, such as polymer brushes or polyelectrolyte multilayers, often takes place at the air/water or solid/liquid interface, where favorable rearrangements can be created in environments with gradients of localized temperatures or connections. Nanoparticles, thick polymer films, or hydrogels also respond to environmental changes, and several-length scale is one of the challenging aspects of the field. Scientists need to design spatial distribution that will facilitate simultaneous rearrangements of molecular arrays in space while maintaining their responsive character. A successful strategy, as demonstrated in several articles, relies on the organization of multi-functional components in which stimuli-responsive objects such as nanoparticles or nanotubes, are distributed in carefully designed microarrays. Applying an external or internal stimulus to such materials triggers the release of active agents serving functions as diverse as self-healing, controlled dosing of bioactive agents, or corrosion inhibitors. Most properties hinge on cooperative interactions, which depend on high- and low-density domain packing, together with interdomain manipulations.

Inspired by nature and bearing in mind the design philosophy of binary cooperative complementary structures, we can definitely succeed in fabricating a variety of novel intelligent materials via use of traditional and modern stimuli-responsive materials, through the organic combination of structure and function. A colorful multifunctional material system with wide applications can be rationally expected.

References

1. Lavine M, Vinson V, Coontz R. *Science*, 2005, **310**: 1131.
2. Fendler J H. *Nanoparticles and Nanostructured Films: Preparation, Characterization, and Applications*, Wiley-VCH, New York, 1998.
3. Jiang L, Wang R, Yang B, Li T J, Tryk D A, Fujishima A, Hashimoto K, Zhu D B. *Pure Appl. Chem.*, 2000, **72**: 73.
4. Hansen M, Anderko K. *Constitution of Binary Alloys*, McGraw-Hill, New York, 1958.
5. Pope C E. *Electronic Processes in Organic Crystals*, Oxford University Press, 1982.
6. Jiang L, Kim Y, Iyoda T, Tryk D A, Li J, Kitazawa K, Fujishima A, Hashimoto K. *Adv. Mater.*, 1999, **11**: 649.
7. Jiang L, Iyoda T, Tryk D A, Li J, Kitazawa K, Fujishima A, Hashimoto K. *J. Phy. Chem.*, 1998, **102**: 6351.
8. Shevchenko E V, Talapin D V, Kotov N A, O'Brien S, Murray C B. *Nature*, 2006, **439**: 55.
9. Kalsin A M, Fialkow M, Paszewski M, Smoukov S K, Bishop K J M, Grzybowski B A. *Science*, 2006, **312**: 420.
10. Do T O, Jiang L, Kitazawa K, Fujishima A, Hashimoto K. *J. Phys. Chem. B.*, 1999, **103**: 3511.
11. Guo P Z, Zhang L, Liu M H. *Adv. Mater.*, 2006, **18**: 177.
12. Tian B Z, Liu X Y, Solovyov L A, Liu Z, Yang H F, Zhang Z D, Xie S H, Zhang F Q, Tu B, Yu C Z, Terasaki O, Zhao D Y. *J. Am. Chem. Soc.*, 2004, **126**: 865.

13. Yamada H, Ogawa Y, Ishii Y, Sato H, Kawasaki M, Akoh H, Tokura Y. *Science*, 2004, **305**: 646.
14. Ohtomo A, Hwang H Y. *Nature*, 2004, **427**: 423.
15. Liu W, Zhang Z D, Liu J P, Chen L J, He L L, Liu Y, Sun X K, Sellmyer D J. *Adv. Mater.*, 2002, **14**: 1832.
16. Sata N, Eberman K, Eberl K, Maier J. *Nature*, 2000, **408**: 304.
17. Altewischer E, van Exter M P, Woerdan J P. *Nature*, 2002, **418**: 304.
18. Lee B, Park Y, Hwang Y, Oh W, Yoon J, Ree M. *Nat Mater*, 2005, **4**: 147.

Index

www.ingramcontent.com/pod-product-compliance
Lightning Source LLC
Chambersburg PA
CBHW061620220326
41598CB00026BA/3827